MOLECULAR WORLD

Transformations: Studies in the History of Science and Technology
Jed Z. Buchwald, general editor

A list of titles in this series appears at the back of the book.

MOLECULAR WORLD

MAKING MODERN CHEMISTRY

CATHERINE M. JACKSON

The MIT Press
Cambridge, Massachusetts
London, England

The MIT Press would like to thank the anonymous peer reviewers who provided comments on drafts of this book. The generous work of academic experts is essential for establishing the authority and quality of our publications. We acknowledge with gratitude the contributions of these otherwise uncredited readers.

This book was set in Stone Serif and Stone Sans by Westchester Publishing Services. Printed and bound in the United States of America.

Library of Congress Cataloging-in-Publication Data

Names: Jackson, Catherine M., author.
Title: Molecular world : making modern chemistry / Catherine M. Jackson.
Description: Cambridge, Massachusetts : The MIT Press, [2023] | Series:
    Transformations: studies in the history of science and technology |
    Includes bibliographical references and index.
Identifiers: LCCN 2022033267 (print) | LCCN 2022033268 (ebook) |
    ISBN 9780262545549 (paperback) | ISBN 9780262374484 (epub) |
    ISBN 9780262374477 (pdf)
Subjects: LCSH: Organic compounds—Synthesis—History.
Classification: LCC QD262 .J33 2023 (print) | LCC QD262 (ebook) |
    DDC 547/.2—dc23/eng20221206
LC record available at https://lccn.loc.gov/2022033267
LC ebook record available at https://lccn.loc.gov/2022033268

To those who enabled my journey into history.

CONTENTS

ACKNOWLEDGMENTS

This is my first book. I hope it won't be my last. As with all such endeavors, the path has been long and bumpy. Perhaps the greatest pleasure of this moment is the opportunity to thank those who helped me along the way.

One person has been there throughout. Hasok Chang guided me through my MSc, expertly supervising my dissertation. He then took the extraordinary step of applying for funding from the Leverhulme Trust to support my PhD, building my research proposal into a larger project. Without Hasok, my journey into the history and philosophy of science would have been short lived indeed. His support has been constant and generous. It has taken a great deal longer than either of us imagined for this book to take shape. I hope, Hasok, that it will prove worth the wait.

Sam Gellman convinced me that chemists are both interested in their history and eager to apply its lessons in the present. His institutional support made a huge difference during my time in Madison. His encouragement and willingness to read and reread drafts kept me going when things were tough, and his wise comments inspired me to want to do better. Through him, I had the opportunity to talk with many chemists in the United States and Europe. I learned a lot from those conversations. Thank you, Sam.

Even a cursory glance at this book will show that several of its central arguments concern the materiality of glass. Glassblowing and lampworking do not typically form part of the historian's skill set. I am no exception—though I have taken some baby steps in that direction, thanks to the patience and generosity of Tracy Drier, master scientific glassblower in the Chemistry Department at the University of Wisconsin-Madison. Time spent in Tracy's glass shop was a revelation. It was a privilege working with you, Tracy. Let's do it again soon.

Academic life necessarily entails movement. My path took me from London to the then Chemical Heritage Foundation and the University of Notre Dame before I was appointed to a tenure-track position at the University of Wisconsin-Madison. These formative years were guided and enabled by many excellent colleagues around me, and I was fortunate to gain good friends as well as experience on my travels. After more than a decade in motion, I write now from the University of Oxford, where colleagues in the Faculty of History, at Harris Manchester College, and in the Department of Chemistry have provided a welcoming environment and supported my scholarly endeavors. As I thank my present and former colleagues, I am also delighted publicly to acknowledge the generosity of those who have written letters for me over the years, lending their support to numerous job and fellowship applications: Katherine Brading, Ron Brashear, Bill Brock, Hasok Chang, Sam Gellman, Michael Gordin, Chris Hamlin, Myles Jackson, Ursula Klein, Christoph Meinel, Alan Rocke, and Simon Schaffer.

It is an honor to list here the many institutions that have supported this project over two decades: the Leverhulme Trust, which funded my doctoral research at University College London through a grant to my then supervisor Hasok Chang; the former Chemical Heritage Foundation (now Science History Institute) in Philadelphia, where I first spent a week funded by one of their travel grants, returning as 2011–2012 Gordon Cain Fellow; the Max Planck Institute for the History of Science in Berlin, where I spent 3 months at the kind invitation of Ursula Klein and Hans-Jörg Rheinberger; the University of Notre Dame for generous research funding associated with my postdoctoral appointment at the John J. Reilly Center for Science, Technology and Values; the Research Committee of the Office of the Vice Chancellor for Research and Graduate Education, University of Wisconsin-Madison for a Fall Competition Award; the Center for the Humanities at the University of Wisconsin-Madison, which funded a manuscript workshop for what was then called *Material World* through its First Book program; and the Research Committee of the University of Oxford's Faculty of History for a grant toward the cost of image reproduction and permissions.

As a material culture history, this book contains a lot of images: more than 50 of them, in fact. A couple are photographs from private collections, reproduced by kind permission of Steve Beare, whom I met at the then Chemical Heritage Foundation in 2012, and Ilia Guzei, a friend and former colleague in the Chemistry Department at the University of Wisconsin-Madison. I owe a particular debt of gratitude to Robin Rider, Curator of Special Collections at Memorial Library, University of Wisconsin-Madison, for her help with images. Always a generous and supportive colleague, as well as a dear friend, Robin went far beyond duty or reason in helping bring this book to completion. The Grenelle's on me, Robin.

Collections professionals on two continents have ably assisted my search for the evidence on which this book is based: the Bancroft Library, University of California-Berkeley; the Bavarian State Library, Munich; the Archive of the Berlin-Brandenburg Academy of Sciences; the Berlin State Library; the Bodleian Library, University of Oxford; the British Library, London; Cambridge University Library; the National Library of the Czech Republic; the Archive of the Deutsches Museum, Munich; the Hathi Trust Digital Library (https://www .hathitrust.org/); the Bibliothèque National de France via its digital library Gallica (https://gallica.bnf.fr/accueil/en/content/accueil-en?mode=desktop); the Hesburgh Library at the University of Notre Dame; the Archive of the Humboldt University, Berlin; the Library and Archive of Imperial College, London; the Donald F. and Mildred Topp Othmer Library of Chemical History, Philadelphia; the Library of the Royal Society of Chemistry, London; the Library and Archive of the Royal Society of London; the Science Museum Library, London; the Library of St John's College, Cambridge; Special Collections at the University of St Andrews; University College Library, London; the Library of the Wellcome Collection, London; and the Department of Special Collections and Memorial Library at the University of Wisconsin-Madison. I thank all of them for their patience and work. Harald Kleims helped me write formal letters to several German libraries and archives, making me appear a better linguist than I am. Danke schön, Harald!

Producing a book manuscript relies just as much on reading as on writing. It is therefore a happy duty record here my debt to those who have read all or part of this book in various stages, several of them on multiple occasions. Some are historians, philosophers, sociologists, and science studies scholars: Carin Berkowitz, Bill Brock, Hasok Chang, Pablo Gomez, Ben Gross, Florence Hsia, David Kaiser, Ursula Klein, Joe Martin, Martijn van der Meer, Gregg Mitman, Nicole Nelson, Mary Jo Nye, Lynn Nyhart, Jennifer

Ratner-Rosenhagen, Robin Rider, and Alan Rocke. Some are chemists: Sam Gellman, Bryan Hanson, Helma Wennemers, and my brother, Richard Jackson. Other—perhaps unwitting or unwilling—readers include the many students I have introduced to the history of chemistry at the University of Notre Dame, the University of Wisconsin-Madison, and the University of Oxford. Finally, I am indebted to those who read my manuscript for the MIT Press, and especially to Jed Buchwald, series editor for Transformations.

One group of people deserves a special mention. Every author knows how much the quality of their published work can be affected—for good and for bad—by the people tasked with ensuring that it is clearly expressed and meets accepted norms of style and presentation. I count myself extremely fortunate to have worked with the MIT Press on this book. Their acquisitions editor for Science and Technology Studies, Katie Helke, has been consistently engaged, responsive, and helpful. Working with Debbie Kuan on the marketing text was enjoyable and instructive. Kate Elwell steered *Molecular World* through production, ably assisted by a team including my excellent copy editor, Cyd Westmoreland. Madhulika Jain was project manager for production at Westchester Publishing Services. She has been charming and efficient throughout. My warm thanks to all those involved for their professionalism and meticulous attention to detail.

All my readers have helped make this a better book. It goes almost without saying that its remaining deficiencies are my responsibility.

A great many individuals have helped me reach this point, some of them perhaps without realizing how much their kindness meant to me. Many are named above, but others are not. I hope they will not be offended. To include all those who warrant it would make for a long read and I'm unwilling to select. The unnamed are not forgotten. On the contrary, they are close to me, encouraging me along the road ahead.

ARCHIVAL SOURCES

The Liebig-Wöhler correspondence is cited from transcriptions made under the direction of Christoph Meinel at the University of Regensburg, Germany from manuscript sources held at the Bavarian State Library, Munich, Germany. I am most grateful to Professor Meinel for permission to use this currently unpublished material.

This study cites material from the following archives:

Archive of the Humboldt University, Berlin: Humboldt-Universität zu Berlin, Universitätsarchiv zu Berlin

Archive of the Berlin-Brandenburg Academy of Sciences and Humanities, Berlin: Archiv der BBAW, PAW (1812–1945)

Bavarian State Library, Munich: Bayerische Staatsbibliothek München

Archives Imperial College London: AICL

INTRODUCTION: MOLECULAR WORLD

This book is the result of a personal voyage of historical discovery. In 2002, I enrolled in a masters' course in history of science, technology, and medicine in London. I set out to understand the history of the science that had beguiled me more than 20 years previously. Training as a synthetic organic chemist in Cambridge in the late 1980s, I had submitted to the field's unique discipline. Now, as a returning student, I was drawn into its history. I wanted to know when, where, and why organic synthesis began. How, I wondered, had my predecessors learned to manipulate, modify, and mimic organic nature? How did they acquire the constructive capability that I, like many synthetic organic chemists, regarded as the defining feature of my field?

As early as 1860, the arch-propagandist for organic synthesis, Marcellin Berthelot, used synthetic organic chemistry's ability to mimic and transcend nature's constructive power to justify its status as a uniquely productive, experimental—rather than observational—science. "Chemistry," Berthelot famously declared, "creates its own object."[1] As this book will show, his claim for what synthesis could then accomplish was somewhat exaggerated. The point, however, remains clear: Synthesis was a philosophically as well as practically significant science.

Fulfilling Berthelot's promise would take organic chemists many decades. Indeed, the "golden era of total synthesis," a period celebrating organic chemistry's productive capability, began in the second half of the twentieth century—not the nineteenth.[2] Twentieth-century synthesis melded material creativity with scientific value. By 1963, Robert Burns Woodward— preeminent exponent of the art and science of synthesis—considered the laboratory preparation of defined molecular targets a "unique challenge" to every aspect of the chemist's skill.[3] Woodward's career epitomized that golden era. Already with a string of complex natural product syntheses to his name, Woodward had recently embarked on the project that culminated in the landmark 1973 synthesis of vitamin B_{12} and his elucidation (with Roald Hoffmann) of the rules governing reactions used in that synthesis.[4]

From the late 1980s, suggestions that synthetic organic chemistry had become a "mature science," in danger of ceding disciplinary identity—and research funding—to the related fields with which it so usefully interacts, provoked outrage.[5] Its practitioners are adamant that organic synthesis continues to be an "advancing art and science" in its own right.[6] Synthetic organic chemistry's vital contributions to material science, biology, and medicine warrant its status as "the central science."[7]

According to K. C. Nicolaou—whose research group synthesized the antibiotic vancomycin and the anticancer drug Taxol® during the 1990s—synthesis "touches upon our everyday lives in myriad ways," by producing new "medicines, . . . and tools for biology and physics."[8] For Nicolaou, chemistry "played a central and decisive role in shaping the twentieth century"—most notably by transforming crude oil into a gamut of useful products.[9] Since the late twentieth century, synthetic organic chemists have been able to make almost any desired target molecule, given adequate time and resources.[10]

My goal has been to understand how this happened and what made it possible. At the outset, I expected to find answers to my questions in the extensive existing historical literature concerned with nineteenth-century organic chemistry.[11] I did not. That is why I wrote this book. It is also why I believe this study is worthy of your attention. In answering the questions that drove me to become a historian of chemistry as well as a chemist, this book explains the origins and the nineteenth-century rise of synthetic organic chemistry, inviting a new appreciation of its nature and significance—past, present, and future.

This book argues that organic synthesis was the ground on which chemists forged a new relationship between experiment and theory, with fundamental consequences for chemistry as a discipline. Before elucidating what terms like "experiment" and "theory" signify in this account, and how they relate to chemical practice, let us pause to consider the scope and import of that claim. Viewed from the disciplinary perspective, it implies that to a considerable degree, the rise of synthetic organic chemistry was constitutive of chemistry's nineteenth-century development.[12] Thus, in addition to answering my original questions, I also intend this book as a disciplinary history of nineteenth-century chemistry.

Such ambition requires qualification. This study does not seek to be comprehensive, and I shall have more to say shortly about how its leading actors and major episodes were chosen, why it focuses almost exclusively on German chemistry (by which I mean chemistry as it was practiced in German-speaking lands and by German-speaking chemists working elsewhere), and where it stands in relation to existing scholarship. Nor is it a disciplinary history in the institutional sense.[13] Yet its purview is reflected in an account that makes visible the intrinsic connections between chemists' ability to advance their science through laboratory work and their decision to move chemistry into lampworked glassware. As later chapters will show, the practice of synthesis in glass ultimately transformed the material environment of the laboratory as well as chemical instrumentation. The laboratory, in this account, was not merely the place where chemistry was done; it was as important to chemists as the test tube or any reagent—and thus an essential component of chemistry's material culture.[14]

Working in glass—a change for which I have previously coined the term "glassware revolution"[15]—altered technique and training throughout the chemical community. The consequences of chemists' adoption of glass and glassblowing pervaded the discipline, but these changes were overwhelmingly rooted in organic chemistry. Although the glassware revolution began in analysis, synthesis expanded its transformational reach. An example from the book will clarify this point. Lampworked glassware was fundamentally implicated in how organic chemists worked with individual, pure substances—a category that was severely challenged by the turn to synthesis. Thus, synthesis prompted two inextricably linked developments, both realized in glass: the standardization of melting and boiling points as criteria of identity and purity, and the introduction of separation techniques capable of producing substances that met this definition. Although these methods and instruments hugely transcended this context, their origins lay in organic synthesis.

This book's historiographical offering reinforces its disciplinary relevance. It introduces several new historical landmarks—the glassware revolution of the 1830s, organic chemists' turn to synthesis a decade later, and their emerging ability to prepare nature-identical substances by laboratory synthesis during the latter part of the century. In doing so, this study helps disentangle an existing historiography of organic synthesis that is full of internal contradictions. Most notably, Friedrich Wöhler's celebrated 1828 production of urea has retained its symbolic importance as an origin moment, despite several studies demonstrating its peripheral role in the development of synthesis and its limited contribution to contemporary vitalist debates.[16] The advent of structural theory in the 1850s and 1860s has provided generations of chemists and historians of chemistry with a seemingly adequate explanation for the productive power of synthesis.[17] At the same time, for reasons examined below, many histories identify instances of organic synthesis going back to the eighteenth century.[18]

This study resolves these contradictions by showing that, although they certainly aspired to the productive power implicit in later understandings of the term, nineteenth-century organic chemists pioneered synthesis as an investigative method. Learning that synthesis in organic chemistry initially denoted a systematic experimental approach intended to elucidate the molecular nature of organic compounds—rather than laboratory production per se—highlights the term's various meanings.[19] In doing so, it alerts us to a problem of definition with consequences for our ability to make sense of the history of synthesis.

I have already alluded to the established practice that recasts earlier preparations as syntheses. This practice has undeniable advantages in chemistry, where many such preparations became part of chemists' synthetic repertoire, and hence of training in synthesis. This neglect of historical terminology is nevertheless problematic if we seek to understand chemistry's past trajectory. Wöhler's preparation of urea is a good example, for this was neither intentional nor thought of as a synthesis by chemists at the time. Nor did Wöhler's serendipitous transformation (of ammonium cyanate into urea) shed much light on how organic substances were constituted, let alone how they might be made. To describe Wöhler's achievement as a synthesis thus misrepresents not only the contemporary language of chemistry but also his methods and goals. In doing so, it misleads modern readers—who are likely to equate synthesis with intentional molecular construction in which vital force plays no part—concerning both the outcome of his work and its disciplinary significance. Far from marking either the death of vitalism or the birth of synthesis, Wöhler's urea mattered at the time because it supported the unity of organic (animal and vegetable) and inorganic (mineral) chemistry.[20]

As we shall see, the term "synthesis" was first used in the context of organic chemistry in the 1840s, as an explicit counterpoint to analysis.[21] The introduction of such a term marks a watershed, identifying the point at which chemists were conscious of doing something so innovative that it required new language. Historians' tendency to project back their current understanding of synthesis has worked to obscure that innovation, as well as the term's subsequent changes in meaning—each change reflecting precisely the developments that this book seeks to recover.[22] If synthesis was not, in its early stages, synonymous with making things, then any claim to understand its history requires an account of its initial goals and immediate outcomes, as well as how it acquired its present-day, inevitably constructive meaning.

This book offers such an account, explaining the rise of synthetic organic chemistry to become both the dominant branch of chemistry and a powerhouse of medicine and industry. Although it includes, and in no sense intends to negate, the contributions of chemists of other nationalities working in other places, this history—especially in its later chapters—focuses on German chemistry (in the sense already defined). This is no whimsical preference. Despite Berthelot's grand philosophical claims, French chemists' role in developing organic synthesis as a theoretically powerful and materially productive discipline was limited by comparison.[23]

The contributions of other European nations were still less significant. From the late eighteenth century to around 1850, British industrial and imperial might in particular had attracted considerable attention, inciting attempts to industrialize formerly agricultural German states.[24] Yet after early success with mauve—the valuable artificial dye first produced by William Henry Perkin in London in 1856—Britain's chemical industry rapidly lost ground to its ever-more dominant German competitor.[25] Academic chemistry in Britain was increasingly seen as impoverished, while other European states apparently fared little better.[26] From about 1865 until the outbreak of the First World War, academic chemistry was largely German, heavily organic, and mainly synthetic.[27] The laboratory syntheses of natural products (chemists' name for naturally occurring substances), including alizarin (1869), indigo (1878), and coniine (1886)—all of which were accomplished by German-speaking chemists working in German laboratories—secured and intensified synthetic organic chemistry's 50-year dominion.

Although I offer this study as a contribution to disciplinary history rather than as a collective biography, I have nevertheless shaped it around the work of three protagonists. The first two are well known, the third much less so. Justus Liebig and August Wilhelm Hofmann are dominant figures in the history of nineteenth-century chemistry, and their central presence in this book perhaps requires little explanation—beyond noting that my interpretation of their disciplinary significance differs considerably from that of previous scholars. Albert Ladenburg has been little studied in comparison, a circumstance that is partly attributable to the relative paucity of his surviving archive but also, I conjecture, to historians' somewhat limited interest in his achievements. Persistently opposed to August Kekulé's theory of aromatic structure, Ladenburg instead devoted himself to laboratory synthesis. Thus, Ladenburg is both much less accessible to the historian and was engaged in a project that is problematic in the context of a historiography based on the assumption that aromatic structural theory was the primary enabler of synthetic success.

This book by no means rectifies the imbalance in historical attention paid to Liebig and Hofmann on one hand and to Ladenburg on the other. However, it does establish commonalities in their approaches and accomplishments, showing how each pioneered instruments and methods that transformed first analysis and then synthesis, ultimately becoming constitutive of the new science of organic chemistry. All three used their personal networks, industrial connections, and disciplinary prominence to lead an extended transformation in chemistry's methods, instruments, material basis, modes of training, and institutional context, shaping the field over three generations. My actors are thus neither typical nor merely illustrative but have been chosen for their disciplinary impact. As explained below, I have likewise selected key episodes to highlight the resources that made organic chemistry possible, thereby providing the basis for a general explanation of how nineteenth-century organic chemists built the science of synthesis.

Organic synthesis was a remarkable achievement that ultimately empowered chemists not only to mimic nature but also to move beyond it, to make compounds that nature never had. It was also—as indicated above—inseparable from the increasingly lucrative manufacture of dyes, drugs, explosives, flavors, and perfumes.[28] Nineteenth-century chemists built a new world on the by-products of coal gas distillation. In doing so, they set the model for how their successors would transform crude oil into the "pharmaceuticals, high-tech materials, polymers, fertilizers, pesticides, cosmetics, and clothing" that epitomize modernity.[29] Chemistry has always been an "impure science."[30]

Yet even as it furthers historical efforts to promote the importance of chemistry's practical utility to its intellectual development, this study aspires to a more radical revision, responding to an existing literature that persistently overemphasizes the power of theoretical innovation to drive practical progress. Rather than merely counterbalancing that bias, this book urges the necessity of prioritizing experiment as the primary force behind theoretical advance, promoting a new approach to elucidating how experiment and theory interact in chemical practice.

This approach is exemplified by my account of Hofmann's introduction of the ammonia type—a major innovation in conceptual understanding of how organic compounds are constituted. Resisting premature theoretical commitment in favor of accumulating vast, systematic experimental knowledge was crucial to Hofmann's achievement. Although theory—in the form of stable concepts, such as composition and molecular formula—was ever-present in Hofmann's work, insecure, speculative theories were never his guide.

In a more general sense, an approach of this kind has the potential to illuminate how past chemists equipped with only simple apparatus obtained stable, useful knowledge about organic nature. From the 1840s onward, synthesis initiated an unprecedented transformation in the material world. It produced increasingly sophisticated theories concerning the nature of organic compounds—ultimately including the now taken-for-granted view of molecules as real, three-dimensional entities.[31] Yet all this was accomplished without the instrumentation that has mediated organic chemists' interactions with their object of study since the mid-twentieth century.[32] Chemists produced the molecular world using test tubes, flasks, and thermometers, not million-dollar nuclear magnetic resonance spectrometers.

Despite its centrality to the history of chemistry, the relationship between localized laboratory manipulation and universal, abstract theories remains a conundrum.[33] Its resolution nevertheless seems to me to be a most pressing and worthwhile goal for anyone interested in chemistry and its history—whether chemist, historian, or philosopher.[34] Studying existing histories alongside the archive of nineteenth-century chemistry has convinced me that one reason this objective has proved so elusive lies in present-day chemistry's dominant association with the visual and the spatial. Being a chemist today entails daily interaction with molecular representations and models as much as with substances and glassware. Modern chemistry demands and develops spatial perception and visual thinking.

But, as this study shows, organic chemistry remained for most of the nine-teenth century a taxonomic and not a visual science.

Since the mid-twentieth century, chemists have relied on three-dimensional models produced by black-box devices. By interpreting stable theories of bonding, structure, and stereochemistry (chemistry in three-dimensional space), sophisticated instruments naturalize molecules and the models and formulae that represent them in ways that make it hard to imagine that it was ever otherwise. Those professionally trained in this world struggle to accept that nineteenth-century chemists accomplished so much with the limited instruments and methods at their disposal, and—above all—with prior theories we now perceive as rudimentary, provisional, and often false.

I have frequently been quizzed—by both chemists and historians—along the following lines: "But surely [*insert name of nineteenth-century chemist*] was working with some notion of [*insert name of cutting-edge theory*], wasn't he?" From our present perspective, the enabling power of novel theories is certainly a tempting proposition. But it is not supported by the archive, and therefore constitutes a real barrier to understanding how chemistry oper-ated in the period before theories of molecular structure and stereochemis-try were established as productive working tools.

In seeking to transcend this barrier, this book uses an extensive historical archive to reconstruct a series of key episodes concerned with the chem-istry of alkaloids—a diverse family of physiologically active substances derived from plants.[35] From about 1820 onward, the medical significance of alkaloids—most notably morphine, quinine, and strychnine—established their commercial and medical value, making them an important testing ground for chemistry's power and utility.[36] What organic chemists recorded in their notebooks, wrote to their mentors and friends, and published for their colleagues helps me elucidate how chemists tackled the alkaloids. Thus, Liebig's analyses of morphine, Hofmann's ammonia type, and Lad-enburg's coniine synthesis serve here to illustrate how nineteenth-century organic chemists learned to relate bubbling liquids and shimmering crystals manipulated in glass to abstract concepts, such as formula and structure.

I have introduced a new term to describe how chemists used wet chem-istry in glass to produce the molecular world we take for granted. I have not done so lightly. "Laboratory reasoning" is how chemists stabilized those cutting-edge theories that are otherwise so easily assumed to be the cause of their developing knowledge and practical capability. It is central to the new relationship forged between experiment and theory in the practice of synthesis, and thus to this study's overarching argument. This term also expresses in the case of organic synthesis something I take to have more general purchase: namely, that the relationship between experiment and theory can only be exemplified within a science as it is practiced. Rather than attempting an abstract definition of either experiment or theory—or even practice—I therefore restrict myself here to offering a brief explanation of laboratory reasoning, a term whose meaning will be most effectively clarified by the examples discussed in this book.[37]

Laboratory reasoning, as I shall show, was nineteenth-century chemists' crucial connection between experiment and theory. Theory's cutting edge mattered to the chemists in this book. But they knew from bitter experience that it constituted a poor guide to experiment. Chemists therefore invoked only established theoretical concepts in their reasoning, working to acquire empirical knowledge made replicable by standard instruments and methods. Useful knowledge required an agreed-on foundation of practice as well as belief. Provided they were embedded in a stabilizing network of this kind, even theories that might later be discarded could be fruitful.

Because chemists' theories had no independent life, they functioned only in the context of shared methods, techniques, and instruments. Standardization was just as big a barrier to disciplinary development as any other. Both Hofmann's ammonia type and Ladenburg's coniine synthesis, for example, depended on the use of glassware and melting point measurements to solve complex problems of chemical identity. Chemists therefore built a disciplinary world in which results and measurements retained their meaning when shared. That world was the institutional chemical laboratory—and it was crucial in laboratory reasoning's ability to be productive in the face of the unstable and highly disputatious nature of much mid-century speculative theorizing.

Because I am a chemist and a historian seeking to engage chemists among my readers, this history is intrinsically technical. I have nevertheless sought to structure and write this book so that its major claims and overarching narrative can be appreciated without becoming more involved in technical detail than the reader desires. Nonchemists may find the detailed accounts of laboratory work presented in chapters 1, 5, 6, and 9 demanding. They should feel free to skim those sections without fear that this will render the book's overall message incoherent. For chemists and others interested in the practice of science and how it produced new knowledge, those accounts are intended to offer a glimpse of how nineteenth-century organic chemists worked that is technically, experimentally, and logically convincing.

Dealing with technical material makes the historian's task harder, not easier. Technical skill is as human as any other form of knowledge. It is equally historically placed. The professional world of chemists like Liebig, Hofmann, and Ladenburg encompassed much that is strange to us, regardless of whether or not we ourselves are chemists. Engaging with it is initially perplexing to the point of being intellectually bruising. But if a present-day chemist could not tell my story, nor could their nineteenth-century predecessors. Taking their world too much for granted, they would have no idea what we would like to know. My task as historian is therefore to bring their world of chemistry alive in ours.

A range of approaches from science studies and the sociology and philosophy of science, as well as the history of science, have helped me reconstruct the work and world of nineteenth-century chemists. The methods I have found valuable suggest an attitude toward the sciences. Yet they in no way dictated this history. Instead, techniques learned during 20 years of historical study have been crucial in enabling me to express my chemist's sensibilities in a history that does justice to its sources by foregrounding experimental work, highlighting laboratory practice rather than cutting-edge theory as chemists' only reliable guide.[38] In seeking to make this book approachable for a diverse readership, I have chosen not to include an explanation of its production. The questions underpinning this study are fundamentally historical and historiographical, not methodological. My method, insofar as I have one, is a continuous presence in the history that follows.

ISSUES, SCOPE, AND STRUCTURE

My overarching historical claim is that synthetic organic chemistry was built on practice, not theory. Practice certainly encompassed established theories as well as experiment. But I shall show that organic chemists' laboratory work was independent of as-yet-unstable theoretical innovations. Previous histories of organic chemistry have offered precisely these cutting-edge but still-insecure theories as the explanation of chemists' developing preparative capability. This account shows instead that synthesis operated without reference to what remained unproven, serving to refine, stabilize, and—in some instances—refute provisional theories as its taxonomic scope and productive power grew. In establishing this claim, I show that practical, commercial goals and access to novel industrial resources were inseparable from chemists' struggle to master organic nature—even though academic chemists frequently deployed the rhetoric of pure, theory-driven science in their search for intellectual recognition and essential financial support.

Loosely structured in three sections, each of three chapters, this book elucidates in chronological sequence how my chosen nineteenth-century chemists—Liebig, Hofmann, and Ladenburg—worked. It explains the experimental basis of their growing practical capability, the processes of laboratory reasoning by which they advanced abstract chemical knowledge, and their central role in building the shared methods, standards, and infrastructure that constituted the new discipline of organic synthesis.

Each section makes a distinct contribution to the book's overall argument. The first (chapters 1–3) reveals both the persistent failure of analysis to produce stable molecular formulae for many organic substances and its inability to resolve fundamental questions concerning the nature and behavior of organic compounds. Building on this, the second section (chapters 4–6) explains how synthesis helped stabilize molecular formulae, providing the crucial basis for hotly contested theories of constitution that chemists introduced in their efforts to conceptualize the arrangement of elements in molecules.

This account of the role of synthesis in advancing constitutional theory has implications for the development of chemical structure. These implications are played out in the final section (chapters 7–9), where the landmark 1886 synthesis of coniine exemplifies how the synthesis of a molecular target established its structure, in turn contributing evidence capable of persuading chemists to accept structural concepts they had hitherto regarded as hypothetical, or even unfounded. Structural theory, far from being the decisive enabler of synthetic success, was in fact established by synthesis.

Ambition, labor, and instrumentation are key themes of the opening section. Chapter 1 relates how, when forced to leave the French capital Paris for rural Giessen, the young Liebig responded in 1830 by tackling one of the hardest research problems of the age. Investigating the chemistry of alkaloids, including morphine and quinine, was Liebig's chosen way of competing with his Parisian rivals. He anticipated alkaloid analysis would establish his reputation by producing new understanding of these commercially important drugs. In fact, this work had a very unexpected outcome.

The crux of the matter was nitrogen, the element chemists held responsible for the alkaloids' chemical properties and physiological activity. But finding out how much nitrogen alkaloids contained was very tricky. Parisian chemists relied on large, expensive instruments made by skilled makers. Chapter 2 explains why and how Liebig turned instead to small glass apparatus that he made himself by glassblowing. Unable to pay for laboratory assistance, geographical displacement also led Liebig to enlist student labor in this research. In the end, Liebig failed to solve the problem of nitrogen. Nor did he discover anything very new about alkaloids. His success was to change how elite chemists worked and were trained.

Moving chemistry into glass—my "glassware revolution"[39]—was a necessary condition for the transformed economy of laboratory labor in Liebig's Giessen research school. It would also, as later chapters explain, prove vital in chemists' ongoing struggle to master organic nature. Just as lampworked glassware was essential to chemistry's disciplinary expansion, so it was integral to chemists' ability to know what substances were and to control how they reacted. Identifying why and how this material transformation began, and with what disciplinary consequences it continued, is therefore an intrinsic requirement of understanding chemistry's scientific development during this period.

During the 1830s, Liebig, aided by his students, learned that ongoing debates about the alkaloids reflected a fundamental problem with analysis—then chemists' leading quantitative method for investigating substances. Analysis was how chemists tried to discover how much of each element was present in molecules of a particular compound—which chemists called "composition," expressed in a molecular formula. But the persistent difficulty of analyzing organic substances left chemists unable to agree on formulae for many simple compounds—never mind more complex substances like alkaloids. Far from being a solved problem, by about 1840, organic analysis was in deadlock—unable to establish the nature of compounds, including the alkaloids morphine and quinine, whose medical importance made them a central testing-ground for chemistry's practical value as well as its scientific worth.

After a decade of intense struggle, as chapter 3 concludes, Liebig had had enough. Without agreed-on formulae, organic chemistry could not progress. Before withdrawing from the field, he therefore guided his protégé Hofmann to develop a new approach to foundational questions about how elements combined in organic molecules. Synthesis began in 1840s Giessen as a response to the limitations of analysis. Rather than equipping chemists to make desired target molecules, organic synthesis was initially a new way of tackling questions that analysis could not answer.

Locating and defining the turn to synthesis resolved several issues that had puzzled me from the start—including the contradiction between the much-repeated claim that Liebig solved the problem of organic analysis, thereby reducing the determination of molecular formulae to mere routine, and the fact that even simple alkaloid formulae were not agreed on until near the end of the nineteenth century.[40] But it also raised new questions. If synthesis was a response to the limitations of analysis, I wondered how early synthesis operated, and how chemists transformed it into a tool for re-creating and expanding nature's molecular repertoire.

Studying Hofmann's early career—the focus of section two—began to answer these questions, showing how synthesis functioned as a tool for advancing theory's cutting edge. It also led me to some unanticipated conclusions regarding the relationship between natural and artificial substances in Hofmann's research and—more broadly—about the balance between commercial and academic motives in guiding both his choice of problem and his decision in 1845 to move from Bonn to London, to become first director of the recently founded Royal College of Chemistry (RCC).

The RCC's role as one of the progenitors of Imperial College, London's elite university for science, technology and medicine, has tended to obscure its roots in commercial pharmacy.[41] Initially unconvinced by the job's prospects, Hofmann was persuaded to accept in order to secure the profitability of a private commercial scheme to supply "amorphous quinine" as an alternative to quinine, a costly drug whose fever-fighting abilities were essential to imperial expansion.[42] As chapter 4 explains, commerce, industry, empire, and romance—rather than his commitment to academic science—were decisive in Hofmann's move to London.

Hofmann's involvement with industry was not new. Indeed, the investigations that had established his chemical credentials were possible only because their subject, coal tar, was then a low-value waste product of rapidly growing coal gas manufacture.[43] By contrast, Hofmann's partnership with Liebig over amorphous quinine drew him into the lucrative, competitive, and already global quinine market—with almost disastrous consequences. Scientific chemistry proved dangerously inadequate in equipping Liebig and Hofmann to navigate professional pharmacists' expert domain. For almost 20 years afterward, I shall argue, this experience directed Hofmann's research toward primarily academic goals.

Organic chemists had not yet mastered natural products. But they were about to transform their field of study by creating a multitude of artificial compounds. Aniline derived from coal tar was the raw material for this change, and Hofmann's "synthetical experiments" its key method.[44] Hofmann's focus was not the new substances he made, however, but the reactions by which these substances formed. Despite the later significance of aniline dyes to the nascent chemical industry, Hofmann's synthetic research program during the 1840s and 1850s—analyzed in chapters 5 and 6—aimed to elucidate the behavior and molecular properties of organic nature. It was not a search for commercially relevant products.

Hofmann used aniline as an experimentally tractable model for natural alkaloids, and synthetical experiments to extend Liebig's original research program. Where his mentor hoped analysis would show how much nitrogen alkaloids contained, Hofmann instead wanted to know how that nitrogen was arranged with respect to the alkaloids' other elements. Molecular arrangement or constitution was chemists' response to observations that distinct compounds could share the same molecular formula. At first, constitution was a somewhat vague concept that prompted much speculative theorizing. Chapters 5 and 6 reveal Hofmann as a very different kind of theory builder, whose work illustrates how meticulous laboratory reasoning led from the results of synthetical experiments to a significant and durable theoretical innovation, the ammonia type.

Understanding early synthesis as an investigative method rather than a way of making things clarifies the history of nineteenth-century chemistry. It refigures haphazard and usually futile early attempts to make natural products as systematic investigations intended to shed light on chemical behavior and reactivity. It also undermines persistent claims that organic chemists of this period operated without and were frequently hostile to theory.[45] It is perhaps a measure of the inadequacy of existing histories that such claims persist alongside the assumption that chemists' synthetic accomplishments were enabled by cutting-edge theory. The present study resolves this contradiction by showing what roles theory did, and did not, play at that time. Hofmann's ammonia type—the subject of chapter 6—exemplifies the power of laboratory reasoning to link experiment and theory, showing how wet chemistry could produce abstract concepts. Hofmann's theoretical contributions—like those of many of the century's greatest practitioners—perhaps appear limited. But, as section 2 demonstrates, this reflects the difficulty of producing an adequate empirical basis for reliable theorizing and the distinct nature of chemical theory in this period rather than any failure to appreciate the value of theory.

Resolving these historical conundrums concerning the experiment/theory relationship in early synthesis left me with a final question, one that found expression in a seeming paradox in Hofmann's later career. Existing histories of organic chemistry either state or imply that novel theoretical developments—specifically, the theory of chemical structure that displaced constitutional theory during the 1850s and 1860s—enabled chemists to make desired natural products in the laboratory (a practice sometimes called "target synthesis"). Yet Hofmann, despite major contributions to constitutional and structural understanding of alkaloids during the 1880s, produced no natural alkaloid in the laboratory.

Instead, it was the much younger and relatively unknown Ladenburg who in 1886 achieved this important landmark by making nature-identical coniine, a poisonous alkaloid found in the hemlock plant. Chapter 9 explains why Ladenburg succeeded where Hofmann—although he assigned coniine's molecular formula and structure during the early 1880s and certainly tried to make it—failed. This explanation concludes a final section that elucidates the many interrelated, discipline-wide developments in practice and instrumentation by which chemists established the resources necessary for productive laboratory reasoning and transformed synthesis into a reliable way of making things. As we shall see, none relied on cutting-edge theory.

Following the turn to synthesis, chemists mobilized glass and glassblowing, melting and boiling points, model compounds, reactions and reagents, student training and labor, standardization, and laboratory infrastructure in their quest to master nature. Crucially, the proliferation of artificial substances during the 1840s and 1850s challenged chemists' ability to know what their reactions made—a question no theory could answer. Chapter 7 explains how chemists and glassblowers working together solved this "chemical identity crisis," developing melting and boiling points measured in standardized glassware as reliable and communicable characteristics of identity and purity.[46]

Chemists often did synthetical experiments by sealing flammable mixtures in glass tubes, which they heated, sometimes for several days. This was dangerous work, especially in inexperienced hands. Synthesis threatened to make accidents, fires, and explosions a regular feature of laboratory life. Protecting chemists from injury and even death was necessary for synthesis to flourish in the context of academic training and research, so that chemistry shaped the laboratory and not the other way around. This transformation of the laboratory—the subject of chapter 8—intensified during the latter part of the century in response to novel reactions, reagents, and techniques.[47] Synthesis changed laboratory infrastructure, driving organic chemistry's expanding scale and increasing disciplinary dominance.

Chapter 9 clarifies the resources required for natural product synthesis by juxtaposing Hofmann's synthetic investigations against Ladenburg's successful synthesis of coniine. For example, Ladenburg's starting material was an aromatic compound derived from coal tar. He developed innovative glass melting point apparatus to prove his synthetic product was identical to natural coniine. Above all, he dedicated himself to making his chosen target. Attaining this goal contributed to a change in what it meant to be a synthetic organic chemist, helping establish the total synthesis of complex molecules as a legitimate disciplinary objective, capable of answering some of organic chemistry's most erudite theoretical questions. By the end of the nineteenth century, organic synthesis had become synonymous with making specified target substances, including the natural products chemists had long desired to produce.

There is one surprising, noteworthy absence from this story. Although his synthesis relied on the chemistry of aromatic compounds, Ladenburg made no reference to August Kekulé's celebrated theory of aromatic structure during this work. He knew this theory better than most, having assisted in its development. Yet Ladenburg resolutely rejected Kekulé's proposal that the structure of aromatic compounds could be conceptualized as atoms linked in hexagonal rings until synthesizing coniine began to convince him of its merits. Like Hofmann before him, Ladenburg was eventually pushed by laboratory reasoning to represent aromatic compounds on paper using hexagonal rings, helping make chemistry newly visual. The reactions Ladenburg developed and used to synthesize coniine provided evidence for aromatic structural theory—not theory driving practice but quite the reverse.

By the late nineteenth century, organic synthesis was changing understanding of the molecular world and the relationship between artificial and natural substances. Industrially powerful, it was also poised to launch a range of new disciplines in the molecular life sciences. This book advances our understanding of what nineteenth-century chemists did, and why and how they did it; of what they knew, and—equally important—what they did not; of how they learned and practiced the art of chemical experimentation; of how laboratory reasoning brokered the fragile, intrinsic connection between experiment and theory; and of how chemists' reliance on laboratory reasoning transformed not only their own science but also—ultimately—the disciplinary landscape of the sciences. I offer it as a model for how history can simultaneously elucidate the nature of chemical theory and inform our understanding of science as a profoundly human endeavor. I dedicate it to those chemists—past, present, and future—who strive to make our world a better place. Most particularly, I hope it will encourage chemistry students all over the world in their struggle to learn—much as Liebig, Hofmann, and Ladenburg did—how laboratory work can make sense of the molecular world.

A NOTE ON NAMES AND FORMULAE

To avoid the problems of approaching past chemistry from a present-day perspective, I have prioritized contemporary, historical names and formulae, providing modern equivalents in parentheses only where this may assist the reader. I explain chemical terms, historical and current, when they first appear—but with one important proviso. Just as the term "synthesis" changed its meaning, so other terms that are crucial in this history, including "isomer(ism)" and "type theory," were modified by organic chemistry's development. I have therefore avoided defining such terms according to their ultimately acquired meaning in favor of explaining their contemporary meaning at relevant points in the narrative. Despite any appearance of repetition, this approach is required to elucidate the historical relationship between experiment and theory in chemical practice.

The absence of modern structural or skeletal formulae may likewise initially seem unhelpful, but it is essential in making apparent the processes
by which nineteenth-century chemists produced composition, constitution, structure, and reactivity. Translating historical formulae into modern
form undoubtedly makes it easier for readers to interpret the results that
past chemists obtained. Such translations, however, have significant disadvantages. They tempt the reader to evaluate the conclusions drawn in ways
that are historically misleading. Even more problematic, modern formulae
embody conceptions of substance that were the outcome of the work being
described. Thus, they conceal the work required to produce this new knowledge, effacing precisely the processes of experimentally driven laboratory
reasoning that this study seeks to recover. Chemists worked for well over
a half century to enable us to visualize molecules as spatial structures. To
understand how they did this, we must see chemistry through their eyes—
not ours.

1 ANALYSIS MANIA

Justus Liebig is widely regarded as the nineteenth century's greatest chemist, famous for developing a school for chemical research and a new piece of chemical glassware he called the *Kaliapparat* (potash apparatus). Iconic from the start, Liebig's Kaliapparat (phonetically, kalɪapara:t) remains familiar to present-day chemists everywhere from the American Chemical Society logo.[1] Triumphalist accounts of Liebig are based on the attractive proposition that when Liebig invented the Kaliapparat in 1830, he solved the problem of analyzing organic compounds, and that this achievement was crucial in propelling his rapid rise to fame. In such accounts, it was always obvious to Liebig that organic analysis would be his route to success, and that he should found a school for research in this area.[2]

This standard account of Liebig is, in large part, derived from what Liebig himself later wrote—autobiographical recollections that entered the historical literature through the first biography of Liebig, written by his devoted pupil Jacob Volhard.[3] Despite numerous more recent studies, the central elements of this hagiography remain unchallenged.[4] Generations of historians and chemists have learned that Liebig solved the problem of organic analysis by inventing the Kaliapparat and building a school in which students learned to use this new analytical device.[5] The self-evidence of Liebig's achievements in turn rendered it largely superfluous to explain his choices and ambitions, leaving us without a clear understanding of the resources that enabled his early success or of why Liebig ultimately became so disillusioned with his chosen field of research that he abandoned it.[6]

This book's opening three chapters answer these questions, clarifying Liebig's trajectory from young unknown to becoming "the greatest chemist of his time."[7] In doing so, these chapters focus attention on other equally important issues, including the mid-century turn to synthesis that is a linchpin of this book. Understanding how and why Liebig's analytical program prompted the introduction of a new investigative approach to organic compounds resolves longstanding historical confusion, locating the origins of organic synthesis in 1840s Giessen.[8]

Taking seriously the contingencies of Liebig's early career, meanwhile, has driven me to seek explanations for why Liebig decided to tackle organic analysis, how he came to invent the Kaliapparat, why he did so using lamp-worked[9] glass, and what he accomplished using his new apparatus. This work was so intense that it drove Liebig into the grip of what his friend and fellow chemist Friedrich Wöhler called "analysis mania."[10] The Kaliapparat—a device that measured carbon content—did not solve the problem of organic analysis. Nor, despite Liebig's efforts, did it fulfill its intended purpose, which was to improve analytical determination of nitrogen. But as we shall see, the Kaliapparat not only established Liebig's career, it also enabled him to set organic chemistry on an entirely new course.

PARIS TO GIESSEN

The young Liebig was poor and unknown. Second son of a hardware merchant in Darmstadt, whose business included making the varnishes, paints, polishes, and pigments sold in his shop, Liebig became interested in chemistry at an early age. In 1822, aged just 19, he won state funds to travel to the French capital Paris on the recommendation of his chemistry professor, Karl Wilhelm Gottlob Kastner.[11] Paris, city of Antoine Laurent Lavoisier and his revolutionary chemistry of the elements, was at that time the undisputed center of chemical excellence in teaching as well as research.[12] While there, Liebig entered the laboratory of Joseph Louis Gay-Lussac, where he learned the state of the art in organic analysis, a method embodying the pinnacle of early nineteenth-century precision science.

Working under Gay-Lussac's direction, and with his apparatus, Liebig provided the first reliable analysis of silver fulminate, a highly explosive substance whose use in gunpowder and percussion caps made it a focus of common interest among chemists throughout Europe. When Gay-Lussac presented Liebig's results to the Parisian Academy in early 1824, they were justly admired.[13] Liebig was ideally positioned to launch a career in chemical research: in Paris, already contributing to chemistry's cutting edge, and rapidly coming to the notice of leaders in his field.

Within months of this Parisian triumph, Liebig traveled to Giessen to take up a position as extraordinary professor of chemistry in the university.[14] Lacking an independent income, Liebig was bound to comply with his patrons' wish that he return to his homeland. He was well aware that Giessen—a small market town in the rural German state of Hessen-Darmstadt—did not offer the advantages of metropolitan Paris, or even of his hometown, Darmstadt. "I have no great desire to go to Giessen," Liebig had written to his parents the previous year, "I would prefer a position in Darmstadt, in the medical college or in some other suitable place."[15]

Having tasted early success amid the intellectual excitement of Paris, Liebig's move to Giessen transformed his situation. This was the start of a period of desperate struggle for resources and for recognition. How to acquire an income, when his professorial appointment provided none? How to do research with only "100 Gulden for the laboratory and for buying instruments, reagents and materials?"[16] How to build his scientific reputation from somewhere that offered such "restricted opportunities" for chemistry?[17] How, in Liebig's own words, to "manage with so little?"[18] No wonder Liebig left Paris with such reluctance.

ADVERSARIAL ANALYSIS

Locked in a struggle to establish himself, Liebig looked for ways to overcome the disadvantages of his Giessen situation. Organic analysis—chemists' method for assigning molecular formulae to organic compounds—was evidently not an obvious way forward, since his research ranged widely during the 1820s.[19] In fact, Liebig's first independent foray into organic analysis ended badly, and he did not devote himself to this sphere until 1830. Learning how and why Liebig came to focus on organic analysis is a necessary precursor to understanding what drove him to invent the Kaliapparat.

Following his arrival in Giessen, Liebig continued to rely on work done in Paris as he attempted to develop an independent career. In 1825, he used his Parisian analysis of silver fulminate to challenge Friedrich Wöhler. Wöhler, a former pupil of Swedish master analyst Jöns Jacob Berzelius, was becoming a successful young chemist in the Prussian capital, Berlin. He had recently published an analysis of a quite different compound, silver cyanate. The problem was that Wöhler's formula for silver cyanate matched the formula Liebig had previously assigned to silver fulminate—something established chemical theory indicated was impossible. Liebig was convinced Wöhler must be wrong, and he published accordingly. A heated, public exchange between the two young men ensued. Their quarrel was noted as far away as Paris. Wöhler's mentor Berzelius intervened to resolve the dispute. In the end, the young chemists agreed it was possible for two distinct organic substances to have the same formula—a phenomenon Berzelius later called "isomerism."[20]

Liebig's challenge to Wöhler seems to me best understood as an ill-fated attempt to establish himself—if not in the eyes of elite Parisian chemists, then at least in Germany's leading chemical venue, Berlin. Liebig could certainly be intemperate.[21] Yet his archive reveals an ambitious and extraordinarily hard-working man, who was capable of friendship and loyalty. Following his reluctant removal from Paris, Liebig struggled to produce valuable results with limited resources. He was also very isolated, lacking a mentor to help him navigate the elite scientific world of his day. Given these circumstances, it is plausible that Liebig's desire for recognition caused him to overreach himself.

Following this episode, Liebig returned to Paris on several occasions. Perhaps initially drawn to seek Gay-Lussac's guidance, Liebig soon began using these trips to try to recreate Paris in Giessen. In 1826, he transported Parisian analytical apparatus to Giessen and began working with it. The next year, he taught his student Heinrich Buff how to use this apparatus so that Buff might aid him in his research.[22] In late 1828, when its glass components needed repair, Liebig traveled to Paris to take lessons in glassblowing in order to do this for himself.[23] And in 1829, he attempted to analyze uric acid (found in human urine) and hippuric acid (its equine analog) using Parisian apparatus.

This work marked an important watershed in Liebig's career and in his relationship to Parisian chemistry.[24] Both uric and hippuric acids contained nitrogen in addition to carbon, hydrogen, and oxygen. The presence of nitrogen considerably increased the difficulty of analysis. Especially in the case of hippuric acid, which contained very little nitrogen, Liebig now confronted some serious technical problems. In fact, hippuric acid contained so little nitrogen that its discoverer, the Parisian pharmacist Hilaire-Marin Rouelle, had first identified it with benzoic acid, which contains no nitrogen at all. Liebig soon concluded from analyses made using the Parisian apparatus that it was possible to achieve "no sharply defined result" for hippuric acid's small nitrogen content. This experience suggested to Liebig that there was a general problem with the Parisian method of measuring nitrogen.[25] As we shall see, this was how organic analysis—and especially the problem of nitrogen—became the focus of Liebig's research.

Just as when challenging Wöhler, Liebig began by drawing upon his Parisian experience. In analyzing silver fulminate (which also contains nitrogen), Liebig and Gay-Lussac had begun using a vacuum pump to remove air from the apparatus, together with atmospheric moisture that interfered with the analysis. Liebig now rationalized the difficulty of analyzing hippuric acid as a consequence of residual atmospheric nitrogen within the analytical apparatus. Because air is mainly nitrogen, any residual air would significantly increase how much nitrogen the analysis measured. The less nitrogen a substance contained, Liebig reasoned, the greater effect this contamination would have on the analysis. The solution was to remove air from the apparatus using a vacuum pump.

Liebig now made a fateful choice, one that would bring him into a new field of analysis and so into direct conflict with Paris—home to the world's most revered and best-equipped chemists. Seeking to demonstrate the importance of using a vacuum pump when analyzing substances containing little nitrogen, Liebig structured his paper on hippuric acid around a new analysis of the opium alkaloid, morphine.[26] Liebig chose morphine, because—like hippuric acid—it contained only a very small amount of nitrogen. This was a rational choice: as one of the most widely used drugs of the day, purified morphine was relatively easy to obtain from pharmacists. But it was also a high-stakes choice. The best previous analysis of morphine had been published in 1823 by Jean Baptiste Dumas and Pierre Joseph Pelletier.[27]

Pelletier was one of the most respected pharmaceutical chemists in Paris, codiscoverer of quinine and several other medicinally valuable alkaloids. Increasingly preoccupied with commercial quinine production, Pelletier had stepped back from cutting-edge research.[28] Dumas, meanwhile, was emerging as the rising star of Parisian chemistry, increasingly viewed as a worthy successor to Lavoisier.[29] Liebig's analysis, published in 1829,

indicated morphine contained about 10% less nitrogen than Dumas and
Pelletier had reported—a huge discrepancy, even in the context of contemporary analytical uncertainty. According to Liebig, Dumas and Pelletier's analysis of morphine was at the very least "unreliable."[30] This was
Liebig's first published alkaloid analysis. As I now show, it started a chain of
events that resulted in alkaloid analysis becoming Liebig's chosen route to
recognition and success.

ANALYZING ALKALOIDS

Liebig was convinced that Dumas and Pelletier's analysis of morphine was
flawed. Spotting an opportunity for advancement, he began gathering
resources to help him persuade other chemists this was so. His first gambit
was to call upon Wöhler. The two young chemists had resolved their previous differences and were now on friendly terms. Earlier that year, Wöhler
had approached Liebig suggesting they collaborate.[31] Liebig agreed to pursue an analytical project.[32] In fact, Wöhler—who found organic analysis
"extremely awkward"—provided only the sample and, when their first
joint publication appeared early in 1830, all the analyses were Liebig's.[33]
    Having helped Wöhler in this way, Liebig now sought reciprocal assis-
tance from his new friend. His initial goal was to undo the damage his
earlier quarrel with Wöhler had done to his reputation by undermining
Parisian chemists' perception that he and Wöhler "live[d] in open feud."
He therefore asked Wöhler to publish "any small piece of work" under their
joint names—a request that produced their second coauthored paper.[34] It
mattered deeply to Liebig what his "friends in Paris" thought.[35]
    Soon afterward, Liebig again approached Wöhler, this time proposing a
joint analytical study of the alkaloids. Now that he and Wöhler were publicly reconciled, Liebig was changing tack, seeking to enhance his chemical
credentials. Liebig was "actively convinced that, if little by little we . . . clarify important points in organic chemistry, we will thereby produce a revolution," and he had already selected a suitable target. "What do you think of
the organic bases [alkaloids]?" he asked Wöhler. "I scarcely doubt [i.e., am
almost certain] that the analyses of D[umas] and P[elletier] are incorrect."[36]
    Liebig failed to convince Wöhler to join him, and there are at least two
plausible reasons why. Wöhler's reply to Liebig indicates he regarded analyzing alkaloids as too difficult, and therefore unlikely to be productive—an

entirely reasonable position, given Wöhler's limited proficiency in organic analysis.[37] In other respects, however, Wöhler was more experienced than Liebig and certainly better connected. A couple of months later, after reading Liebig's investigation of hippuric acid, Wöhler not only privately agreed with his mentor Berzelius that Liebig deserved the motto "fast and sloppy," he thought such "negligent" work deserved a slap on the wrist (*handplagga*).[38] Thus, Liebig's hasty dismissal of Parisian chemists cast doubt on his analytical capability, leaving Wöhler unconvinced of Liebig's ability to tackle the problem of alkaloid analysis. To Wöhler and Berzelius, Liebig still had a great deal to learn.

Events were about to overtake Liebig in the form of a published rebuttal from Dumas. Dumas had spent the months since Liebig's hippuric acid paper appeared preparing his response. By May 1830, he was ready. Dumas's vigorous critique rested on a claim to superior analytical skill and experience, and it drew upon new analytical results. Rather than analyzing morphine again, however, Dumas instead used analyses of a quite different compound called oxamide to refute Liebig. The move to oxamide was significant. Whereas Liebig's criticism referred to the difficulty of determining morphine's small nitrogen content, oxamide contained nitrogen in much greater proportion. Dumas was using oxamide to shift the terms of the dispute.

Dumas claimed superior analytical experience, declaring with evident condescension that it was his "duty to explain the facts."[39] Dumas defended both the method he and Pelletier had employed and the results they had obtained. Where their results differed from Liebig's, Liebig was at fault. Dumas also disagreed with Liebig about what caused unreliable nitrogen determinations for hippuric acid and morphine. According to Dumas, the real problem was that analysis indicated too much carbon dioxide rather than too much nitrogen. Only inexperience led Liebig to conclude the opposite. As this example illustrates, the technical difficulties of organic analysis, especially where nitrogen was concerned, meant differences in practice and interpretation easily transmuted into attacks on personal credibility. This was a thorough, and very public, dismissal.

Dumas's censure targeted Liebig's vulnerability. A rough contemporary of Liebig's, Dumas had also traveled to Paris in the early 1820s to learn chemistry.[40] Since then, however, their paths had diverged. Where Liebig was struggling in isolation to build a career, Dumas had married a wealthy woman and was established at the heart of the Parisian chemical elite.[41]

Criticism from such a source was both a personal affront and a serious threat to Liebig's ambitions. Liebig responded by tackling the might of Paris head on. Only better alkaloid analyses could save his reputation.

TURNING POINT

Liebig was soon hard at work, preparing to analyze alkaloids—even without Wöhler. Liebig tested new analytical procedures, including some developed by Berzelius, and by August 1830, he was searching for novel instruments in hopes of solving what he called the "fatal" problem of determining the alkaloids' nitrogen content.[42] Whether Liebig had in mind the technical difficulty of this problem or the seriousness of his own situation, or both, we cannot be sure. In any case, he now—with Wöhler's mediation—took another supremely important step.[43] Liebig had recently learned from Wöhler that Berzelius planned to attend a conference in Hamburg in September 1830.[44] He had long wished to meet Berzelius. Now Liebig's mind was made up: Despite the cost to a poor Giessen professor, he would travel to Hamburg to meet Berzelius.[45]

Although Liebig may not have realized it, he was about to find a way of tackling the problem of alkaloid analysis and responding to Dumas—as well as acquiring a new mentor. The evidence is insufficient to show exactly why Liebig chose that moment to go to Hamburg to meet Berzelius—though Berzelius's reputation as a master of organic analysis is suggestive. Nor can we know the details of their conversation. We do know, however, that while in Hamburg, they discussed alkaloid analysis and that Liebig afterward declared himself "literally at peace."[46]

Equally suggestive reasoning indicates why Berzelius was able and willing to help the young Liebig. Three decades spent working in Sweden, on the periphery of northern Europe, meant that no one was better placed than Berzelius to understand Liebig's isolation and struggle. Like Liebig, Berzelius was used making do with limited resources. He was also no friend of Parisian chemistry.[47] He, too, had sought to engage Parisian analysts and been rebuffed.[48] His work introduced several important analytical innovations but nevertheless attracted stinging criticism from Paris.[49] Despite this, at least one of the innovations Berzelius introduced had been adopted by Gay-Lussac, apparently without acknowledgment.[50] Recruiting Liebig to his methods offered Berzelius a means to propagate his analytical approach in a new generation beyond Paris. Perhaps his protégé Wöhler's reluctance to work in this area helped convince Berzelius—despite his earlier criticism of Liebig's analytical practices—to give his support.[51]

One thing is certain. On returning to Giessen, Liebig developed a series of analytical techniques that were closely modeled on methods that Berzelius had previously pioneered. Within weeks of meeting Berzelius, Liebig's analytical approach was transformed. Until now, Parisian chemistry had been Liebig's touchstone. From here on, he would take his lead from Berzelius in challenging Paris.

THE STATE OF THE FIELD

Before rejoining Liebig in Giessen, it will be useful to outline the state of organic analysis, and particularly alkaloid analysis, in 1830. Learning how chemists used analytically derived composition and formula in their attempts to understand organic compounds like the alkaloids will explain why Liebig found the analyses by Dumas and Pelletier so problematic, and why—as we have seen—nitrogen was the crux of the matter. By the middle of 1830, most likely prompted by Dumas's recent rebuttal of his earlier analysis of morphine, Liebig urged nitrogen determination as the single most important problem confronting organic analysis.[52]

For roughly 40 years, chemists had been struggling to determine the composition of organic compounds—that is, how much of which elements they contained. Analysis showed that organic compounds were composed of just a handful of elements (carbon, hydrogen, oxygen, and sometimes nitrogen). Elemental analysis relied on combustion, which converted carbon to carbon dioxide gas and hydrogen to water vapor. Nitrogen, where present, also escaped in gaseous form. In principle, burning a sample of known mass and measuring these combustion products allowed the analyst to determine how much of each element the sample contained, expressed as percentage composition.[53] But in practice, the more organic compounds chemists studied, the harder they recognized this type of analysis to be.

Organic compounds could not be distinguished simply by which elements they contained, as was often the case in inorganic chemistry. Rather, the same few elements combine in minutely different proportions to make distinct organic compounds.[54] For analytical results to be useful, they had not only to be consistent for the same compound but also to differentiate reliably between distinct substances in which the same elements were combined in almost identical proportions. To put this important issue in the context of alkaloid chemistry, consider that Dumas and Pelletier found a difference of less than 1% between the carbon content of two alkaloids, each derived from tree bark: brucine from *nux vomica* and quinine from the *cinchona* tree.[55] The carbon content they assigned to morphine, meanwhile, was the average of two values separated by well over 1%.[56] Differentiating between alkaloids based on their percentage composition was pushing analysis to its limits.[57]

At the same time, Berzelius's demonstration that atomic theory applied to organic compounds was changing how analysis was done, reported, and interpreted. Atomic theory implied that organic compounds, like their inorganic counterparts, consisted of molecules containing a fixed number of atoms of each kind, expressed in a molecular formula (e.g., C_2H_6O for alcohol). Although many chemists continued to prioritize percentage composition over formula during the 1820s, it became increasingly common to publish formulae from about 1830 onward.[58]

Assigning formulae required measuring atomic and combining weights (today's atomic and molecular masses, respectively) as well as percentage composition. With this information, chemists could in principle calculate how many atoms of each kind a molecule contained.[59] In practice, however, such calculations were made highly insecure by the uncertainty of the underlying experimental data. Producing a formula was far from being a routine deductive process at this time. In 1823, Dumas and Pelletier reported the percentage composition and formula of every alkaloid they studied. These were the first published alkaloid formulae, including the formula for morphine that prompted Liebig's analytical project.[60]

Assigning a formula in addition to establishing composition had particularly significant consequences for alkaloid analysis. Because alkaloids like morphine are large molecules composed of relatively light atoms, small differences in percentage composition determined by analysis could have a noticeable effect on how many atoms the analyst assigned. When Dumas and Pelletier measured morphine's nitrogen content, for example, they found 5.53% nitrogen and assigned two nitrogen atoms in the formula ($C_{30}H_{40}N_2O_5$).[61] If Liebig's 1829 analysis was right and their nitrogen determination was 10% too high, it was entirely possible that Dumas and Pelletier had assigned the wrong number of nitrogen atoms to morphine.[62]

It mattered a great deal to chemists around 1830 how many atoms of nitrogen alkaloids like morphine contained. This was because, in addition to their powerful physiological action, many alkaloids also neutralized acids—something chemists attributed to the presence of nitrogen in the molecule. This led the Parisian pharmacist Pierre Jean Robiquet to speculate that nitrogen existed within alkaloid molecules in the form of ammonia, a simple compound of nitrogen and hydrogen that neutralized acids. Thus, the quantity of acid an alkaloid neutralized should be proportional to the number of nitrogen atoms it contained.[63]

Robiquet's idea was attractive, because it opened the possibility of explaining the alkaloids' medicinal activity as well as their chemical behavior. The problem was that chemists could not agree on the numerical relationship between number of nitrogen atoms and quantity of acid neutralized. Dumas and Pelletier rejected Robiquet's proposal outright, because their analyses refuted the simple proportionality it implied.[64] But Liebig, because he doubted Dumas and Pelletier's results, saw this matter as unresolved and therefore worthy of further study.

By late summer 1830, Liebig's technical dispute with Dumas over the alkaloids' nitrogen content had brought him to the heart of disciplinary debates concerning the methods, apparatus, results, and interpretation of organic analysis. On one side was Paris, the dominant tradition in which both Liebig and Dumas had trained. On the other was Berzelius, whose preference for simple methods and reliance on homemade apparatus had led him to develop a substantially different analytical approach.

Parisian chemists prioritized the pursuit of maximum precision in measuring percentage composition—which, in Paris at this time, meant measuring gases by volume.[65] This was why Gay-Lussac and Thenard made organic analysis into a high-precision science based on measuring the volume of gaseous combustion products collected over mercury and under glass. Only precise measurements of gas volumes, they believed, would permit chemists to determine reliable composition and formulae for organic compounds. This was the Parisian analytical tradition, and it was decidedly volumetric.

This legacy explains why Dumas and Pelletier measured the alkaloids' carbon and nitrogen content by volume. To do this, they collected carbon dioxide and nitrogen gases together and measured the volume of this mixture before introducing a substance (potassium hydroxide) that absorbed the carbon dioxide, leaving nitrogen as a residue. Measuring this residual volume of nitrogen enabled them to calculate the volume of carbon dioxide removed, and hence to calculate the ratio between the volumes of nitrogen and carbon dioxide. For morphine, this ratio was roughly 1 part nitrogen to 30 parts carbon dioxide, which represented the proportions of nitrogen and carbon in morphine.[66]

In fact, this bare ratio was the only experimental result Dumas and Pelletier published for morphine—something that greatly annoyed Liebig, because it made it difficult to evaluate their analyses.[67] Already in 1829, Liebig realized the extremely small proportion of nitrogen in morphine meant that any inaccuracy had a very much larger effect on the percentage nitrogen determined by analysis than it did on carbon content.[68] As we shall see, this differential effect on carbon and nitrogen would be crucial to Liebig's 1830 alkaloid analyses. He would also learn that errors in the percentage of nitrogen determined by analysis had much less effect on the number of nitrogen atoms assigned in the formula than at first appeared to be the case.

Berzelius, meanwhile, had for well over a decade been a solitary challenger of Parisian analytical supremacy based on volumetric measurement. To Berzelius, consistent measurements were more important than absolute analytical precision—provided, of course, the results enabled the analyst to assign the number of atoms with confidence, and hence to produce a reliable formula. This commitment to replicability meant that the results should be very similar, regardless of who did the experiment, and it led Berzelius to prefer simple experiments and apparatus. Because he found it easier to weigh things than to measure gas volumes, Berzelius measured the products of combustion by mass and not by volume—called "gravimetric measurement."[69]

Berzelius's endorsement of gravimetric rather than volumetric measurement would be crucial to Liebig. So, too, was another strategy Berzelius developed to improve the reliability of measuring the hydrogen content of organic compounds containing no nitrogen. Previously, analysts had measured hydrogen (as water) in connection with carbon dioxide. In 1814, Berzelius pioneered a direct, independent measurement of each element.[70] His innovation produced more reliable results, especially for hydrogen, and it was soon adopted by Parisian analysts.

Chemists in Paris acknowledged the advantage of measuring carbon and hydrogen separately. When it came to compounds containing nitrogen, however, they persisted in measuring nitrogen and carbon dioxide in a single experiment. This approach necessarily linked the results for nitrogen and carbon. This was how Dumas and Pelletier measured nitrogen, and it was the method that Liebig had learned from Gay-Lussac. By the time he met Berzelius in Hamburg, however, Liebig's early attempts to analyze alkaloids meant that he was looking for something new.

IN LIEBIG'S LABORATORY

Let us now rejoin Liebig in Giessen on his return from Hamburg in late September 1830. Liebig immediately resumed analyzing alkaloids. In addition to morphine, he also analyzed brucine, strychnine, quinine, cinchonine, and narcotine—all medicinally important alkaloids that Dumas and Pelletier had studied in 1823.[71] Liebig's goal was to improve nitrogen determination and hence to produce new formulae for the alkaloids that were better than those of Dumas and Pelletier. Adopting the strategy Berzelius had introduced for hydrogen and carbon, Liebig planned to separate the measurement of carbon and nitrogen, launching a two-pronged initiative involving the search for new methods for determining each element.[72] Although his overall purpose concerned nitrogen, Liebig almost simultaneously sought new ways of measuring both carbon and nitrogen. For the sake of clarity, however, I begin with carbon and the story of the Kaliapparat before continuing to Liebig's attempts to develop a direct, independent measure of nitrogen based on using Parisian volumetric apparatus. In both cases, Liebig's efforts were fraught with difficulty.

The remainder of this chapter follows Liebig inside the laboratory in order to understand how he approached this demanding work. Liebig's 1830 notebook was crucial to the new insights on which this account is built. Although historians have known about Liebig's notebooks for decades, they have been little used—despite increasing recognition that such rare manuscript sources provide invaluable access to scientists at work.[73] Somewhat ironically, one of the leading proponents of notebook studies, historian of chemistry Frederic Holmes, may be partly responsible for this neglect. In 1987, Holmes concluded that Liebig's few, "rather unsystematic notebooks" probably did not provide "sufficient detail and coverage to reconstruct an unbroken research trail."[74]

Holmes urged a completeness that Liebig's surviving archive could not support. The only other study to use these sources—by chemist Melvyn Usselman and historian Alan Rocke—had different objectives. Usselman and Rocke used their survey of Liebig's 1830 notebook, focused on Liebig's sketch of a prototype Kaliapparat (folio 55a verso: hereafter 55av; see figure 1.1), to outline a timeline for the invention of the Kaliapparat.[75] As we shall see, Liebig's sketch was one of the notebook's few readily comprehensible entries.[76]

I sought to make sense of Liebig's alkaloid analyses through a detailed reading of his 1830 notebook that elucidated exactly how he tackled just one alkaloid, morphine. Morphine was the first alkaloid Liebig analyzed and it was the easiest to obtain.[77] It was also the alkaloid to which he repeatedly returned during the final months of 1830. Highlighting a series of key moments in Liebig's struggle with analyzing morphine allows me to explain what was at stake for Liebig in this work. My study of Liebig's notebook changes almost everything we think we know about Liebig's goals during this period, and what he in fact accomplished using the Kaliapparat. Exposing Liebig's struggle and the origins of his "analysis mania" significantly alters our understanding of this crucial early phase in Liebig's career.

DEVELOPING THE KALIAPPARAT

Almost immediately on his return from Hamburg, Liebig invented the Kaliapparat, a small piece of glassware that trapped the carbon dioxide gas released when the sample burned (figure 1.1). By weighing the Kaliapparat before and after analysis, Liebig obtained a direct measure of the mass of carbon dioxide produced by combustion, and hence of the mass of carbon in the sample—much as Berzelius had previously pioneered direct measurement of hydrogen by mass.[78] Measuring carbon by mass meant that Liebig could burn a much larger sample than possible in Parisian analysis (about 1 g compared to 100 mg). Crucially, because the Kaliapparat measured only carbon dioxide, it enabled Liebig to determine carbon independently of nitrogen—something volumetric analysis could not do.[79]

Liebig's initial goal was to establish that the Kaliapparat produced similarly reliable values for carbon as the existing volumetric method. But when Liebig began using a prototype Kaliapparat in early October 1830, his first recorded measurement of morphine's carbon content (55%, 43v) was much smaller than the best existing values—which, in 1830, were still Dumas and Pelletier's (72.02%).[80] Since this result was clearly far too low, Liebig now

(a)

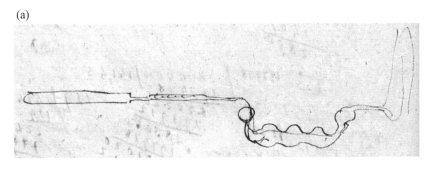

(b)

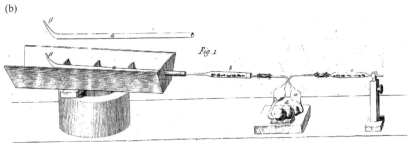

FIGURE 1.1

(a) Sketch of a prototype Kaliapparat from Liebig's 1830 notebook; (b) an illustration of the final apparatus from Liebig's 1831 published paper. Both diagrams show the Kaliapparat within the combustion train. The organic substance under investigation was introduced into the combustion tube (labeled "a" in figure 1.1(b)) together with an oxidizing agent to ensure complete combustion. When the sample burned, it produced water vapor and carbon dioxide gas. A glass tube containing calcium chloride (labeled "b") trapped the water vapor. Acidic carbon dioxide, meanwhile, passed into a Kaliapparat containing alkaline potassium hydroxide solution, where it dissolved. By weighing the Kaliapparat before and after combustion, the analyst could calculate the mass of carbon dioxide. The difference between prototype and final versions had two benefits. Most importantly, the final triangular shape slowed gas flow through the liquid, maximizing absorption of carbon dioxide. The triangular Kaliapparat could also be suspended from the hook of an analytical balance, making it much easier to weigh.

Source: (a) Bayerische Staatsbibliothek München, Signatur, Liebigiana I. C. 1, fol. 55av. Courtesy of the Bavarian State Library, Munich; (b) Justus Liebig, "Ueber einen neuen Apparat zur Analyse organischer Körper, und über die Zusammensetzung einiger organischen Substanzen," *Annalen der Physik und Chemie* 97 (1831): Tafel I, Fig. 1. Courtesy of the Bibliothèque National de France (https://gallica.bnf.fr/ark:/12148 /bpt6k15106q/f637.item).

worked to improve his prototype Kaliapparat, obtaining rapidly increasing values for carbon.

Because Liebig believed Dumas and Pelletier's method produced reliable measures of carbon, he had a valuable indicator of when his device was working properly. In 1823, Dumas and Pelletier had calculated the percentage carbon in morphine as the average (72.02%) of two distinct values (71.38% and 72.68%).[81] Their results therefore defined a range of acceptable values for morphine's carbon content. As he continued experimenting with the Kaliapparat, Liebig soon recorded a result that fell within this range (71.4%, 44v). This moment was surely important for Liebig. For the first time, his Kaliapparat had produced a reliable value for carbon.

We may be fairly certain that this is what Liebig was doing because of what happened next. Immediately after obtaining these results, Liebig recorded in his notebook two markedly lower values for carbon (68.5%, 69.1%, 45v) produced using the Kaliapparat (figure 1.2). This change in the direction of his results seems to have unsettled Liebig. Perhaps wondering whether his sample of morphine in fact contained less carbon than Dumas and Pelletier's, Liebig analyzed morphine again, this time using the established Parisian volumetric method. The result (72.0%, 46v) was right in the middle of Dumas and Pelletier's given values, which confirmed the Kaliapparat was still recording values for carbon that were too low.[82]

Liebig set to work with the Kaliapparat again. Since his three further analyses increased the carbon content, we must assume that he modified his method and, possibly, the apparatus. Of this most recent set of results, the last two (56v and 68v) produced remarkably similar values for carbon (calculated on 63v and 69r as 72.38% and 72.3%, respectively). Both these values lay in the upper part of the range previously defined by Dumas and Pelletier. For reasons examined later in this chapter under the section "Skating on Thin Ice," Liebig published only these two results for morphine's carbon content in January 1831. The average of these figures somewhat exceeded that of Dumas and Pelletier, but Liebig's results nevertheless lay well within the range defined by their 1823 analyses. Although Liebig had not surpassed them, his new apparatus had at last produced carbon determinations that he considered to be as good as those of Dumas and Pelletier. Following weeks of work, Liebig's struggle to measure carbon independently of nitrogen was at an end.

LIEBIG'S STRUGGLE WITH NITROGEN

Liebig, meanwhile, had also been searching for a direct, independent measure of morphine's nitrogen content. Following Berzelius's strategy of measuring each element separately, he intended to design a procedure that measured only nitrogen.[83] As with carbon, Liebig anticipated that measuring morphine's small nitrogen content independently of carbon would improve reliability. But Liebig did not develop another new apparatus, nor did he attempt to measure nitrogen by mass. In fact, his innovations were largely procedural modifications of the Parisian volumetric approach described above. Close attention to exactly how Liebig measured nitrogen is therefore necessary if we are to understand the trajectory of this work.

The key distinction between Liebig's method and the Parisian approach lay in the results he recorded. He followed the Parisians in collecting nitrogen and carbon dioxide gases together. Rather than measuring this combined volume, however, Liebig removed carbon dioxide before recording the residual volume of nitrogen. Although he measured nitrogen by volume rather than by mass, Liebig measured nitrogen separately from carbon in a process he made as similar as possible to measuring carbon dioxide using the Kaliapparat.

Liebig used this method—which he termed "direct," because the measured value related only to nitrogen—to tackle suspected problems with nitrogen measurement.[84] He believed some nitrogen eluded measurement because it did not reach the receiver, remaining elsewhere inside the analytical apparatus. In the standard experiment, which measured both carbon and nitrogen, the potassium hydroxide used to absorb carbon dioxide had to be introduced into the collected mixture of carbon dioxide and nitrogen gases after combustion. Since Liebig was not measuring carbon, potassium hydroxide could be present at the outset. He therefore incorporated a small glass vessel containing potassium hydroxide solution inside the apparatus, using this liquid to flush the interior once combustion was complete, in an attempt to force nitrogen gas into the collecting vessel.[85]

Ultimately, Liebig recommended using potassium hydroxide in this way when measuring nitrogen. But he also experimented with a variation of this method that warrants particular mention here, because it produced results he chose to publish (4.51%, 44v; 4.99%, 45v).[86] Instead of potassium hydroxide, Liebig substituted ammonium hydroxide, introduced before

FIGURE 1.2

(a) This page from Liebig's 1830 notebook records the raw results of some of his analyses of morphine, showing how he matched raw results to percentage composition. It reports results of standard gravimetric hydrogen determination; two determinations of carbon using the Kaliapparat; a volumetric determination of nitrogen. The form and units (cc vs. g) of Liebig's results for nitrogen indicate that he obtained the volume of nitrogen by using ammonia to absorb carbon dioxide. This is one of three published results for morphine's nitrogen content (discussed later in this chapter under the section "Skating on Thin Ice") and the only one Liebig used to fix his 1831 formula for morphine: $C_{34}H_{36}N_2O_6$.

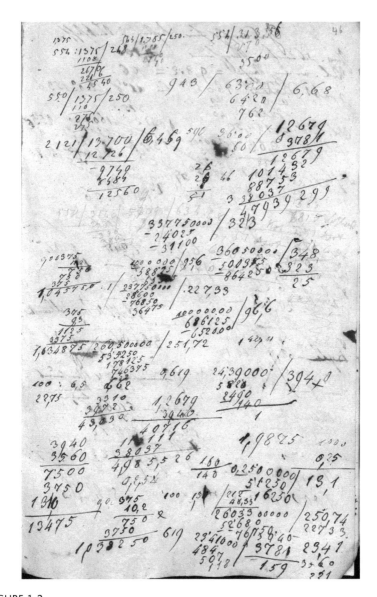

FIGURE 1.2

(b) The facing page shows the detailed calculations that Liebig performed in order to produce percentage composition.

Source: Bayerische Staatsbibliothek München, Signatur, Liebigiana I. C. 1, fol. 45v/46r. Courtesy of the Bavarian State Library, Munich.

combustion in just the same way. He soon learned that this method gave unpredictable results, dismissing it as "fruitless."[87] But as we shall see below, this did not prevent Liebig from using the results it produced in fixing morphine's formula.

These morphine analyses showed Liebig the significance of another problem with nitrogen determination—one previously recognized in Paris and which applied to any method based on combustion.[88] Having highlighted a role for residual atmospheric nitrogen in producing results for nitrogen that were too high, Liebig now acknowledged a distinct problem with exactly the opposite effect. One of the underlying chemical assumptions of organic analysis was that combustion released nitrogen in elemental form. As Liebig now realized, this was far from being the case.

In fact, the conditions of combustion analysis promoted unwanted side reactions of nitrogen. These side reactions produced gaseous products that were chemically similar to carbon dioxide. Potassium hydroxide therefore removed them (together with carbon dioxide) from the gas mixture collected on combustion. As a result, the measured residual volume of nitrogen was falsely low.[89] Although Liebig's observations were based on the modified method in which he measured nitrogen separately from carbon, the same problem affected standard Parisian analysis in a very similar way.[90]

Where side reactions occurred, combustion analysis always produced values for nitrogen that were too low.[91] The analyst could sometimes detect these side reactions, because the products appeared as brown fumes inside the analytical apparatus. But especially for compounds like morphine that contained so little nitrogen, this telltale sign was easily overlooked. Precisely where its effects were most damaging, the problem of side reactions was hardest to establish.[92]

Liebig responded to this problem in three ways. One involved adjusting the conditions of analysis to try to prevent the problematic side reactions of nitrogen from occurring in the first place. Liebig also recommended many repetitions of the nitrogen determination. Finally, he insisted that the analyst should not simply average these results—as was then standard analytical practice. Rather, Liebig advised using only those results "in which a completely colorless gas was obtained." Liebig claimed data selection enabled him to determine nitrogen "with greater accuracy" than could be attained using the standard method.[93] In fact, however, he obtained only one publishable result this way (3.97%, 69v) and did not use this when assigning morphine's formula.[94]

Although Liebig recommended this method of nitrogen determination in his 1831 paper, his endorsement was highly equivocal. It was, he said, merely "the least bad" among all those methods he had tried.[95] In private, Liebig was even more scathing. As he admitted to Berzelius, his method of nitrogen determination was "tiresome, time-consuming and, in a word, quite unbearable," while nitrogen's side reactions had driven him to "despair."[96] By the end of 1830, Liebig knew he had failed to come up with a method of nitrogen determination that surpassed Dumas's. After all his efforts, this outcome was surely dispiriting.

CHALLENGING DUMAS

Having reached this impasse with nitrogen, Liebig made another strategic move that combined his approaches to carbon and nitrogen. He had already begun using the independent measure of carbon provided by the Kaliapparat to evaluate results for nitrogen produced by Dumas and Pelletier's volumetric method (56v, 68v). Having failed to find a better method for nitrogen determination, Liebig now set out to use the Kaliapparat to expose the flaws in their method. Dumas's response to Liebig's previous criticisms indicated that Dumas did not see anything problematic in the way he and Pelletier had measured morphine's nitrogen content. But Liebig certainly did, and he was determined to convince others that he was right.

As we have seen, Liebig's fundamental problem with Dumas and Pelletier's volumetric method was that they determined both nitrogen and carbon from linked measurements made in a single experiment. Thus—for a given combined volume of carbon dioxide and nitrogen, as measured in the volumetric analysis—any erroneous increase in the volume of one gas inevitably decreased the volume of the other. Because the volume of nitrogen was so much smaller than that of carbon dioxide, any error in how the total volume of the mixture was split between carbon dioxide and nitrogen had a very much greater effect on the nitrogen measurement than on that for carbon dioxide. This gave rise to a significant problem with Dumas and Pelletier's nitrogen determinations that did not noticeably affect carbon.[97]

Liebig wished to demonstrate that the link between nitrogen and carbon, combined with the nitrogen's tendency to undergo side reactions during combustion analysis, invalidated Dumas and Pelletier's nitrogen determinations. His strategy was to compare the total gas volume measured by their standard volumetric method with the same volume calculated from a carbon determination made using the Kaliapparat. His notebook contains more than one such calculation (beginning at 57r), but these entries are more than usually difficult to sequence reliably because of extensive overwriting and revision. For the sake of clarity, what follows is an explanation of the version of this calculation Liebig published in January 1831, linked to the relevant notebook entries.[98]

This is how Liebig's calculation worked. Using his largest carbon determination made with the Kaliapparat (72.38%, 56v), Liebig converted the mass of carbon dioxide to a volume, which he then scaled to take account of the much smaller sample burned in volumetric analysis. In this way, Liebig used a result obtained with the Kaliapparat to calculate the volume of carbon dioxide that volumetric analysis *should* measure (131.86 cc, 69r). By adding this calculated volume of carbon dioxide to an average volume of nitrogen determined by his method of direct, volumetric measurement (3.608 cc), Liebig produced a calculated combined volume of carbon dioxide plus nitrogen (135.468 cc).[99]

When Liebig compared this calculated volume to the total volume of carbon dioxide and nitrogen gases actually measured in the standard volumetric analysis (Liebig averaged three results to give 132.071 cc), however, he found a significant discrepancy.[100] According to Liebig's calculation, volumetric analysis measured a lot less gas than it should. Liebig used this outcome to argue that the existing volumetric method routinely produced a total gas volume that was too low, and hence a residual volume of nitrogen gas that was far too small. As a result, Liebig claimed, "all the mistakes that are made in the qualitative [volumetric, Parisian] analysis, as well as all observational errors, fall on the calculation of nitrogen."[101] As discussed above, this concern prompted Liebig to propose several modifications to the standard method of nitrogen determination.

As Liebig presented it in January 1831, this argument was empirically somewhat insecure, and it certainly relied on an intrinsic circularity. Liebig did not make clear how he had obtained the experimental results manipulated in this calculation. Nor did he explain that the average volume of nitrogen was itself an outcome of a method very similar to the one he was criticizing. By using the Kaliapparat in this way, however, Liebig developed his technical dispute with Dumas along lines that capitalized on his own growing expertise. By the end of 1830, Liebig had more experience analyzing alkaloids like morphine than any other chemist had. Even though he was unable to produce better results for nitrogen than Dumas and Pelletier, Liebig used measurements of carbon made using the Kaliapparat to claim superiority in this field.

SKATING ON THIN ICE

To complete and publish this work, Liebig needed to assign formulae to the alkaloids he had analyzed. By 1830, formulae were a required output of analysis, essential to Liebig's goal of superseding Dumas and Pelletier's analyses and also necessary for him to comment on Robiquet's proposal concerning nitrogen's role in the ability of alkaloids to react with acids. Understanding the process by which he assigned morphine's formula reveals that by this time, Liebig prioritized formula over composition as the most significant output of analysis, a shift that changed his style of laboratory reasoning. Comparing private and published records also indicates that producing a formula for morphine that would compare favorably with Dumas and Pelletier's pushed Liebig to the edge of—and perhaps somewhat beyond—the limits of contemporary analytical possibility. Even with the Kaliapparat, Liebig struggled to improve on previous formulae.

It is important to recall that assigning formulae in 1830 required skillful interpretation of uncertain experimental results. Liebig's Kaliapparat did not remove the need for this expertise. Liebig used atomic weights published by other chemists, but even these remained unstable during this period.[102] He measured morphine's molecular mass himself, recording four such measurements in 1830 (beginning at 50v). As soon as he had all the results he needed, Liebig began the process of fitting a formula for morphine (54r). Heavy revisions in Liebig's notebook show that he frequently adjusted analytical calculations, and hence the formulae he assigned, in accordance with variations in the underlying data.[103]

Liebig's early attempts to produce a formula for morphine failed, because he was unable to reconcile his analytical results with the measured molecular mass (50v, 53v, 54r).[104] Once Liebig measured carbon at over 72.3% (56v, 68v), however, he succeeded in producing several possible formulae for morphine, one of which he would publish ($C_{34}N_2H_{36}O_6$, equivalent to

$C_{17}H_{18}NO_3$ today).[105] In tracing how Liebig produced this formula, I shall not address how he measured hydrogen and calculated oxygen. Nor do I examine how Liebig evaluated the relative stability of the various kinds of data on which this formula rested.[106] Instead, what follows continues to focus on how Liebig used his analytical results for carbon and nitrogen to arrive at the number of atoms of each element in the morphine molecule.

Liebig began by selecting for publication only the most reliable results from among all the analyses he had done.[107] In the case of carbon, for example, Liebig knew the Kaliapparat tended to produce values for carbon that were too low, rather than too high. Liebig therefore published only two results (72.3%, 56v and 72.38%, 68v)—those that had produced the largest percentage of carbon.[108] Although these two results for carbon were very similar, Liebig did not average them to produce a mean value for use in fixing the number of carbon atoms in morphine. Instead, having used his raw data to assign the most plausible formula and so to identify the required number of carbon atoms in the molecule, Liebig then selected only the result that most closely agreed with the theoretical percentage of carbon this formula required. This iterative interplay between experiment and established theory is exemplary of what I mean by laboratory reasoning. In this case, it led Liebig to base a new morphine formula containing 34 carbon atoms (equivalent to 17 atoms today) on only the higher value (72.38%).

Laboratory reasoning guided similar selections when it came to nitrogen. As we have seen, he learned to discard nitrogen determinations in which telltale brown fumes indicated unwanted side reactions had taken place.[109] It seems probable that this was why Liebig rejected all but three measurements of morphine's nitrogen content (4.51%, 44v; 4.99%, 45v; and 3.97%, 69v). Just as with carbon, Liebig selected a single nitrogen determination (4.99%, 45v) from among these published values as the basis for his proposed formula. As Liebig explained, this value was significantly higher than the other two published results.[110] It also gave by far the best fit with the theoretical value that his assigned formula required. In choosing this result, however, Liebig's desire to present his analytical results in the best possible light caused him to mislead his readers.

When Liebig published these three results for nitrogen in January 1831, he gave no indication of how they had been obtained—despite having used several different methods for measuring nitrogen during this work. Anyone reading his paper would therefore make the reasonable inference that all three results had been produced by Liebig's recommended method. In fact, this was not the case. Only one of these results was produced by Liebig's recommended method (3.97%, 69v), while the other two (4.51%, 44v; 4.99%, 45v) had been obtained using the method he claimed to have discarded as unreliable. The value Liebig used in fixing morphine's formula was therefore almost 1% higher than the only published value produced by his recommended method (3.97%, 69v).[111]

It is no surprise to find that Liebig's approach to assigning a formula for morphine relied heavily on data selection.[112] Indeed, it is perhaps a mark of the extreme difficulty of determining morphine's nitrogen content that Liebig entirely suppressed the highly dubious empirical foundations on which his new formula rested—at the same time as he extolled the virtues of reporting raw analytical data. This account suggests that both experiment and established theory can be justifiable grounds for selecting data. Liebig's lack of candor in this instance nevertheless seems beyond the pale.[113]

In the case of morphine, it is also worth noting that although Liebig's nitrogen determination differed significantly from Dumas and Pelletier's (4.99% cf. 5.53%), the formula he ultimately assigned contained an identical number of nitrogen atoms (two). Despite his conviction that Dumas and Pelletier's method of measuring the alkaloids' nitrogen content was flawed, Liebig had learned that the number of nitrogen atoms in the formula depended on more than the analytically determined percentage composition.

This distinction between analytical results and formulae had important consequences for Liebig. Although in general his nitrogen determinations were rather similar to Dumas and Pelletier's, Liebig nevertheless assigned different numbers of nitrogen atoms than Dumas and Pelletier had in the formulae of several other alkaloids. Yet, in the end, Liebig was dismayed to realize he could come to no decisive conclusion concerning Robiquet's proposal. Although the alkaloids' ability to react with acids seemed "linked to the presence of nitrogen," it was impossible to say whether nitrogen was present as ammonia or in some other form.[114]

Toward the end of 1830, Liebig was in a tricky situation. He had already criticized Dumas and Pelletier's morphine analysis in print, and he had discussed his intention to pursue this criticism with both Wöhler and Berzelius. Yet Liebig had in fact failed to resolve the problem of measuring the alkaloids' small nitrogen content and, as we have seen, his formula for morphine was based on rather insecure experimental evidence. Despite intensive investigation and all his work in developing the Kaliapparat, Liebig had not accomplished his original purpose.[115] At the outset, Liebig had been convinced that Dumas and Pelletier's alkaloid analyses were deeply flawed. In fact, his own analytical results were remarkably similar.[116] Nor, finally, could Liebig point to any decisive theoretical intervention as justification for publication.

Far from being poised on the brink of success, Liebig now confronted the difficult task of managing a catalog of disappointments. Liebig's career ambitions meant he was nevertheless under considerable pressure to publish. It was surely unthinkable that so much effort might not lead to a significant publication. Reading Liebig's notebook, however, indicates only one genuinely successful aspect of this work: Liebig had invented a new piece of analytical apparatus that produced results for carbon every bit as good as those of Dumas and Pelletier in a fraction of the time, and with far less effort and expense. In reviewing the outcome of 3 months' exhausting and frustrating struggle, Liebig evidently arrived at the same conclusion as this study has. When he published these alkaloid analyses in January 1831, Liebig called his paper "On a New Apparatus."

LIEBIG'S ACCOMPLISHMENT

Even after leaving the French capital for Giessen in 1824, Liebig continued at first to define his career in relation to Paris, seeking to build his reputation by adopting Parisian analytical methods. That situation changed in 1829, when analyzing morphine prompted Liebig to challenge Paris. Understanding this reversal explains how Liebig became embroiled in conflict with Dumas, and why their early disputes centered on analyzing alkaloids. Liebig's struggle with Dumas—and with Paris—was fundamental to his ambitions, and it provides the crucial and hitherto missing context for Liebig's meeting with Berzelius. This was Liebig's first direct exposure to a different chemical tradition, in which small, cheap apparatus displaced the costly equipment that enabled Parisian chemists' achievements. Immediately following this meeting in August 1830, Liebig invented the Kaliapparat.

One issue in particular has obscured Liebig's reasons for inventing the Kaliapparat. Historical focus on the ultimate consequences of Liebig's alkaloid analyses rather than on his original purpose in tackling this problem has made it difficult to appreciate what role Liebig originally intended the

Kaliapparat to fulfil. Although his new device measured carbon, and this eventually became its primary function, Liebig introduced the Kaliapparat as part of an analytical strategy intended initially to supersede—and later to undermine—the Parisian approach to *nitrogen*.

This conflation of function and purpose has resulted in neglect of the disputatious nature of Liebig's work on the alkaloids. By emphasizing a retrospectively assumed success, scholars have done less than justice to Liebig's struggle—with Dumas and with the technical difficulties of organic analysis ca. 1830—making it impossible to grasp what was at stake in this work. Those stakes were high, not just for Liebig personally, but also in the larger contest between the established resources of elite Parisian chemists and a newer, more materially accessible vision of how chemistry should be done.

This study of Liebig's 1830 notebook suggests several revisions in the historiography of nineteenth-century chemistry. The empirical insecurity surrounding Liebig's development and use of the Kaliapparat undermines the prevalent historical assumption that the Kaliapparat solved both the specific problem of analyzing alkaloids and the more general problem of organic analysis. As we have seen, Liebig relied on skillful negotiation between experiment and theory—his approach to laboratory reasoning—to produce useful analytical results, even with his new device. In fact, Liebig would continue to grapple with the problems of organic analysis for at least another decade, and he regarded the alkaloids as beyond the reliable limits of analysis throughout this period.[117]

In January 1831, Liebig published new formulae for morphine and the other alkaloids he had studied. Notwithstanding the fragility of his evidence, this was a significant achievement that helped enable a molecular understanding of these medicinally important and chemically interesting substances. Liebig was pushing the boundaries of applicability of Berzelian formulae and the atomistic conception of matter such formulae embodied. This makes it particularly important to appreciate that Liebig's formulae were plausible rather than definitive, becoming merely some among the many possibilities whose correctness was debated by mid-nineteenth-century chemists.[118]

In fact, although recent work has praised the accuracy of Liebig's formulae for morphine and other alkaloids, this aspect of Liebig's work was largely ignored by his contemporaries.[119] Even Berzelius, who—having helped Liebig pursue this project—took far more interest than did anyone else, was not persuaded by Liebig's formulae. On the contrary, Berzelius privately pointed out several errors in Liebig's published analytical results, suggesting alternative formulae he found preferable. Over the coming months, Berzelius repeatedly encouraged Liebig to seek more convincing evidence concerning the alkaloids' composition and formulae.[120] It is surely a mark of his failure to solve the problem of analyzing alkaloids that Liebig could not be induced to do so. "These experiments," he complained to Berzelius, "were so laborious that they drove me to despair."[121] Unable by now to decipher his own laboratory notebook, Liebig would not return to alkaloid analysis for almost a decade.[122]

Nitrogen was largely responsible for Liebig's despair. He certainly had not solved the problem of measuring the alkaloids' nitrogen content.[123] Nor was the method he proposed adopted by other chemists. In fact, it was Dumas—rather than Liebig—who introduced what became for the remainder of the decade the accepted method of measuring nitrogen.[124] Throughout the 1830s, even as their professional situations relative to each other changed, nitrogen determination continued to be a focus of ongoing, heated analytical disputes between Liebig and Dumas.

Although it reported work that was far from consistently successful, Liebig's 1831 paper marked a turning point in his career. The Kaliapparat indeed attracted the attention and admiration of other chemists. But this was not because it measured carbon by mass.[125] Nor did it improve the accuracy of carbon determination.[126] As Liebig himself explained in 1831, "Nothing is new about this apparatus apart from its simplicity, and the complete reliability it offers."[127]

This reference to reliability—together with Liebig's assertion, elsewhere in the same paper, of "the higher degree of precision" his device enabled—helps explain the persistent assumption that Liebig's Kaliapparat improved the accuracy of analytical results.[128] Reliability and precision, however, do not imply accuracy. Instead, Liebig's claims indicate that the Kaliapparat produced more consistent results than the volumetric method it displaced—an improvement that would prove crucial in enabling Liebig to build a school around his new device.

At this stage, the Kaliapparat's most significant virtues were speed and simplicity. Berzelius, despite his other concerns, was full of admiration for Liebig's productivity. "I find it utterly inconceivable," he marveled, "how you have been able to carry out all these things in such a short time."[129]

Liebig's good friend Wöhler was finally persuaded to give organic analysis a try, becoming a competent organic analyst during 2 weeks in Giessen in late 1831.[130] Even Liebig's rival Dumas conceded that the Kaliapparat was "destined without any doubt to change the state of organic chemistry within a very short time."[131] By early 1832, presumably persuaded by its ease of use, Dumas was using Liebig's Kaliapparat regularly and "with full success."[132]

Like Dumas, other chemists recognized the utility of Liebig's new analytical device while remaining indifferent to his alkaloid formulae and skeptical about his method of measuring nitrogen.[133] The Kaliapparat changed the course of Liebig's career. It succeeded because, by transforming skill and scale, it made possible a new kind of chemical experimentation. Adopted by chemists across Europe over the coming few years, Liebig himself no longer mentioned it in papers published after 1833. As we shall see, the Kaliapparat also introduced a revolutionary new relationship between maker and instrument, teacher and student, and chemist and chemical community that would ultimately liberate chemistry from the dominance of Paris. How it did so is the subject of the next chapter.

2   SURE REAGENT

The 1830s were an exciting decade for Justus Liebig. As chemists across Europe heard about the Kaliapparat, they flocked to Giessen to learn how to use it. Liebig was quickly recognized as one of the greatest chemists of his day, and by about 1840, his laboratory became the world's foremost center of chemical training. Liebig's Giessen research school is widely credited with putting organic chemistry on a proper scientific footing. According to standard accounts, two connected developments were responsible: first, the Kaliapparat; and second, the reorganization of chemical training and laboratory labor this instrument enabled. The Kaliapparat made analysis so routine that even students could easily assign reliable formulae to organic compounds, while the new laboratory economy enormously multiplied Liebig's research productivity.[1]

This received view is attractive—partly because it helps reconcile Liebig's status as one of the nineteenth century's greatest chemists with his rather meager contributions to chemical theory. In the absence of decisive theoretical advances, historians have instead stressed Liebig's pedagogical achievement—leading to considerable debate concerning the extent to which his school was, in fact, innovative and successful.[2] Whatever their differences, existing accounts assume Liebig's school was a natural consequence of his introduction of a device that "perfected" organic analysis.[3]

As chapter 1 showed, however, in 1831 Liebig was far from solving the problem of analyzing organic compounds. This understanding of Liebig's situation reopens two important questions. First, why did Liebig found a school for organic analysis? Second, where did his research program actually lead? By shifting focus from the Kaliapparat to experimental practice, this chapter explains why analysis was Liebig's "only sure reagent," and what his school was intended to accomplish.[4]

The Giessen school was not an inevitable outcome of the Kaliapparat's success. Instead, like the Kaliapparat, Liebig's school was a product of his twin ambitions: to reform organic analysis and so to forge a career. Both school and instrument were simultaneously essential tools in, and consequences of, Liebig's efforts to develop and propagate his analytical methods. What I have called the "Giessen approach" to organic analysis was constitutive of Liebig's mode of laboratory reasoning, central to his vision for disciplinary development.[5] This vision, as we shall see, was destined to be thwarted. By about 1840, Liebig's research program was in deadlock—a reinterpretation that clarifies his decision to abandon cutting-edge organic chemistry. Of even greater significance for this book, identifying the limits of Giessen analysis sheds new light on the origins of organic synthesis.

FRIENDS AND RIVALS

If Liebig imagined his 1831 paper "On a New Apparatus" would cause other chemists to adopt his new apparatus, he was wrong. Although it appeared in J. C. Poggendorff's *Annals of Physics and Chemistry*—then the leading scientific journal of the German-speaking world—neither Liebig's new alkaloid formulae nor his new apparatus initially attracted much interest. Despite illustrating the Kaliapparat (see figure 1.1), Liebig offered no directions for how to make or use it—leaving other chemists ill equipped to adopt his new device. As explained in chapter 1, Liebig's alkaloid formulae were neither strikingly different from nor obviously superior to Dumas and Pelletier's, published in Paris almost a decade earlier. Parisian chemists' analytical domination made it hard for other chemists to see anything useful in Liebig's work.

This situation was a serious threat to Liebig's ambitions. He therefore began actively to promote his apparatus, beginning with his friends before seeking to engage the wider chemical community.[6] This work, as we shall see, required enormous effort for limited returns. First, Liebig approached Berzelius, seeking to open a correspondence with the man whose guidance had been decisive in his invention of the Kaliapparat. Berzelius responded positively, becoming Liebig's mentor. But despite Liebig's explicit incitement, Berzelius did not immediately begin analyzing organic compounds using the Kaliapparat.

Berzelius was impressed by how much Liebig had accomplished since their meeting in September 1830 and he was delighted to see Liebig undermine Parisian analysts. He urged Liebig to extend his study of the alkaloids in order to resolve problems he had identified with Liebig's results and data handling—a proposal Liebig rejected, because he found alkaloid analysis too difficult. As far as we know, Berzelius did not use the Kaliapparat until 1832, when Liebig sent him an exemplar device. Even the skilled glass-blower Berzelius seems not to have made his own Kaliapparat until he had a model to work from. Berzelius was never a routine user of the Kaliapparat.[7]

Most chemists required much greater contact with Liebig than Berzelius did before being able to make and use the Kaliapparat—indicating they found it far from straightforward to replicate analysis using Liebig's new device. This is exactly what his friend Wöhler traveled to Giessen in November 1831 to learn. During this visit, Liebig and Wöhler produced results leading to a significant contribution to chemical theory, the benzoyl radical.[8] Even this experience, however, did not persuade Wöhler to focus his own research on organic analysis.[9]

Paris was similarly high on Liebig's agenda. But persuading chemists there to use the Kaliapparat proved far harder than getting his mentor and his friend to try it. Parisian analysts were heavily invested in their own methods. These techniques, pioneered in Paris, relied on different apparatus (examined in chapter 3), and they embodied ideals of precision and elite philosophy that were a far cry from Liebig's homemade glassware, punishing schedule, and relentless quest to establish the superiority of his methods.[10] Spreading the Kaliapparat to Paris was a major undertaking, and—as we shall see—it required substantial personal contact.

Immediately after sending his paper to Poggendorff, Liebig submitted a French translation to his former teacher Gay-Lussac, seeking publication in the Parisian *Annals of Chemistry*. He was perhaps disconcerted to learn Gay-Lussac had commissioned a review from Liebig's rival Dumas. When Liebig's paper appeared in late summer 1831, it was followed by Dumas's commentary. Dumas declared that the Kaliapparat was "destined without any doubt to change the state of organic chemistry within a very short time." This view, however, was not based on personal experience of using Liebig's device. In fact, Dumas had focused on critiquing Liebig's method of measuring nitrogen—certainly one of the paper's weakest aspects and also an area where Dumas was eager to defend his own expertise.[11]

By October 1831, seeking Liebig's support for his election to the Parisian Academy of Science, Dumas became more receptive toward Liebig's work. It was nevertheless not until the following spring that he began using the Kaliapparat. Meanwhile, in December 1831, Liebig's former student, the Alsatian Charles Oppermann, arrived in Paris. Oppermann had mastered the Kaliapparat during 2 years in Giessen, expertise that led Liebig to recommend him as assistant to Théophile-Jules Pelouze. Pelouze, in turn, had recently been appointed to a position at the prestigious École Polytechnique—where his laboratory bench was next to Dumas's. Physical proximity and timing strongly indicate Dumas learned to use the Kaliapparat from Oppermann or Pelouze, rather than from Liebig's paper or any other source.[12] Indeed, he may well have helped bring Oppermann to Paris for just this purpose. Even an expert like Dumas required face-to-face instruction before he could use the Kaliapparat.

Once it had been drawn to his attention, Gay-Lussac seems to have been interested in Liebig's invention. He did not, however, attempt using the Kaliapparat himself. Instead, in the fall of 1831, he dispatched his son Jules to Giessen. Jules Gay-Lussac already had experience of Parisian volumetric analysis. Prior to his sojourn in Liebig's laboratory, working with Pelouze, he had analyzed salicin, a substance derived from willow bark.[13] While in Giessen, Jules Gay-Lussac repeated this analysis using "Liebig's apparatus and under his gaze." Toward the end of 1831, he and Pelouze published this new analysis in the Parisian *Annals of Chemistry,* using it to correct their previous formula for salicin. This was Parisian chemists' first published analysis using the Kaliapparat. Far from indicating the Kaliapparat's arrival in Paris, however, this work emanated from Giessen.[14]

This publication nevertheless signaled a major reversal in Liebig's professional standing relative to Paris. It was surely gratifying for Liebig when Jules Gay-Lussac celebrated obtaining results that "agree perfectly with those from the same analysis performed by M[onsieur] Liebig." Where once Liebig had defined his career aspirations in relation to Parisian chemistry, by 1831, he had received his former teacher's son into his laboratory and his home. Jules Gay-Lussac was the first Parisian student in Giessen, and his arrival no doubt helped establish Liebig's laboratory as the new destination of choice for ambitious young chemists from across Europe and beyond.[15]

These successes, however, should not be taken to imply widespread adoption of Liebig's new device.[16] The Kaliapparat's early users were few and almost exclusively drawn from Liebig's immediate circle. Dumas, Berzelius, and Wöhler—Liebig's rival, his mentor, and his closest friend—began using the Kaliapparat within a year or so of its invention. Apart from Berzelius, these chemists learned to use the Kaliapparat through personal contact with Liebig and others trained in Giessen. There is no evidence chemists unconnected to Liebig began using his new device during the early 1830s.[17]

Berzelius in 1833 considered his protégé "the greatest master in the art of carrying out precise organic analyses."[18] But this did not mean other chemists agreed or were ready to adopt Liebig's apparatus and methods. The Kaliapparat had not spread beyond Giessen except when actively promoted by Liebig. This experience taught Liebig how hard it would be to gain recognition for his invention. He now began a concerted campaign to teach others how to make and use the Kaliapparat and to persuade them of what he regarded as its proper use.

GIESSEN VS. PARIS

After publishing his paper "On a New Apparatus" in January 1831, Liebig committed himself to organic analysis. He knew the Kaliapparat had not solved the problem of organic analysis, especially for complex substances like alkaloids. In fact, he would not return to alkaloid analysis for many years. The difficulties of analyzing alkaloids were nevertheless crucial in shaping Liebig's strategy at this point, highlighting the Kaliapparat's improved "simplicity" and "reliability." Experience persuaded Liebig that analysis *could* elucidate the composition and formulae of many organic compounds—if it was done right.[19]

Liebig began by directing his student Oppermann to compare the Kaliapparat directly with Parisian volumetric analysis—something no paper of Liebig's ever did.[20] Oppermann analyzed oil of turpentine, a substance similar to camphor—the subject of Liebig's final publication before inventing the Kaliapparat.[21] Even though it contained no nitrogen, Liebig had struggled to analyze camphor using Parisian apparatus, making it plausible that he selected oil of turpentine as a way of demonstrating both Oppermann's abilities and the Kaliapparat's superiority over Parisian methods. Whatever Liebig's motives, oil of turpentine, camphor, and their related compounds were about to reignite his disagreement with Dumas, prompting a fresh crisis in his research program.

Liebig's difficulty stemmed from camphor's volatility (which made it hard to ensure complete combustion) and high molecular mass (which meant any error in measuring the volume of carbon dioxide had a large effect on the number of carbon atoms assigned in the formula). His paper explicitly referred to both these problems and the consequent uncertainty of his results. But this had not prevented Liebig from using conclusions about camphor's composition, formula, and relationship to camphoric acid to cast doubt on previous analyses.[22]

Liebig's criticisms of three prior analysts confirm his particularly adversarial attitude to Parisian chemists ca. 1830. According to Liebig, German pharmacist Rudolph Brandes had described preparing camphoric acid "with great accuracy and thoroughness," while Swiss chemist and physiologist Nicolas-Théodore de Saussure's analysis was merely "incompatible" with Liebig's. Parisian chemist Edme-Jean Baptiste Bouillon-Lagrange, meanwhile, had not analyzed camphoric acid at all but a different compound of camphor and camphoric acid.

These criticisms also highlight fundamental issues that Liebig no doubt hoped Oppermann's analyses would help resolve. Ongoing analytical uncertainty made it hard to identify organic compounds reliably. Two analyses of what was believed to be the same substance could be equally valid, even if they did not lead to the same formula. It was easy to mistake one substance for another, hard to obtain pure samples, and not uncommon to observe differences between samples that were in fact the same substance. Oppermann indeed made valuable contributions in these areas. But, as we shall see, his work had additional, unwelcome consequences.

In 1831, working with the "support" and "direction" of his "most esteemed teacher, Prof. Liebig," Oppermann published his "most successful" analyses of oil of turpentine. His stated goal was to unravel inconsistencies in analyses published by de Saussure and François Joseph Houtou de Labillardière. Each chemist's results were "so consistent, they appeared to rule out any doubt of their correctness," yet their formulae and composition were quite different. Saussure considered that oil of turpentine contained nitrogen, while Labillardière's analysis indicated it was a hydrocarbon, containing carbon and hydrogen only. Neither chemist believed oil of turpentine contained oxygen, which Oppermann found hard to reconcile with its ability to oxidize calcium.[23]

As with camphor, oil of turpentine's volatility was a major problem. Oppermann found the Kaliapparat "most helpful" in this regard. Because it relied on measuring mass rather than gas volume, Liebig's apparatus could handle a much larger sample, which enabled Oppermann to develop a special procedure for introducing oil of turpentine into the analytical apparatus. He analyzed 0.338 g of oil of turpentine using the Kaliapparat— roughly 10 times what was possible with Parisian apparatus. Two analyses using Parisian apparatus produced results for carbon that were consistent to within 1% (83.1677% and 83.9828%) but nevertheless differed by more than 0.5% from that obtained using the Kaliapparat (84.5923%). All Oppermann's analyses indicated that oil of turpentine contained oxygen—which explained the oxidation of calcium. But using the Kaliapparat suggested a quite different composition and formula than did the Parisian method.[24]

Disturbed by these findings, Oppermann now adopted what would become a leading strategy of Giessen analysis. Reacting carefully purified oil of turpentine with hydrogen chloride gas (a standard inorganic reagent), he isolated "artificial camphor"—a substance that smelled similar to natural camphor but (unlike its natural counterpart) contained chlorine. Since he believed artificial camphor and oil of turpentine were related by a simple chemical reaction, Oppermann expected their formulae would be similar—valuable additional context for interpreting his analyses. Analyzing artificial camphor, however, suggested a formula that was impossible to reconcile with its formation from oil of turpentine.[25]

Noticing that the yield of artificial camphor depended crucially on how oil of turpentine was sourced, Oppermann—no doubt guided by Liebig— now reasoned that artificial camphor was not produced from oil of turpentine at all but from a distinct hydrocarbon (compound of hydrogen and carbon only) in his starting material.[26] He obtained this hydrocarbon by decomposing artificial camphor and converted it back to artificial camphor. Having successfully completed this cycle of decomposition and reconstitution, Oppermann concluded—despite significant gaps in the evidence— that this new substance was "the basis of camphor."[27]

Neither Oppermann nor Liebig seem to have anticipated Dumas's likely interpretation of this work. But for Dumas, Oppermann's identification of a core atomic grouping found in oil of turpentine and related compounds corroborated his earlier proposal (with Félix-Polydore Boullay) regarding the arrangement of atoms in organic ethers. Dumas therefore appropriated Oppermann's new substance to his theory, renaming it "camphogène" (literally, "maker of camphor") and studying its properties. Camphogène reacted with hydrogen chloride to produce artificial camphor and with oxygen to give natural camphor.

Liebig, who vehemently rejected Dumas and Boullay's ideas, was furious.[28] Worse still, Dumas's new analyses and formulae for oil of turpentine, camphor, and camphogène differed from Liebig's and Oppermann's. Although he diplomatically acknowledged Liebig's "well-known talent and exactness," Dumas found that camphor contained more hydrogen and "a little less" carbon than Liebig had measured. He agreed with Oppermann that oil of turpentine's composition varied according to its source but, unlike Oppermann, Dumas found it contained no oxygen.[29]

Dumas's findings were a serious threat to Liebig. Despite his tact, Dumas had measured 2.5% less carbon than Liebig had—a significant discrepancy that undermined Liebig's analytical expertise and, even more important, the reliability of his new apparatus. Oppermann's work indicated that students using the Kaliapparat could produce results as good as, if not better, than Liebig's. But Dumas's response called Liebig's methods into question. Analytical disagreement with Dumas had prompted Liebig's introduction of the Kaliapparat; it now precipitated a crisis that incited Liebig's next step in the development of his research school.

RESEARCH SCHOOL

Liebig had first opened a school in Giessen in 1826 in response to the financial limitations of his unsalaried university position. The school initially offered instruction in return for fees that helped fund Liebig's personal research, remaining a private venture until 1833.[30] Liebig established systematic laboratory training in Giessen well before organic analysis became the focus of his research.[31] In 1827, a curricular revision required students to spend the "entire winter semester . . . in the chemical laboratory, whereby they must occupy themselves with analytical work from morning until evening."[32]

Liebig also began training students in organic analysis, albeit in an initially limited way. Between 1827 and 1830, two young chemists in his laboratory—Heinrich Buff and Friedrich Kodweiss—published analytical studies performed using Parisian apparatus.[33] No longer merely a source of income, students henceforth made vital contributions to Liebig's research. From now on, Liebig was well placed to involve students in organic analysis, and his reliance on their labor increased following his invention of the Kaliapparat.

Between 1831 and 1834, Charles Oppermann, Karl Ettling, Jules Gay-Lussac, Rodolphe Blanchet, and Ernst Sell joined Liebig's nascent research school.[34] These early students stabilized the use of the Kaliapparat and aided Liebig's quest to demonstrate its superiority and persuade others to use it. Understanding students' role in developing Giessen analysis explains the relatively slow growth of Liebig's school during this period.[35] It also highlights the difficulties confronting Liebig as he strove to transform his personal research methods into a reliable analytical approach suitable for use by others.

Producing comparable results required every analyst to work the same way using a standard device. As introduced in January 1831, however, Liebig's Kaliapparat was anything but standard. At first, he provided only limited guidance for its use. Starting in his own laboratory, Liebig sought to improve and standardize the Kaliapparat and to integrate its use in stable, teachable analytical methods. Chapter 3 pursues the material aspects of standardization, including the contributions of Ettling, a skilled glassblower. By 1833, the Kaliapparat was smaller and lighter than Liebig's original, and it produced significantly better results.[36] Meanwhile in 1832, Liebig instructed Blanchet and Sell to repeat Dumas's analyses of camphor and oil of turpentine. As we shall see, their investigations marked a watershed in Giessen research, in the development of Liebig's school, and in the dissemination of his methods beyond Giessen.

GIESSEN APPROACH

Liebig's decision to delegate responding to Dumas over camphor to his students Blanchet and Sell privileged the emergent school over his personal reputation. Disciplinary advance, based on establishing a community steeped in his analytical methods, now took priority over Liebig's individual success. One plausible motive was Liebig's growing realization that the camphor analyses he and Oppermann had previously published were flawed—which Blanchet and Sell's new analyses confirmed.[37] Up to this point, it had been impossible to adjudicate the analytical disagreements between Liebig and Dumas. But now, as Liebig certainly recognized, there was no way of moving forward without admitting Dumas's analysis of camphor was superior to his own.

Liebig's chosen strategy effectively deflected attention from this difficult admission. Although Blanchet and Sell's 1833 paper conceded Liebig and Oppermann's earlier errors, they nevertheless rejected Dumas's methods and interpretation. Working under Liebig's "guidance and direction," they reported many analyses, camouflaging their discussion of camphor in a vast body of new evidence.[38] Moreover, this evidence enabled them to develop their critique of Dumas into an authoritative commentary on organic chemistry's present state and goals. This was how students—rather than Liebig himself—came to publish the first description of the Giessen approach to organic analysis. For Liebig and his followers, adherence to proper procedure now took precedence over assigning any valid formula in isolation.

According to Blanchet and Sell, determining the laws governing combining proportions[39] in organic bodies by analysis had been "the exclusive object of the work of chemists in recent times." They argued for a "true system of chemistry" based on "an exact knowledge of the composition of organic compounds." Recent developments—no doubt Liebig's work was meant—had rendered organic analysis reliable in any hands, given care and practice. Echoing Liebig, Blanchet and Sell claimed that discovering the "true composition of organic bodies" relied not on eliminating errors but on "containing them within narrow limits" that defined formula, and they explained how to accomplish this using the Kaliapparat. Material improvements in the Kaliapparat's design since 1831 were not the sole cause of improvements in analytical reliability. Instead, Blanchet and Sell described a series of procedural controls introduced to maximize confidence in assigned formulae.[40]

Blanchet and Sell had taken turns repeating the analysis of each substance until any variation in results had no effect on formula. Thus, their combined effort unambiguously established how many atoms of each kind these relatively simple molecules contained. Much as Jules Gay-Lussac had compared his analysis of salicin with Liebig's in 1831, Blanchet and Sell now claimed consistency at the level of molecular formula was an effective way of securing the validity of the underlying analytical results.[41] There was no expectation that two analysts—even when they had been trained and were working in the same laboratory, with the same apparatus—would measure identical percentage composition.[42]

It is also worth noting that—just as with Liebig's 1830 alkaloid analyses—Blanchet and Sell did not use an average percentage composition to arrive at the correct molecular formula. Instead, they reported two sets of analytical results for each substance, both of which were consistent with the same molecular formula. Calculating the theoretical percentage composition from this formula showed which set of raw results fitted the formula better. This "calculated analysis" also provided a baseline against which to compare analyses published by chemists elsewhere.[43]

Blanchet and Sell investigated whether oil of turpentine's formula depended on its source—as Dumas had proposed, following Oppermann's previous claim. Analyzing samples purified from three different sources, they assigned the same composition and formula to each. Their formula agreed with Dumas's, which meant that his results were reliable. But they strongly contested his speculation that oil of turpentine was none other than camphogène, the radical Dumas asserted as the common constituent of both artificial and natural camphor. For Blanchet and Sell, Dumas's proposal could not be satisfactorily reconciled with their analyses of other, related substances and was incompatible with camphor's chemical properties. Although Dumas had assigned a correct formula in this instance, Blanchet and Sell now rejected his analytical approach as inferior to their own.[44] Ensuring that the formulae assigned to related substances were compatible with their reactivity and relationships was crucial to Giessen analysis.

Blanchet and Sell's discovery that multiple samples of oil of turpentine shared the same formula allowed them to develop Oppermann's analyses of related compounds into an experimental and interpretative framework. Reiterating Liebig's emphasis on the paramount importance of molecular formulae,[45] they now defined Giessen analysis. Blanchet and Sell claimed multiple analyses of related substances, collaborative confirmation, and careful balancing of experiment and theory—rather than isolated analyses and speculative interpretations of the kind they attributed to Dumas—were essential to reliable organic analysis. Where previously "different methods produced different results," any analysts following these procedures should assign the same formula—even if their raw analytical results were not identical. This was the first published description of the Giessen approach, and it was a landmark in the development of Liebig's school.

Equipped with this investigative method, the Giessen school now entered a period of more rapid growth, soon supplying Liebig with a community of skilled analysts under his direction. More than a dozen students joined Liebig's research group between 1834 and 1837, including Hermann Fehling, Charles Gerhardt, William Gregory, Friedrich Knapp, Victor Regnault, and Heinrich Will. Several had significant prior chemical experience—with important consequences, negative as well as positive.[46] From 1833 on, students in Liebig's laboratory learned the Giessen approach. In other words, they were trained in Liebig's method of laboratory reasoning—not merely in how to use the Kaliapparat. Once published, their work enhanced Liebig's reputation, making Giessen the global center for research in organic chemistry. Students were crucial in developing the Giessen approach, advancing organic analysis, responding to Liebig's critics, and promoting his methods. This strategy, as Liebig evidently recognized, also required access to publication.

PUBLISHING AND PROMOTING

Right from the start, Liebig encouraged his students to publish under their own names, even when—as was clearly the case with Oppermann, Blanchet, and Sell—their work was performed under his close supervision, serving his ends and propagating his views. Such publications enabled students to demonstrate analytical skills acquired in Giessen. For Liebig, they highlighted the superiority of his approach by advertising what chemists in his laboratory were doing. By 1834, the six original members of Liebig's early research school had published seven papers in Poggendorff's *Annals*. In 1832, meanwhile, opportunity had come Liebig's way, in the form of an invitation to join the ageing Philipp Lorenz Geiger as editor of the *Magazine of Pharmacy*.[47]

When Liebig sought Berzelius's advice about Geiger's offer, his mentor advised against taking on this additional workload. Liebig was under intense physical and mental strain at this time, and Berzelius feared for his health.[48] He may also have questioned the value to Liebig's career of editing a journal concerned primarily with pharmacy rather than chemistry as an academic science: Geiger's *Magazine* was certainly no match for Poggendorff's *Annals*. But Liebig saw potential that Berzelius did not and accepted Geiger's invitation.

Subsequent events confirm that Liebig viewed Geiger's journal as far more than a source of additional income, and that he intended from the start to develop it as an outlet for Giessen analysis and a competitor to Poggendorff's *Annals*. He twice renamed it, first the *Annals of Pharmacy* and then the *Annals of Chemistry and Pharmacy*, changes that reflected his use of editorial power to reshape the journal's goals and content. During the early 1830s, Liebig fashioned the new journal into a venue for Giessen students' work and a forum for his own views on organic analysis. In this way, Liebig transformed the new *Annals* into a leading chemical journal that promoted the Giessen approach, defining the standards of method and instrumentation that its editor was determined others should follow.

From 1832 on, Liebig published both original articles and reprints in his new journal, including much work done outside Giessen. This meant Liebig's *Annals* not only served to disseminate his students' research, it also allowed him to bring selected publications to a wide German-reading audience, frequently in translations from French and English (most likely produced by students). Publishing work that he endorsed gave it Liebig's stamp of approval. And when reprinting the work of chemists with whom he disagreed, Liebig did not hesitate to exercise his editorial right to comment.[49] Thus, Liebig's *Annals* became a major venue for the analytical disagreements that tore through organic chemistry during the 1830s, playing a vital role in his attempts to direct disciplinary development. The journal's contents also offer historical insight into the difficulties confronting Liebig, and the unexpected origins of some of his most vocal critics.

CUTTING EDGE

Having explained the rise of Liebig's school and journal, let us now turn to his personal research between 1831 and 1834. In 1832, working with Wöhler, Liebig used the Kaliapparat to identify the benzoyl radical—a major contribution to constitutional theory (concerning the arrangement of elements in organic molecules) and Liebig's most significant theoretical accomplishment.[50] While his students developed the Giessen approach for compounds containing no nitrogen, Liebig also worked tirelessly to advance organic chemistry's cutting edge by extending Giessen analysis to nitrogen.

Nitrogen's widespread occurrence in plants and animals made this element crucial to chemical understanding of living phenomena. According to Liebig, "step by step" analysis of nitrogen-containing compounds was the only way of acquiring "insight into the secret processes of nutrition etc. of the animal organism."[51] But measuring how much nitrogen organic compounds contained remained very difficult. This problem had drawn Liebig into organic analysis, and it was why he invented the Kaliapparat during his 1830 alkaloid investigations. Although the Kaliapparat measured carbon, Liebig introduced it to tackle the problem of nitrogen. His initial attempts failed, but Liebig continued working after 1831 to incorporate the Kaliapparat into an improved method of nitrogen determination.

Liebig's most successful strategy involved developing his earlier use of the Kaliapparat to criticize Dumas's volumetric measurement of morphine's nitrogen content. The discrepancy between the Kaliapparat's direct measurement of carbon and the same value calculated from volumetric analysis (which measured carbon and nitrogen together in a single experiment) enabled Liebig to argue that volumetric analysis significantly underestimated morphine's nitrogen content. Liebig now proposed this comparison as a routine evaluation of the reliability of volumetric analysis, central to the Giessen approach to nitrogen.[52] If volumetric analysis produced results for carbon that matched the value measured using the Kaliapparat, then the volumetric analysis was valid, and the nitrogen determination could be considered reliable. But if the two values for carbon differed, then the nitrogen determination must be discarded as "incorrect."[53]

In early 1834, Liebig published a new formula for uric acid, produced by comparing values for carbon as just described. This, according to Liebig, was the first time that analysis had produced consistent composition data for uric acid. Liebig claimed errors in the volumetric analysis of uric acid were "unavoidable," because its nitrogen tended to undergo unwanted side reactions during analysis. But using the Kaliapparat's "direct" measure of carbon to validate volumetric nitrogen determinations had enabled Liebig to assign uric acid a new formula ($C_5H_4N_4O_3$) that he believed was reliable. Based on this success, he offered comparing carbon values as a fully fledged "control" for nitrogen.[54]

Improvements in the Kaliapparat also enhanced the effectiveness of Liebig's approach to nitrogen, especially for substances containing very little nitrogen, including hippuric acid. Within months of publishing his new formula for uric acid, Liebig followed it with a fresh analysis of hippuric acid. After more than 5 years, Liebig and Dumas were still vying to determine a stable formula for hippuric acid. Liebig's newly revised formula for hippuric acid ($C_{18}H_{18}N_2O_6$) appeared in Poggendorff's *Annals* shortly before Dumas's most recent analysis of the same compound. According to Poggendorff, Liebig's analysis had been made possible by his "now perfected apparatus"—almost certainly a reference to the Kaliapparat's recently refined form (see chapter 3).[55]

Liebig's 1834 analyses of uric and hippuric acids further embedded the Kaliapparat in his analytical strategy. Abandoning his search for a superior method of measuring nitrogen, Liebig now recommended using the Kaliapparat to identify and discard erroneous nitrogen determinations made with existing methods. In fact, Liebig at this time used a range of volumetric techniques and apparatus to measure nitrogen, depending on how much nitrogen the substance under examination contained. At least two of these methods emanated from Paris.[56] But in every case, Liebig relied on a match between carbon measured using the Kaliapparat and the same element determined by volumetric analysis to identify reliable nitrogen determinations, making the Kaliapparat central to the Giessen approach to nitrogen.

Collective analytical experience shaped Liebig's understanding of research. This led him to place reaction-based taxonomy at the core of his laboratory reasoning, central to the Giessen approach. Merely describing a compound's reactions was "pointless" unless the chemist also knew "its composition and that of two or three of its products," in which case, its nature became "obvious of its own accord."[57] Only analyzing related compounds—as Blanchet and Sell had done—produced secure composition and formula. Liebig learned by running a school that investigating networks of organic compounds connected by reaction—rather than individual substances in isolation—was essential in managing the uncertainties of analysis and hence was necessary for useful research.

Liebig's explicitly taxonomic approach aligned organic with inorganic chemistry, whose taxonomy was by this point well established. Taxonomy and transformation were the framework for Giessen analysis. This framework, as chapters 4, 5, and 6 explain, would transcend the turn to synthesis. A decade later, taxonomic relationships and well-understood reactions remained central to Hofmann's synthetic study of organic bases. In 1834, meanwhile, Liebig drew on the language of inorganic chemistry when he identified analysis as organic chemistry's "only sure reagent" and a core component of the Giessen approach.[58]

LIEBIG'S *INSTRUCTIONS*

From the mid-1830s, more chemists recognized the Kaliapparat's advantages. This progress nevertheless confronted Liebig with new problems. Chemists elsewhere often used modified versions of Liebig's apparatus, and they rarely adopted the Giessen approach. Unable to achieve the standardization that his disciplinary vision required, Liebig apparently recognized the need for a new strategy. Liebig first published his approach to organic analysis in the first part of Liebig and Poggendorff's *Handbook of Pure and Applied Chemistry* in 1836, issuing a standalone volume under the title *Instructions for the Analysis of Organic Bodies* (*Anleitung zur Analyse organischer Körper*) the following year. As I now show, this manual—which appeared in English translation in 1839—was a manifesto for the Giessen approach and for Liebig's views on research.[59]

Liebig's *Instructions* did not merely supplement his paper "On a New Apparatus" with additional detail.[60] His manual described how to make and use a standard Kaliapparat that was smaller, lighter, and more regular in shape than Liebig's original. Especially where nitrogen was concerned, but also in other respects, Liebig's methods differed from those he had published in 1831. The *Instructions* described procedures and controls developed by Liebig and his students during the 1830s. They explained how Liebig and his students worked, assembling previously published material into a pedagogically oriented guide to the mature Giessen approach.

It was essential that the analyst select targets appropriate to their skill and experience. Because volatile liquids were easiest and gave the most accurate results, "beginners will do well to occupy themselves first with the combustion of such substances."[61] High-molecular-weight substances and most compounds containing nitrogen, by contrast, were unsuitable. As Oppermann had done, the *Instructions* explicitly prioritized formula as the crucial analytical output, urging molecular weight measurement as a "control" on raw analytical data.[62] Collaborative confirmation and analyzing related compounds—first described by Blanchet and Sell in 1833—were key recommendations.[63]

Liebig also used the *Instructions* to promote his approach to nitrogen, confirming that this manual was primarily intended for experienced chemists working at the cutting edge. Liebig still believed that students lacked the necessary skill to tackle nitrogen. Two factors indicated the most appropriate way of measuring nitrogen: the sample's tendency to undergo side reactions (e.g., uric acid), and how little nitrogen it contained (e.g., hippuric acid). Measuring small nitrogen content required a "special operation," remaining subject to "a constant error, which is unavoidable."[64] In every case, the final arbiter of a reliable nitrogen determination was a match between carbon measured using the Kaliapparat and determined by volumetric analysis—the control for nitrogen that Liebig had introduced in 1834.[65]

Ongoing analytical uncertainty was evident in the requirement for data selection based on both empirical and theoretical considerations. Empirical factors, including visual criteria, were essential in distinguishing "trustworthy" from "erroneous" analyses.[66] For example, if the sample, once ignited, "continues to burn spontaneously, . . . the analysis is good for nothing."[67] Calculations, corrections, and adjustments incorporating theoretical knowledge, meanwhile, were still needed to "discover" a convincing formula.[68] Much as with Blanchet and Sell's work, reconciling raw analytical results with combining weight entailed more than routine numerical processing.

In 1837, Liebig continued seeking converts to his methods. He had persuaded some of Europe's most eminent chemists to adopt his new device during the early 1830s, and others had learned to use it by studying in Giessen or knowing someone who had. But not everyone was using Liebig's Kaliapparat, and even those who were frequently did not use it according to his recommendations. The Kaliapparat had not reduced organic analysis to routine manipulation. Standard methods were therefore as vital as standard apparatus in producing the reliable analytical data and agreed-on formulae necessary to expand chemical knowledge. Assigning formula still relied on skill and judgment. Liebig's *Instructions* therefore sought to promote the Giessen approach—including the apparatus, procedures, and essential forms of data selection at the heart of research in Liebig's laboratory.

ANALYTICAL DISCORD

The research agenda laid out in Liebig's *Instructions* significantly altered the terms of his disputes with other chemists. Emphasizing procedures, controls, and standards—rather than analytical results in themselves—redefined the crucial barrier to disciplinary progress. Whereas previously, Liebig's analytical arguments had focused on Paris, beginning in 1837, he became embroiled in a wider series of debates about analytical practice, theory, and interpretation. Beginning with his mentor Berzelius and later involving one of his most promising former students, Regnault, these increasingly heated disagreements would ultimately strip Liebig of support exactly where he might otherwise have had most reason to expect it.

Organic analysis in 1837 remained far from routine, especially for large molecules, compounds containing nitrogen, and substances that were hard to purify. Chemists frequently failed to agree on composition and formula, preventing the development of a shared understanding of organic compounds—a situation that Liebig attributed to other chemists' rejection of the Giessen approach. The stakes were high, leading Liebig to dismiss those who disagreed with him. Even Berzelius and Regnault became targets of his intemperate criticism. Liebig's passionate commitment to the superiority of his own methods polarized opinion in the wider community, making cutting-edge analysis more fraught than ever.

These disputes originated in Berzelius's claim that cork—which Liebig used to connect the glassware components comprising his analytical apparatus—released moisture during analysis. Berzelius preferred *caoutchouc* (natural rubber) but Liebig was convinced caoutchouc absorbed water—and said as much in his *Instructions*. Berzelius, Liebig declared, could only prefer caoutchouc because he had never tested cork. Worse, Liebig dismissed Berzelius's apparatus as inaccurate and unreliable in the hands of most analysts. This was an especially barbed criticism in light of Berzelius's efforts to improve organic analysis and make it more accessible—not to mention his mentorship of Liebig. Liebig and Berzelius later patched up this quarrel, but their relationship never fully recovered, failing altogether following a second disagreement about the interpretation of analytical results.[69]

The cork-caoutchouc debate focused on an element yet to feature in this account of organic analysis: hydrogen. Because water contains hydrogen, experiments to measure hydrogen would be disrupted whether cork gave up water or caoutchouc absorbed it. In addition, hydrogen's low atomic weight meant that even small experimental errors had a big impact on the number of hydrogen atoms and hence on the assigned formula. Any moisture—whether present in sample or reagents, air in the apparatus, or introduced by any other means—could have a disastrous effect on analysis. Despite striving to exclude moisture, Giessen chemists believed they consistently measured slightly more hydrogen than was actually present in the sample. According to Liebig, this systematic error occurred whether cork or caoutchouc was used and should be handled by procedural means.[70]

Liebig's response to Berzelius—perhaps because of its public and ungenerous nature—attracted the interest of others, including Hermann Hess in St Petersburg. In 1838, Hess published experiments that confirmed cork gave up moisture when heated. Siding with Berzelius, he argued that the errors in measuring hydrogen were not systematic, as Liebig claimed, but were caused by using cork. This was a fundamental challenge to Liebig's expertise.[71] Liebig—despite then attempting a reconciliation with Berzelius—brusquely rejected Hess's criticism.[72] Liebig "had made the experiment described by my distinguished friend [Hess] a great many times," but had "never observed either a gain or loss of weight." When Liebig heated cork, he found no evidence that it absorbed or released moisture.

Liebig now used editorial privilege to restate the Giessen approach in an extensive commentary on Hess's work. Not only did he believe Hess was wrong about cork, Liebig also rejected other aspects of Hess's approach that he considered undermined disciplinary development. "Progress in organic chemistry," Liebig declared, was "out of the question, without researches." The "analysis of a single substance"—Hess's approach—did not constitute research, while in Giessen, "60-70-100 or more analyses [we]re not rare in the course of one investigation." Only multiple analyses converging on a coherent outcome enabled the analyst to "arrive at the knowledge of a substance, whose properties are uniform, and whose composition explains the discordance of the first results. The last analysis alone is published; the others were merely tests."

The "advantage of our methods," Liebig argued, "does not consist in greater accuracy, for . . . the old method was susceptible of the greatest exactness," but instead in their "greater simplicity and security, with the same degree of exactness." The superiority of Giessen analysis did not depend on producing "mathematically accurate results" in a single experiment, but on the coherence of a very large number of analyses—more than 400 analyses in an average year.[73] Despite this evidence, Liebig still struggled to convince others that consistency was more important than accuracy in analytical research.

Liebig's claim to superior experience was certainly justified, since Hess had only lately learned how to analyze organic compounds. The most likely source of Hess's newfound analytical expertise was Aleksandr Voskresensky, who had recently returned to St Petersburg after studying with Liebig.[74] Hess's connection to Giessen, though indirect, is significant. It indicates that chemists' movement from place to place continued throughout the 1830s to be important in spreading the Kaliapparat. It also suggests that even close association with Liebig did not result in adherence to the Giessen approach.

Once provoked, Liebig was a tenacious opponent. His second public clash with Hess places data selection at the core of their disagreement. When Hess published an analysis of saccharic acid, Liebig was evidently unpersuaded by his findings. Reprinting Hess's article in his *Annals,* Liebig set one of his students, the Norwegian M. C. J. Thaulow, to repeat Hess's analyses.[75] In 1838, Thaulow published a new formula for saccharic acid. Hess responded by accusing Thaulow of suppressing results that undermined this formula.[76]

Liebig again reprinted Hess's article, following it with a strong defense of Thaulow and a withering critique of Hess's work. Hess had failed to address the "all-important question" of "where saccharic acid came from and what its relatives were." The mere act of analyzing a substance such as saccharic acid was easy; the problem was "to make it speak"—Liebig's expression for situating it within a framework of taxonomy and reaction, and a key component of the laboratory reasoning associated with analysis. In Liebig's view, such a framework was required to stabilize a substance's assigned formula and rationalize its chemical reactions and behavior. Hess had neglected crucial features of the Giessen approach.[77]

The sequel to this exchange confirms that Hess rejected the Giessen approach's reliance on data selection to produce formulae. In 1840, having repeated his investigation, Hess claimed that Liebig was wrong to criticize his saccharic acid analysis.[78] Liebig could not have disagreed more strongly. The "true scientific value of a chemical investigation," he countered, "does not consist of discovering numerical results and placing them next to one another." What mattered was that "these figures imply a mental concept" consistent with observed behavior.

Such conceptual work, Liebig argued, entailed identifying and excluding failed experiments; and it implied data selection on theoretical as well as empirical grounds: "The correctness of a number depends on chance occurrences, the discovery of the truth which arises from it is the task of science, but it will not be found using copper oxide, spirit lamps, and oxygen gas alone." In Liebig's view, "If Thaulow's explanation is right, then his analyses are true; if it is false, then his analyses may nevertheless be reliable, but they require a different interpretation." Only a poor chemist would become preoccupied with technicalities at the expense of skillful interpretation: "The worse the chemist . . . the sharper the evidence!!" Exasperated by Hess's rejection of the Giessen approach, Liebig dismissed him as a mere "dilettante," outside the community of scientific chemists.[79]

During this period, Liebig was also in dispute with William C. Zeise, professor of chemistry in Copenhagen.[80] Zeise, like Hess, used the Kaliapparat but neglected key features of the Giessen approach. Data selection was again the crucial issue, confirming this as a primary cause of Liebig's mounting frustration. In 1837, Zeise discovered and analyzed a novel substance.[81] His work prompted familiar, strong objections from Liebig. Because Zeise had not investigated the new substance's decomposition products, Liebig dismissed his proposals concerning its formula and chemical nature as mere opinion. This, Liebig wrote, was "not the course of chemistry today."[82]

Liebig's criticisms focused on hydrogen—whose determination was becoming ever more central to his analytical concerns. Wöhler and Liebig had recently established the "limit of accuracy" in hydrogen determination as around 0.2%, even when following the Giessen approach.[83] An error of this magnitude had no effect on the number of hydrogen atoms in the formulae of small molecules. But it was a serious obstacle to assigning reliable formulae for larger molecules.[84] Liebig's suspicions were aroused because Zeise's hydrogen determinations varied widely. Overall, they indicated less hydrogen than Zeise's formula required—which was problematic, because "hundreds of organic analyses from the most varied chemists" suggested measured values for hydrogen were always slightly higher than formula required.[85]

Liebig inferred that Zeise had published all his results, "even those whose correctness one might have reason to doubt," and including "those hydrogen determinations that far exceeded his theoretical formula."[86] To Liebig, such divergent results were unhelpful, because only the analyst knew "which analyses to trust and which not; if some kind of error has occurred during the operation, he must discard the result." It was his duty to "separate the wheat from the chaff" as "only the analyst himself can."[87] Without appropriate selectivity, analysis might occasionally produce results leading serendipitously to a reliable formula—as indeed would prove to be the case here: Ultimately, other chemists accepted Zeise's formula for what became known as Zeise's salt.[88] But, Liebig argued, such chance occurrences did not constitute systematic, progressive, scientific chemistry.

A final exchange seems to have brought these issues to a head, underlining chemists' inability to determine reliably how many hydrogen atoms large molecules contained and the challenges Liebig faced in disseminating the Giessen approach. This dispute, with his former pupil Victor Regnault, prompted Liebig's first return to alkaloid analysis in almost a decade, and it confronted him with two unwelcome facts. Not only was alkaloid analysis still an unsolved problem in 1838, but—despite Liebig's fame and his school's global reputation—even his best students sometimes rejected the Giessen approach.

In 1838, Regnault—who had recently taken up an appointment in Lyon—published new analyses of alkaloids, including morphine. Regnault's analyses and formulae differed significantly from those Liebig had published in 1831. He was consistently critical of Liebig's work, dismissing the notion that the alkaloids' basicity was directly related to their nitrogen content. His former teacher was understandably unhappy and responded at length with new analytical evidence.[89]

According to Liebig, Regnault had abandoned essential aspects of his Giessen training. Although Liebig conceded some of Regnault's formulae were superior to his own, he regarded others as completely arbitrary.[90] Overall, Regnault had failed to present a coherent, taxonomically informative view of the alkaloids' composition. Analysis still could not decide "whether [morphine] contains two atoms of hydrogen more or less."[91] Regnault's manipulative skill was beyond doubt, but his "limited experience" had led him to place too much reliance on "mere numerical results." By choosing to work without the framework of taxonomy and transformation at the heart of Giessen analysis, Liebig considered that Regnault had made it impossible "to distinguish the true analysis from the false" and that his laboratory reasoning was flawed.[92]

Ultimately, Liebig lamented, "with all these difficult and time-consuming experiments we have moved no closer to the underlying question" concerning "the nature of the organic bases [alkaloids] and their capacity to neutralize acids."[93] In other words, Regnault in 1839 was no more able to answer this question than Liebig had been in 1830. Nor had Liebig convinced one of his school's most successful alumni to adopt his methods. By 1840, following a decade spent striving to impose order through his school, journal, and manual, Liebig confronted widespread, mounting analytical discord.

STUDENT SOLUTION

Liebig still could not answer the fundamental questions about alkaloids that had first drawn him to specialize in organic analysis.[94] He was surely frustrated by this lack of progress. Yet Liebig remained as creative as ever in making the best use of available resources. As we have seen, experienced, capable, and ambitious students were attracted to Giessen in rapidly increasing numbers from 1835 onward. For example, in addition to having the temerity to disagree publicly with Liebig shortly after leaving Giessen, Regnault went on to notable accomplishments in both academic and applied science.[95]

Liebig now capitalized on the unique opportunity such a student body afforded, steering a new generation of chemists toward innovations in analytical apparatus and method while simultaneously shifting his own research program toward physiological chemistry. Despite its importance to physiology, Liebig had abandoned the search for a better way of measuring nitrogen in 1834. In late 1840, a letter from Wöhler encouraged him to revisit this problem.[96] It would nevertheless be Giessen students rather than Liebig himself who offered the first durable solution to the problem of nitrogen.

Wöhler's letter outlined an innovative approach to measuring uric acid's nitrogen content. Developed in response to an investigation of human urine published jointly with Liebig in 1838, Wöhler's method applied an earlier method for measuring metallic platinum—due to Berlin's master inorganic analyst, Heinrich Rose—to organic nitrogen determination.[97] Wöhler converted uric acid's nitrogen to ammonia gas, which he trapped in a modified Kaliapparat—a device Wöhler illustrated in an enclosed sketch (see figure 3.8).[98] Based on entirely different chemistry, this method appeared to avoid the most serious obstacle confronting existing methods: nitrogen's unwanted side reactions. It worked well for uric acid, but Wöhler had not tested it on other substances. Much as Berzelius had done for carbon almost a decade before, Wöhler offered Liebig a potentially reliable method of measuring nitrogen by mass.

It is clear from what Liebig did next that he recognized as much potential in Wöhler's suggestion as in Berzelius's earlier guidance. Liebig's circumstances, however, were very different. Where Berzelius had prompted months of solitary work developing the original Kaliapparat, Liebig now deployed the resources of his Giessen school to develop a new approach to nitrogen. By 1840, the enhanced capabilities of his students and assistants meant Liebig could lift previous strictures excluding junior chemists from measuring nitrogen. Soon after receiving Wöhler's letter, Liebig set two young Giessen chemists, Heinrich Will and Franz Varrentrapp, to work on his friend's exciting proposal.

It is a mark of the Giessen school's developed state in 1841 that Will and Varrentrapp between them commanded the entire range of skills needed to transform Wöhler's proposal into a reliable analytical method. Will, a recent PhD in Liebig's laboratory, had stayed in Giessen as one of Liebig's assistants. A highly trained analyst, he was also a skilled glassblower, capable of making and improving Wöhler's proposed apparatus. Varrentrapp was a PhD student fresh from Rose's Berlin laboratory, where he had mastered the practical intricacies of quantitative inorganic analysis—including techniques that would be vital to the new method's success and reliability.

Even so, Liebig did not report significant progress to Wöhler until the end of June 1841.[99] Published later the same year, Will and Varrentrapp's paper detailed how to make and use a modified Kaliapparat that reliably measured nitrogen by mass for the first time.[100] Liebig did not appear as coauthor of Will and Varrentrapp's paper—even though they had worked in Giessen, under his direction, using the full range of Giessen expertise. Nor did Will and Varrentrapp thank Liebig, as Blanchet and Sell had done in 1833. A range of explicit and implicit references nevertheless indicate Liebig's virtual presence in this work. Liebig used editorial privilege to endorse Will and Varrentrapp's method, calling it "one of the most important improvements in organic analysis, because it gives the determination of nitrogen a hitherto lacking reliability, simplicity, and accuracy."[101]

Throughout, Will and Varrentrapp explicitly referenced Liebig's views and methods, as well as his original Kaliapparat. They began by rehearsing concerns, readily identified as Liebig's, about existing volumetric methods of nitrogen determination, particularly when applied to substances containing little nitrogen. For example, they asserted continuing uncertainty about the alkaloids' composition as one of the primary motivations for their work—a clear allusion to Liebig's ongoing dispute with Regnault.

Will and Varrentrapp claimed that their new device—made by Will to function in the same way as Wöhler's prototype—made nitrogen determination "just as simple and secure" as measuring carbon using Liebig's standard Kaliapparat. The significance of their work for the development of chemical glassware is pursued in chapter 3. For now, the important point is that Will and Varrentrapp's modified device measured nitrogen quite separately from any other element, just as Liebig's original Kaliapparat had done for carbon.[102]

Will and Varrentrapp's method offered what Liebig had sought for more than a decade: a simple, direct (i.e., independent) method of measuring nitrogen that was universally reliable. They presented analyses of uric and hippuric acids—Liebig's test substances while developing the Giessen approach to nitrogen. They had also analyzed the alkaloids narcotine and brucine, together with a multitude of other compounds containing widely varying amounts of nitrogen. Their method proved vastly superior to any existing technique, remaining chemists' standard approach to measuring nitrogen for almost half a century. Thus, Will and Varrentrapp at last brought Liebig's original project to a conclusion by offering a durable solution to the problem of nitrogen determination that helped advance chemical understanding of alkaloids. The Giessen school was beginning to fulfill its purpose.

ANALYTICAL IMPASSE

This account clarifies aspects of Liebig's work during the 1830s that have remained puzzling until now—including the fact that the Giessen school did not grow rapidly until 1835.[103] Despite the Kaliapparat's relative simplicity and eventual success, it took Liebig and his students until 1834 to incorporate the new device into a stable, teachable analytical practice. Recognizing students' contribution to the Giessen approach highlights 1835 as an inflection point, explaining both the small size of Liebig's early research school prior to this date and its rapid growth thereafter.[104]

Liebig's laboratory was not fruitful because of mere multiplication of routine work. The Giessen approach entailed many analyses to produce each new formula, but the laboratory's productivity lay in students' ability to assign reliable formulae at a time when analysis remained anything but routine. Local collaborative confirmation improved analytical consistency, while studying substances related by simple reactions constrained the formulae that could reasonably be assigned. Transformation and taxonomy were crucial to the Giessen approach.

Research in Giessen certainly implied proficiency in using the Kaliapparat. It also encompassed the range of procedures—theoretical as well as empirical—that Liebig believed necessary to make sense of uncertain analytical results. The Giessen approach embodied Liebig's form of laboratory reasoning. Convinced that this was the only way that analysis could become scientific, Liebig taught his students how to negotiate between experiment and theory, attempting to codify these guidelines for laboratory reasoning in published papers written by students as well as by himself.

Liebig's editorship of the *Annals* was an important component of his plan to develop and control the new discipline of organic chemistry.[105] His strategy was nevertheless only partly successful. Although many chemists were using the Kaliapparat by 1834, it proved difficult to convince them to adopt the Giessen approach. Liebig responded in 1837 by publishing a manual incorporating significant developments in analytical practice post-1831. The *Instructions* were a manifesto for the Giessen approach, intended to disseminate research methods developed in Liebig's laboratory over several years.[106] Where formerly it was hard to see why Liebig delayed writing this manual for so long, this study shows he could hardly have produced it any sooner.

Liebig had turned the tables on Paris, but by 1837 he was once again becoming isolated. French students studied in Giessen, no doubt enhancing his school's global reputation. Liebig was established as Dumas's peer, despite their ongoing analytical differences. Theoretical developments, however, multiplied the opportunities for disagreement, and a public reconciliation between Liebig and Dumas proved short-lived. Disputes over experiment and theory, meanwhile, undermined Liebig's relationship with Berzelius, leaving Wöhler as his closest ally.[107]

Liebig's efforts to promote the Giessen approach reflected a relentless quest for scientific advance that required the resolution of widespread disagreements about analytical practice and interpretation, and what it meant to do research. Inside his laboratory, Liebig could require compliance. But his attempts to force consistency elsewhere were resisted by those who considered philosophical chemistry a more individualistic pursuit. Instead of establishing agreed-on standards, Liebig's actions generated discord that—as later chapters explain—would not be overcome until organic chemistry was established as a professional discipline.

Liebig's ambitions were thwarted by other chemists' persistent rejection of the Giessen approach. Nitrogen was particularly significant in light of his growing interest in agriculture and physiology. But the solution that Liebig proposed in 1834 was almost entirely ignored, Dumas's method remaining in favor until it was superseded by Will and Varrentrapp's new approach in the 1840s. It was surely an especially bitter blow that even Giessen students did not always appreciate the superiority of his methods. Liebig's personal involvement with cutting-edge analysis—and alkaloid analysis in particular—dwindled rapidly following his dispute with Regnault, his final published contribution appearing in 1840.[108]

If Liebig by now was unwilling to continue pushing the boundaries of organic chemistry, he was ideally placed to shape the future of the field through the work of his students, Will and Varrentrapp. Their development of a modified Kaliapparat at last solved the problem of nitrogen, repaying Liebig for a decade of intense labor. In fact, Will and Varrentrapp's solution—created from glass tubes by glassblowing—seems so obvious it is curious that Liebig did not think of it himself. One explanation is that Wöhler—notwithstanding his early reluctance to tackle organic analysis— was the better experimentalist. It is also plausible that, not burdened by running a large laboratory, Wöhler was at this time more immersed in laboratory work than was Liebig.[109] Liebig, meanwhile, was locked in a futile struggle for disciplinary control.

By January 1844, Liebig abandoned organic analysis altogether.[110] Previous studies have identified several reasons for this behavior—including the seemingly irresolvable disputes about the theoretical interpretation of analytical results discussed above.[111] This study contributes an additional explanation that is also the first to account for subsequent disciplinary developments. By the early 1840s, Liebig was convinced that organic analysis was approaching its limit, incapable of supporting sound conceptual progress in the study of alkaloids, substances whose medical efficacy, commercial value, and scientific interest placed them at the forefront of chemical enquiry. Although chemists could now assign reliable formulae for many simple organic substances, continuing disagreements concerning the composition of natural alkaloids severely hampered disciplinary progress. It was in hopes of transcending this impasse, as chapter 4 explains, that Liebig directed his student August Wilhelm Hofmann to develop synthesis as a new method for investigating organic compounds.

When he dubbed Liebig "the chemical gatekeeper," William Brock was referring to Liebig's diverse entrepreneurial activity at the intersection of chemistry and society.[112] This study indicates that the term might be applied with at least equal justification to Liebig's pioneering contributions to academic chemistry—accomplishments that, as we have seen, enormously transcended the Kaliapparat plus research school pair. As chapter 3 explains, Liebig had also unleashed a powerful material force that would shape the discipline for decades to come. In learning to make and use Liebig's Kaliapparat, chemists came to appreciate new and remarkable possibilities of glass and glassblowing.

3   GLASSWARE REVOLUTION

Justus Liebig's decision in the fall of 1830 to analyze alkaloids using a small piece of home-blown glassware had momentous consequences for chemistry. As chapters 1 and 2 have shown, however, this was neither because Liebig's new device—the Kaliapparat—solved the problem of organic analysis, nor because it powered the development of the Giessen school. In fact, using the Kaliapparat rather than Parisian apparatus did not improve the accuracy with which chemists measured the carbon content of organic compounds. The Kaliapparat's primary significance lay in its speed and simplicity, and its vastly reduced cost. These were the qualities that made it possible for Liebig to challenge and ultimately overthrow Parisian analytical domination.

Small, cheap glassware was essential not only to the transformation of scale and practice at the heart of Liebig's analytical project, but also in driving chemistry's broader development. In launching the Kaliapparat, as we shall see, Liebig played a major role in persuading other chemists to use flameworked glassware—a change that I have called elsewhere the "glassware revolution."[1] From the mid-century, professional chemists in Europe and beyond increasingly relied on apparatus fashioned from glass tubing "in the flame of a proper lamp."[2]

The distinction between lamp- or flameworking and furnace-based glassblowing is vital in recognizing this historical landmark. Although both are commonly referred to as glassblowing, lampworking does not involve the use of a glassmaking furnace. Instead, lampwork entails manipulating glass tubing that is heated in the flame of a lamp or, more recently, a glassblower's torch.[3] The term "glassware revolution" describes how chemical apparatus made by lampworking became a vital supplement to items produced by furnace glassblowing, such as the flasks, retorts, and phials that were already in widespread use by chemists, apothecaries, and pharmacists.

Apparatus they made themselves from glass tubing using simple equipment minimized chemists' reliance on professional instrument makers. Between the mid-nineteenth and mid-twentieth century, trainee chemists routinely acquired at least basic glassblowing skill. Lampworked glassware, as this chapter illustrates, greatly improved experimental flexibility and control. For chemists at the cutting edge, its advantages were decisive. As a result, the glassware revolution opened a new phase in organic chemistry's development, one that would be crucial to disciplinary advance. The very flexibility of working in glass, however, introduced problems as well as opportunities, unforeseen tensions that threatened the nascent science at its core.

Understanding how the Kaliapparat could have such far-reaching consequences requires situating the device in its proper experimental tradition. We saw in chapter 1 that Liebig learned organic analysis in Paris, and that his subsequent isolation in Giessen drove him to make a series of innovations, including the Kaliapparat. Ambition and geography were crucial factors motivating Liebig to take on the might of Paris. As this chapter explains, they also shaped the material dimensions of his project, prompting a radical change in apparatus and ways of working. Essential to Liebig's ability to analyze alkaloids in Giessen, moving chemistry into lampworked glass initiated a much wider disciplinary transformation.

COMPETING TRADITIONS

Wealthy Parisian chemists relied on the skills of professional scientific instrument makers, such as Nicolas Fortin and Charles Félix Collardeau. When Antoine Lavoisier first analyzed organic compounds in the 1780s, he used large, complex apparatus made for him by Fortin (figure 3.1). Despite its sophistication and expense, this apparatus did not enable Lavoisier to produce useful quantitative analyses of organic compounds. By about 1790, Lavoisier abandoned quantitative organic analysis as an unsolved problem.[4]

Twenty-five years later, Lavoisier's protégés Joseph Louis Gay-Lussac and Louis Jacques Thenard took up the same problem, developing new analytical methods based on the precision measurement of gas volumes—then a mainstay of Parisian science (figure 3.2). Collardeau made the accurately

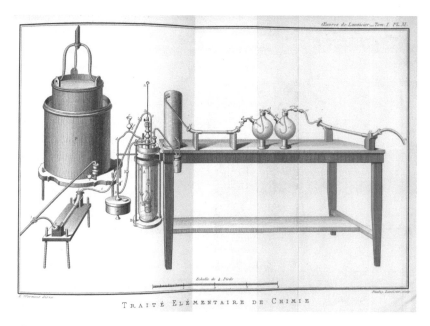

FIGURE 3.1

Lavoisier described this large, complex and expensive apparatus for organic analysis and its use in analyzing oils. The illustrator included a 4-foot ruler to emphasize
the scale of this apparatus. Two glass bulbs, connected by brass tubing and partially
filled with sodium hydroxide solution, absorbed the carbon dioxide produced on
combustion. Lavoisier subsequently used up to nine such bulbs in the attempt to
ensure complete absorption (see Antoine L. Lavoisier, *Oeuvres de Lavoisier,* vol. 3
(Paris: Imprimerie Impériale, 1865), 774). A similar apparatus including five glass
bulbs—attributed to the celebrated Parisian instrument maker Nicolas Fortin—is displayed among Lavoisier's laboratory equipment in the Conservatoire National des
Arts et Métiers, Paris.

Source: Antoine L. Lavoisier, *Oeuvres de Lavoisier,* vol. 1 (Paris: Imprimerie Impériale,
1864), plate XI. Courtesy of the University of Wisconsin-Madison Libraries.

graduated glass vessel Gay-Lussac and Thenard used to measure how much
gas was produced by burning an organic compound. Fortin, meanwhile, is
plausible as maker of the specialized ground glass tap that—according to
the two chemists—"constitute[d] the entire merit of the apparatus."[5] Only
the assistance of some of the world's leading instrument makers made it
possible for Gay-Lussac and Thenard to produce the first convincing quantitative analyses of organic compounds.

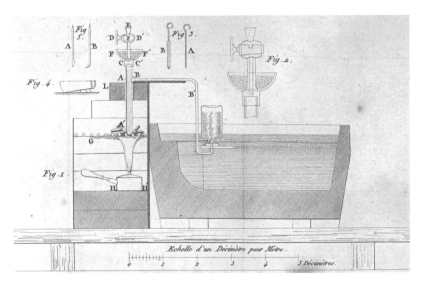

FIGURE 3.2

Note the large size and complexity of Gay-Lussac and Thenard's apparatus for organic analysis, illustrated here with meter scale below. Their method relied on volumetric measurements made using the graduated glass receiver made by Collardeau. This apparatus also incorporated an unusual vertical combustion tube (A′A) fitted with a 3-way ground glass tap at the top (illustrated at higher magnification and labeled as "Fig. 2," top right). Each component of this apparatus was made by skilled instrument makers. As a result, it was costly to produce and impossible to replicate beyond the Parisian metropolis.

Source: Joseph-Louis Gay-Lussac and Louis-Jacques Thenard, *Recherches Physico-Chimiques,* vol. 2 (Paris: Deterville, 1811), plate 6. Courtesy of the Cole Collection, University of Wisconsin-Madison Libraries.

But no chemist outside Paris had access to such resources. In Stockholm, Jöns Jacob Berzelius lamented the 3-month wait for a simple glass retort made by working molten glass drawn from a furnace.[6] The need for self-reliance prompted Berzelius to experiment using small, cheap apparatus that he fashioned himself by working glass tubing in a flame. Expert in mineralogical analysis using the blowpipe—another practice that relied on small, portable apparatus—Berzelius had learned how to blow glass as a young man.[7] He also believed simple, easy-to-use apparatus was fundamental to reliable science.[8]

When Berzelius challenged Gay-Lussac and Thenard's analyses (as discussed in chapter 1), he did so using a completely different experimental approach. Berzelius measured substances by mass instead of volume, a shift that transformed the apparatus required. Because a large volume converts to a tiny mass, Berzelius needed only table-top apparatus composed of small glassware components and a beam balance whose accuracy belied its simple construction (figure 3.3). Parisian analysis was an elite practice based on precision measurement of gas volumes using large, complex apparatus made by expert artisans. The reliability of Berzelius's analyses, meanwhile, depended only on good experimental technique and simple equipment. Whereas Parisian methods and apparatus excluded all but a few, Berzelius's approach was open to anyone with the necessary manipulative skill.

Chapter 1 explained that when Liebig turned to organic analysis in 1826, he did so using apparatus acquired from Paris. Although by this time, Parisian

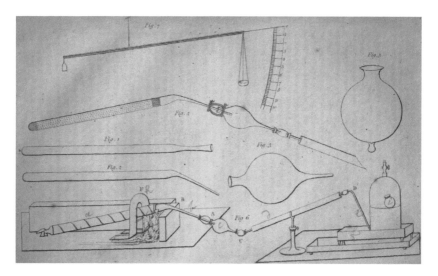

FIGURE 3.3

Berzelius's illustration of his table-top apparatus marked a watershed in both chemical practice and apparatus. Whereas Parisian chemists' elegant, expensive apparatus excluded most other chemists, Berzelius's glassware components and beam balance invited participation. Life-sized drawings of the condenser (labeled "Fig. 3") and the small glass vessel containing solid potassium hydroxide (used to absorb carbon dioxide) provided a template for anyone wishing to replicate Berzelius's experiments.

Source: Jöns Jacob Berzelius, "Experiments to Determine the Definite Proportions in which the Elements of Organic Nature Are Combined," Part 2, *Annals of Philosophy* 4 (1814), plate XXV. Courtesy of Hathi Trust: https://hdl.handle.net/2027/umn .31951000744434n?urlappend=%3Bseq=429%3Bownerid=13510798902441358-449.

analytical apparatus incorporated some aspects of Berzelius's approach, it still relied on volumetric measurement and hence on an accurately graduated receiver that could only be made by a skilled instrument maker (figure 3.4). But Liebig's Giessen location meant that he—like Berzelius—had limited access to such expert makers. In any case, Liebig lacked the financial resources to enlist skilled professional labor to his cause. Even keeping his Parisian apparatus in working order in Giessen was problematic, leading Liebig to learn basic glassblowing skill during a visit to Paris in late 1828. On

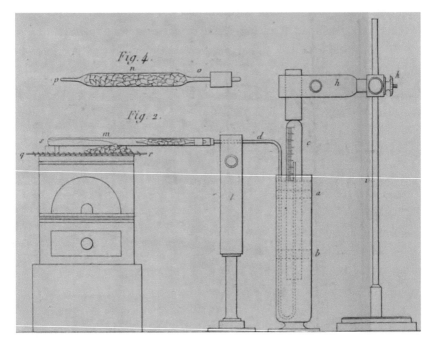

FIGURE 3.4

Liebig used this apparatus to analyze silver fulminate in Paris, bringing it to Giessen in 1826. The horizontal combustion tube, which reflects the influence of Berzelius's work, contained a small calcium chloride tube to trap water vapor, so that only carbon dioxide gas was collected in the graduated glass receiver. As with Gay-Lussac and Thenard's apparatus (see figure 3.2), only a skilled instrument maker such as Collardeau could make this precision measurement device.

Source: Justus Liebig and Joseph-Louis Gay-Lussac, "Analyse du Fulminate d'Argent," *Annales de Chimie et de Physique* 25 (1824): fig. 2. Courtesy of Hathi Trust: https://hdl.handle.net/2027/iau.31858046217851?urlappend=%3Bseq=459%3Bownerid=102385406-3.

his return, Liebig installed a French glassblowing table in his laboratory so that he could make his own repairs.[9] In this way, Liebig was able to continue working in the Parisian analytical tradition in Giessen.

When Liebig began analyzing alkaloids, however, he confronted problems Parisian analytical apparatus could not solve. The resulting pressure to innovate exposed the severe limitations of his circumstances. We saw in chapter 1 that Liebig traveled to Hamburg in August 1830 to meet Berzelius— then one of Europe's leading chemists—and soon afterward began to enjoy his patronage. In meeting Berzelius, however, Liebig not only benefited from the advice of a master analyst. He also encountered a consummate experimenter whose exceptional capabilities were inseparable from his glassblowing skill. Meeting Berzelius exposed Liebig to a chemist who operated in an entirely different way from his Parisian teachers. Immediately on his return from Hamburg, Liebig invented the Kaliapparat. In so doing, he allied himself to the Berzelian mode of working. Chapter 2 explained why this step was essential to the subsequent development of Liebig's Giessen school. I now show how it set the glassware revolution in motion.

LIEBIG'S KALIAPPARAT AND THE GLASSWARE REVOLUTION

The evidence indicates that Liebig himself made the original Kaliapparat, developing the new device through a series of prototypes during the fall of 1830. By the time he published "On a New Apparatus" in January 1831, Liebig's Kaliapparat was beginning to resemble the now-familiar triangular piece of glassware incorporating five glass bulbs (figure 3.5). Having announced his new apparatus, Liebig embarked on a campaign to spread and stabilize its use. Aided by his students, Liebig soon convinced others to use the Kaliapparat. But, as chapter 2 explained, it proved much harder to persuade chemists elsewhere to adopt the procedures that would enable them to replicate results obtained in Giessen. There were also significant material barriers to the widespread uptake and appropriate use of Liebig's new device. Overcoming these obstacles, as we shall see, affected much more than organic analysis.

Just as Liebig's innovation relied on lampworking, so other chemists' access to his new device depended on the availability of glassblowing skill. We can be sure that Liebig recognized this because of what he did immediately after inventing the Kaliapparat. Even as he published "On a New Apparatus," Liebig sent his new device to two people. The first was

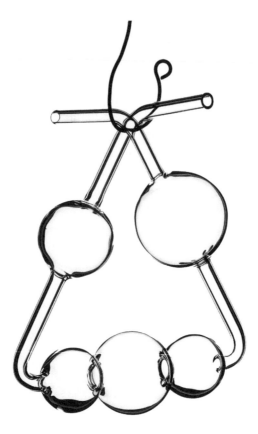

FIGURE 3.5

Liebig's Kaliapparat measured carbon by trapping carbon dioxide gas produced when an organic substance burned in potassium hydroxide (potash) solution. This was how the Kaliapparat got its English name: potash apparatus or potash bulbs. Weighing the Kaliapparat before and after combustion gave the mass of carbon dioxide, and hence the mass of carbon in the sample. The photograph shows a replica Kaliapparat made in 2017 by Tracy Drier, master scientific glassblower in the Chemistry Department, University of Wisconsin-Madison.

Source: Photograph courtesy of Ilia Guzei.

Berzelius, Liebig's inspiration and new mentor. Berzelius, a skilled glass-blower, made Liebig's Kaliapparat for himself with only the finished object as a guide. The other was Prague pharmacist Johann Baptist Batka, who had come to Liebig's notice during his life-changing visit to Hamburg.[10]

Batka was proprietor of the firm founded by and named after his grand-father Wenzel Batka. In addition to chemical and pharmaceutical prepara-tions, Wenzel Batka sold apparatus, including some items of furnace-blown glassware, such as the pharmacists' separatory funnel (forerunner of a similar item of chemical apparatus, discussed further in chapter 6).[11] Thus Batka, whether or not he was himself a glassblower, employed people who were. For Liebig, Batka offered a route to glassblowing skill, a resource he hoped to exploit by sending Batka his new device. For Batka, the connection to Liebig held the promise of commercial gain. When "On a New Appara-tus" appeared, it indicated Batka as a possible supplier of Liebig's analytical apparatus.[12]

If Liebig hoped its availability from Batka would promote immediate, widespread uptake of the Kaliapparat, he was to be disappointed. There are several plausible explanations for this outcome. The community of those working at the cutting edge of organic analysis was small, an elite for whom Batka was an unlikely supplier—especially given the dubious reputation of his products.[13] When Liebig's Kaliapparat first appeared in Batka's catalog in 1832, it was poorly formed, of unspecified dimensions, and unlikely to produce reliable results.[14] Indeed, it is quite possible, given Liebig's level of glassblowing skill, that the Kaliapparat he sent to Batka was somewhat asymmetrical—leading Batka to produce a similarly imperfect device.[15]

Chemists beyond the community of elite analysts as yet had no persua-sive reason to choose Liebig's apparatus—even supposing they wished to enter this contentious field. Thus, Liebig instead proceeded by introducing those in his immediate professional circle to the Kaliapparat—a strategy that maintained control over his new device during these crucial early years.[16] As we have seen, these chemists could access the Kaliapparat without dif-ficulty. Some—like Berzelius—were themselves skilled glassblowers, while others—including Liebig's Parisian analytical rival, Jean Baptiste Dumas—could call on this skill in others.

In Giessen, Liebig continued working to develop the use of the Kaliap-parat. In 1832, as discussed in chapter 2, he guided his students Rodolphe Blanchet and Ernst Sell in developing a reliable, collaborative analytical method based on the Kaliapparat. Published the following year, Blanchet and Sell's paper sought to establish and distribute the standard procedures necessary for chemists beyond Giessen to produce reliable analytical results using Liebig's apparatus.[17]

Their work no doubt also exposed the limitations of a nonstandard device, such as then used in Giessen and advertised by Batka. As he sought to extend the use of his new analytical apparatus, Liebig surely learned that standardizing the Kaliapparat would be vital in its transformation from personal innovation to widely used research tool. Even the most skillful glassblower would require precise dimensions to produce a device capable of replicating Liebig's results.

Thus, standardization of instrumentation as well as technique emerged as a critical barrier to Liebig's goal of reforming organic analysis. Liebig's actions show that he recognized both facets of this problem. His solution was typically ambitious. As he campaigned to distribute his analytical methods more widely, Liebig defined a standard Kaliapparat and promoted chemists' acquisition of lampworking skill. Until now, it had been far from obvious to most chemists that they had anything to gain from this artisanal skill. Glassblowing therefore did not form part of standard chemical training.[18]

Beginning in 1833, Liebig sought to change this status quo. If chemists beyond the elite were to use the Kaliapparat successfully, they had to be able to make it in standard form, which in turn gave them a compelling reason to work in glass.[19] This was how the Kaliapparat promoted chemistry's move into flameworked glassware, becoming a significant driver of the glassware revolution. An important step in this process took place in 1833, when Liebig's *Annals of Pharmacy* published an article titled "On the Art of Blowing Glass" that included the first printed instructions for making a standard Kaliapparat.[20]

Before going further, it is vital to appreciate that very few lampworking instructions intended for chemists were then available, either as standalone books or sections of manuals and textbooks.[21] Hands-on instruction was then almost entirely restricted to metropolitan centers, including London and Paris. Such texts were therefore potentially valuable pedagogical tools in propagating glassblowing skill.[22] They also served as vehicles for raising chemists' awareness of the possibilities of glassblowing. "On the Art of Blowing Glass" provided some of the earliest lampworking instructions to be made widely available to chemists. Liebig's decision to incorporate instructions for making a standard Kaliapparat into this article was thus a highly significant move.

Written by a young Parisian glassblower called Lafond, the text Liebig republished in the *Annals* had first appeared in 1832 in the French *Journal of Everyday Knowledge*. Subsequently translated into German for Johann Gottfried Dingler's *Polytechnic Journal,* in its original form, "On the Art of Blowing Glass" provided basic instructions in lampworking and enameling, describing the lampworker's basic tools and most important skills.[23] When it appeared in German translation, Lafond's article incorporated numerous "improvements" due to Lafond's teacher, the Parisian master glassblower Ferdinand Danger.[24] This German translation presented Liebig with a valuable opportunity. Reprinting Lafond's lampworking instructions (with Danger's improvements) in the *Annals,* Liebig made notable further alterations of his own.

Liebig replaced Lafond's description of how to make simple apparatus including a funnel with detailed instructions for making two items of chemical glassware that he clearly viewed as of far greater utility and relevance to practicing research chemists, one of which—as mentioned above—was his own Kaliapparat.[25] He supplemented the original illustrations of the glassblower's lamp and other tools with diagrams showing each step in the construction of these devices, as well as the exact dimensions of the finished object (figure 3.6). The order in which the Kaliapparat's multiple bulbs should be blown was crucial, since this was particularly difficult to infer from the apparatus's final form.[26] In this way, Liebig did as much as he could using text and illustration to enable other chemists to make a standard Kaliapparat.

"On the Art of Blowing Glass" indicates that spreading the use of the Kaliapparat depended on chemists learning lampworking. It fixes important new milestones in standardizing the Kaliapparat, and in Liebig's campaign to distribute his analytical methods that accord well with chapter 2's account of the development of Giessen analysis.[27] By 1833, Liebig's Kaliapparat—although it had not yet acquired its final form—was a standard device. In addition to using the *Annals* to propagate the standard procedures at the heart of Giessen analysis, Liebig also used his journal to disseminate the material skills and standards required for others to replicate his methods.

From 1833, Liebig no longer referred explicitly to his device when publishing analytical results.[28] But if Liebig imagined other chemists were now making the standard Kaliapparat and using it according to his procedures, he was wrong. Chapter 2 explained how other chemists' reluctance to adopt

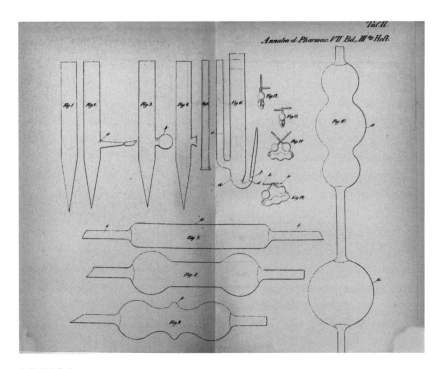

FIGURE 3.6

Liebig published the first instructions for making a Kaliapparat in his *Annals of Pharmacy* in 1833, alongside an introduction to basic glassblowing. As with Berzelius's analytical apparatus, this plate included a life-sized drawing (labeled "Fig. 10") that defined the Kaliapparat's size. But where Berzelius had illustrated only the finished object, Liebig also showed the sequence of steps by which this instrument was best constructed.

Source: Lafond, "Ueber die Kunst Glas zu blasen, mit Verbesserungen von Danger," *Annalen der Pharmacie* 7 (1833): Tafel II. Courtesy of Hathi Trust: https://hdl.handle.net/2027/uva.x002457886?urlappend=%3Bseq=375%3Bownerid=27021597768474033-419.

his analytical standards led Liebig to publish a new manual. In addition to defining methods and procedures, his (1837) *Instructions for the Analysis of Organic Bodies* (*Anleitung zur Analyse organischer Körper*) specified in words and images how to construct a standard Kaliapparat, showing the actual size of the finished object (figure 3.7).

Small but significant differences distinguished Liebig's 1833 Kaliapparat from its final 1837 form. The size and shape of the device were almost

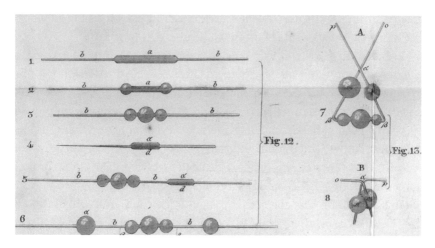

FIGURE 3.7

By 1837, when Liebig's *Instructions for the Analysis of Organic Bodies* (*Anleitung zur Analyse organischer Körper*) appeared, general glassblowing instructions were no longer necessary. But chemists still needed to know how big Liebig's Kaliapparat should be and how best to make it. Liebig therefore illustrated the sequence of glassblowing steps chemists should follow to recreate his apparatus. The bent tubes at α ("Fig. 13") were important, because they made it possible to hang the Kaliapparat from the hook of an analytical balance.

Source: Justus Liebig, *Anleitung zur Analyse organischer Körper* (Braunschweig: Vieweg, 1837), Tafel II, Figs. 12 and 13. Courtesy of the Hanna Holborn Gray Special Collections Research Center, University of Chicago Library.

identical, and the recommended construction sequence remained substantially the same (compare figures 3.6 and 3.7). But by 1837, the internal tube diameter connecting the three bulbs on the base was significantly narrowed. This change improved absorption of carbon dioxide and hence analytical accuracy. It also required significantly greater glassblowing skill and may well have involved the use of specialized tools.[29]

Moreover, the 1837 instructions for making the apparatus were published in a quite different context. When "On the Art of Blowing Glass" appeared in 1833, the Kaliapparat was not yet available from commercial sources, and few chemists were skilled glassblowers. Four years later, many chemists across Europe had learned to blow glass—making it superfluous to explain basic lampworking techniques. Such relatively inexperienced glassblowers continued, however, to need detailed instructions for making the

device, including the exact sequence of operations by which the Kaliapparat was best formed from lengths of glass tubing (see figure 3.7).

Liebig turned to lampworking in response to the limitations of his
Giessen situation. But by developing a cheap, effective analytical tool, he
changed much more than organic analysis. Although the Kaliapparat did
not initially produce superior results, it made carbon determination simpler
and faster than previous analytical methods. Even more important, its low
cost meant any chemist could afford to use it—provided they knew how to
make it. This new economy enabled Liebig to involve students in his project
on a hitherto impracticable scale. But it was equally important to Liebig's
disciplinary vision that chemists throughout the community should adopt
the methods of Giessen analysis. While the Kaliapparat remained a self-
made item, chemists had to be able to blow glass in order to make it. For
this reason, Liebig promoted the fundamental transformation in chemists'
skill set and the material basis of chemical experimentation that constitutes
the glassware revolution.

Glassblowing skill had the potential to liberate chemists—as it had freed
Berzelius and Liebig—from reliance on wealth and instrument makers. By
spreading lampworking skill, Liebig simultaneously drew chemists everywhere into the emerging field of organic analysis and equipped them to
capitalize on the advantages of apparatus made from glass tubing. In this
way, the glassware revolution enabled cutting-edge chemistry's expansion
beyond metropolitan centers like Paris and its rapidly increasing scale during
the remainder of the century. As the balance of this book shows, chemistry's move into glass was vital to the development of the science. Chemists'
growing ability to produce their own apparatus nevertheless proved to be
both a threat to and an opportunity for Liebig's analytical project.

GLASSBLOWING AS OPPORTUNITY:
SOLVING THE PROBLEM OF NITROGEN

Chapter 1 explained how measuring morphine's nitrogen content initially drew Liebig to specialize in organic analysis, and how he strove to
resolve this difficult problem. Despite his efforts—and those of his rival
Dumas—throughout the 1830s, nitrogen determination continued to rely
on complex, custom-made volumetric apparatus that produced results of
dubious reliability. Few chemists could measure nitrogen, and those who

did struggled to agree on how much nitrogen compounds contained. This was especially true for alkaloids such as morphine. As a result, the formulae of these chemically interesting and medicinally important substances remained a matter of debate and uncertainty.[30]

We saw in chapter 2 that this situation did not begin to change until 1841, when two young Giessen chemists, Franz Varrentrapp and Heinrich Will, published a new method of nitrogen determination. Inspired by a letter from Liebig's good friend Friedrich Wöhler that proposed completely changing the chemistry underpinning nitrogen determination, their method was based on a modified Kaliapparat. As we shall see, Will and Varrentrapp brought Liebig's original analytical project to a conclusion by capitalizing on the unprecedented possibilities lampworking offered for designing experiments and controlling their outcomes.[31]

Wöhler's innovation relied on a new piece of glassware he had made himself by glassblowing (figure 3.8). Exactly as Berzelius had done for hydrogen, and Liebig's Kaliapparat accomplished for carbon, Wöhler

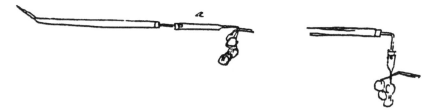

FIGURE 3.8

Wöhler's modified Kaliapparat incorporated an additional tube (a) joined to a Kaliapparat. By warming tube (a) during the analysis, Wöhler ensured ammonia produced by combustion remained in the form of a gas and so passed through a Kaliapparat containing dilute hydrochloric acid, in which alkaline ammonia readily dissolved. Wöhler did not explain when and why the tube connecting (a) to the combustion tube should be bent (as shown on the right), but this seems likely to be intended to ensure that even liquefied ammonia reached the Kaliapparat, and perhaps also to reduce the likelihood that the Kaliapparat's cold contents might be drawn back toward the hot combustion tube. Writing to an experienced glassblower and leading analyst, Wöhler clearly felt it unnecessary to provide details of either analytical or glassblowing procedures.

Source: Wöhler to Liebig, 30 October 1840. Bayerische Staatsbibliothek München, Liebigiana II. B. Wöhler, Friedrich, Nr. 194 vom 30.11.1840, S. 1r. Courtesy of the Bavarian State Library, Munich.

measured nitrogen separately from any other element. Wöhler converted organic nitrogen to gaseous ammonia that he trapped in a modified Kaliapparat containing hydrochloric acid. Developed while analyzing uric acid, his new device functioned well in this instance. But—as Wöhler clearly recognized—a single analysis was far from sufficient to establish the device's wider validity.

It may seem strange that Wöhler should offer such an important innovation to Liebig. It should not. Although Wöhler was by this time a skillful analyst of organic compounds, organic chemistry was not his specialty. Even more important, Wöhler's situation was very different from Liebig's—beyond as well as inside the laboratory. Since Wöhler did not share Liebig's community-building ambitions, young chemists in Göttingen stood in a quite dissimilar relationship to their mentor's research from those in Giessen. Although they sometimes aided Wöhler, their work did not contribute systematically to a defined research program in the way that junior chemists advanced Giessen analysis. But nitrogen determination was immediately relevant to Liebig's research, with the potential to shape organic chemistry as a field. Liebig was therefore both more motivated and vastly better placed than his friend Wöhler to make use of this important innovation.[32]

Guided by Liebig, in 6 months Will and Varrentrapp accomplished for nitrogen what had taken Liebig 3 years for carbon. Their work focused in three main areas: ensuring the chemistry Wöhler proposed worked smoothly; demonstrating the new method's general applicability to a wide range of organic compounds containing nitrogen; and, finally, standardizing apparatus and procedures that rendered measuring nitrogen as reliable as possible. The first two relied solely on chemical understanding. For example, Will and Varrentrapp found, contrary to Wöhler's original report, that not all the nitrogen contained in uric acid was easily converted to ammonia—a problem they overcame by adjusting the conditions of analysis.[33] The last, however, combined knowledge of chemical theory and practice with considerable glassblowing skill.[34]

Where Liebig had struggled to make a device that reliably trapped all the carbon dioxide produced when an organic substance burns, Will and Varrentrapp contended with exactly the opposite problem. In contrast to carbon dioxide's reluctance to dissolve in potassium hydroxide solution, ammonia reacted all too readily with hydrochloric acid—so much so that the vigor of the reaction could easily lead to the production of a vacuum, drawing liquid back into the combustion tube. This phenomenon—known as "suck back" (*Zurückspritzen*)—had to be strenuously avoided. Its occurrence certainly invalidated the analysis, and—though Will and Varrentrapp did not feel it necessary to explain this to their chemist readers—contact with cool liquid would shatter the red-hot glass combustion tube, with potentially dangerous consequences.[35]

Will and Varrentrapp detailed procedures to improve the reliability of their method, including innovations that helped prevent suck back by controlling the complex flow processes taking place in apparatus composed of multiple glassware components. Some of these were chemical. For example, mixing the sample with a roughly equal weight of an organic substance containing no nitrogen ensured there would be enough hydrogen to convert elemental nitrogen to ammonia.[36] It also produced additional carbon dioxide that passed through hydrochloric acid entirely without dissolving. This visible stream of bubbles provided a clear indication of how fast gas was being produced, allowing the chemist to control the analysis and avoid suck back.[37]

Working with small, self-made glassware offered important possibilities for innovation, allowing Will and Varrentrapp to optimize the shape of the glass vessel used to absorb ammonia. Although it implemented similar chemistry, their "chosen form" of modified Kaliapparat differed significantly from both Wöhler's prototype and Liebig's original.[38] After trapping all the ammonia in the absorption device, its contents were transferred to a porcelain dish for the final stage of the analysis.[39] The method's accuracy relied on the completeness of this transfer, achieved by repeated washing of the absorption device to ensure no trace of ammonia was left behind. The ease with which hydrochloric acid absorbed ammonia, meanwhile, rendered superfluous the iconic triangular shape of Liebig's original Kaliapparat—a feature designed to maximize contact between gas and liquid. These factors led Will and Varrentrapp to give their device a more open U-shape that optimized transfer without compromising absorption (figure 3.9).[40]

By early summer 1841, some 6 months after receiving his friend's letter, Liebig reported to Wöhler that the new method of nitrogen determination met all expectations.[41] As with Liebig's original Kaliapparat, this was not because Will and Varrentrapp's device produced more accurate results than did existing methods. It was, however, the first to measure nitrogen by mass, and the first to produce reliable results in a relatively straightforward

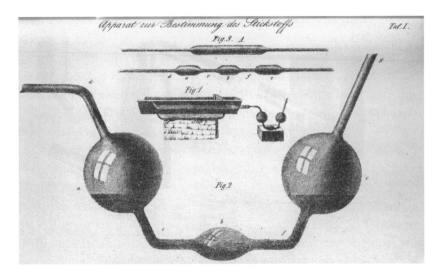

FIGURE 3.9

Will and Varrentrapp provided a life-sized diagram of their modified Kaliapparat for measuring organic nitrogen. Although their device differed from Liebig's original, it nevertheless belonged to a similar generic form. Will and Varrentrapp described and illustrated the series of glassblowing operations by which their device should be made. By 1841, these instructions were sufficient to enable other chemists to make this apparatus "easily for themselves" (Varrentrapp and Will, "Neue Methode," 268–269).

Source: Franz Varrentrapp and Heinrich Will, "Neue Methode zur Bestimmung des Stickstoffs in organischen Verbindungen," *Annalen der Chemie und Pharmacie* 39 (1841): Tafel I, figure 3. Courtesy of the University of Wisconsin-Madison Libraries.

manner using small, cheap apparatus. By developing Wöhler's innovation into a standard and widely applicable analytical method, Will and Varrentrapp transformed nitrogen determination from a tricky activity whose results were often hotly disputed to a routine technique that any competent student could be expected to learn.

When Varrentrapp and Will's paper appeared, it provided details of their analytical procedure. It also specified and illustrated the form and dimensions of their new device, as well as outlining how it was best produced from glass tubing—thereby standardizing the device as well as the method. Will and Varrentrapp assumed that readers would be able to construct the new device "easily for themselves" by following a sequence of bending, blowing, and joining operations.[42] Thus, their paper confirms that chemists across Europe possessed—or had access to—at least basic glassblowing skill by 1841.

In the wake of the glassware revolution, Will and Varrentrapp's method soon spread throughout the chemical community. Equipping chemists everywhere to measure nitrogen reliably made it possible to assign stable formulae to nitrogen-containing substances, including the alkaloids. In solving Liebig's original problem, they also enabled the new project that drew Liebig away from organic chemistry. Many physiologically important compounds contain nitrogen. From 1842, disillusioned by seemingly irresolvable theoretical disputes in organic chemistry, Liebig instead began using organic analysis—including Will and Varrentrapp's new method—as a stable tool of research in physiology.

GLASSBLOWING AS THREAT: LOSING CONTROL OF THE KALIAPPARAT

Will and Varrentrapp's work shows how glassblowing made it possible to improve and diversify apparatus in response to new experimental contexts. Within the confines of Liebig's laboratory, where he retained control, students could develop an important new piece of apparatus without the resources of Paris. By 1840, largely thanks to Liebig, lampworking skill was becoming more widely distributed among the chemical community—which meant that Will and Varrentrapp's device spread more rapidly than did Liebig's original Kaliapparat. But the distribution of glassblowing skill introduced a fundamental tension between Liebig's desire to impose disciplinary standards and the innovative possibilities of working in glass.

This tension confronted Liebig with serious problems. As he launched new glassware and new methods into the chemical community, Liebig also sought to impose the procedures and standards of Giessen analysis. We have already seen that it proved difficult throughout the 1830s to propagate the procedures Liebig associated with using the Kaliapparat. To some extent, as we shall see, other chemists' objections were legitimate. The diversity of the organic world meant that Liebig's methods were not always adequate. But elsewhere, chemists' failure to adopt the standards of Giessen analysis point to a fundamental lack of consensus concerning what it meant for chemistry to be a science.

These problems go to the heart of organic chemistry in its crucial mid-century period of disciplinary formation. As we shall see, Liebig was trapped by his attempts to change chemistry from elite natural philosophy to an industrial science. Depicted by some historians as overly sensitive to criticism and even idly disputatious, Liebig's goals meant that he could not afford to overlook what he regarded as blatant attempts to plagiarize and undermine his work.[43] For example, the Berlin chemist Eilhard Mitscherlich tried in the 1830s to appropriate and modify Liebig's device without acknowledging the latter's priority, even attempting to use his political connections to control admissions to Liebig's laboratory. By such behavior, Mitscherlich stoked the potential for dispute in a science already fraught with disagreement. As his remarkable correspondence with and about Mitscherlich makes clear, Liebig's ambitions necessitated a public response.[44]

This train of events began in May 1832, when Mitscherlich visited Liebig in Giessen. A decade Liebig's senior, Mitscherlich—like Wöhler— had trained with Berzelius and was a skilled glassblower. Following an early career devoted to mineral analysis and crystallography, he was professor of chemistry in Berlin and the author of a textbook Liebig greatly admired. Primarily intended for those seeking a "general scientific training" rather than would-be professional chemists, Mitscherlich's multivolume *Textbook of Chemistry* was comprehensive in scope and positioned chemistry in the disciplinary spectrum of the sciences. It also emphasized chemistry's educational virtues, expressing ideas that Liebig would later develop in his passionate polemics for chemistry as the ideal basis for a liberal education.[45]

Mitscherlich had remained close to his former teacher Berzelius, who had recently become Liebig's mentor. On the face of it, Mitscherlich must have seemed a potentially powerful ally in Liebig's campaign to spread the Giessen approach to organic analysis. Liebig welcomed his more established colleague, and Mitscherlich returned to Giessen later that year to learn how to make and use the Kaliapparat. By October, Liebig was in Berlin. While there, he and Mitscherlich collaborated on the analysis of lactic acid, during which they "tested the operation of an apparatus for organic analysis." This work appeared in Liebig's *Annals* under their joint names at the start of 1833.[46]

What Liebig perhaps did not know is that Mitscherlich was at this time both seeking a new research project and preparing to produce a new edition of the textbook Liebig admired so much.[47] Within weeks of Liebig's departure from Berlin, Mitscherlich was performing independent analyses using the Kaliapparat, and he was already beginning privately to question Liebig's procedures and apparatus.[48] However, Mitscherlich's doubts did not prevent his cultivation of an apparently friendly relationship with Liebig. Mitscherlich spent the summer of 1833 working on his textbook, visiting

Giessen again in early October.[49] Indeed, when the new edition of his text-book appeared, it contained a lengthy discussion of organic analysis—a subject that had not featured in the original.[50] Mitscherlich's method of organic analysis differed in some respects from Liebig's but nevertheless relied on using Liebig's triangular Kaliapparat. Despite this, Mitscherlich neither cited Liebig's work nor thanked him for instruction in how to make and use the Kaliapparat in the first place.

Just as Liebig's work to establish and distribute Giessen analysis was gaining traction, Mitscherlich both denied Liebig's contributions and actively undermined his ambition to enable progress in organic analysis by standardizing procedure and instrumentation. Mitscherlich's textbook incorporated several of his own modifications in a description of what was basically Liebig's method of organic analysis. In light of all he had learned from Liebig, these actions were certainly ungenerous, and as we shall see, Liebig saw them as a betrayal.[51] Most likely on Berzelius's recommendation, he had received Mitscherlich into his laboratory as a friend—only to be challenged at the core of his project by a relative newcomer to the field.[52] By February 1834, Liebig saw Mitscherlich as a wolf "in sheep's clothing," who had used the "mask of friendship" to "disarm" him—presumably by putting him off his guard and damaging his attempts to control the Kaliapparat.[53]

Mitscherlich's behavior also placed Liebig in a tricky situation. Despite the appearance of Blanchet and Sell's foundational paper and Lafond's glassblowing instructions in Liebig's *Annals* during 1833, Liebig had yet to publish full details of his analytical method—making it hard to establish the extent of Mitscherlich's plagiarism. Without definitive documentary proof, challenging Mitscherlich was a risky strategy. Mitscherlich was known to be a touchy, disputatious man, unpopular with many of his Berlin colleagues.[54] Given his seniority and his close friendship with Berzelius, however, it must have been difficult for Liebig to anticipate how the chemical community would react to a direct accusation. In the end, Liebig chose to tackle Mitscherlich indirectly on grounds that were less personal and more clear cut—wisely, it turned out. If Liebig hoped his friends and mentors would rally to his cause, he was to be disappointed.

Keeping a close eye on Mitscherlich's future publications, Liebig soon spotted what he regarded as demonstrable malpractice. According to Liebig, Mitscherlich's recent paper on benzene not only expressed "completely topsy turvy theoretical opinions," it also "borrowed" work properly due to the young Parisian chemist Eugène Péligot.[55] On top of everything else, Liebig also discovered that Mitscherlich had sided against him during an earlier dispute. In 1833, Liebig had accused the Swiss chemist Carl Löwig of plagiarizing the first edition of Mitscherlich's textbook. Far from being grateful, however, Mitscherlich in fact defended Löwig—even after Löwig published a rebuttal incorporating a vicious personal attack on Liebig.[56] Little wonder that Liebig now declared "open war" on Mitscherlich, planning to publish a note claiming Péligot's priority and exposing the weakness of Mitscherlich's science.[57]

Liebig's intentions caused consternation among his friends and advisers. Berzelius wrote to him, urging him to reconsider: Mitscherlich had supplied Berzelius with a sample of benzene the previous summer and was therefore entitled to credit for this work.[58] Wöhler, encouraged by Poggendorff, hoped to avert "scandal" by persuading Liebig to leave his accusations against Mitscherlich unpublished. Liebig "might be perfectly correct, might personally have cause, might do science some service," but challenging Mitscherlich would nevertheless exclude Liebig from the scientific elite, leaving him easily dismissed as possessing "a petty mind." Liebig published anyway.[59]

Liebig's actions show clearly that much more was at stake here than a priority dispute affecting a third party. For the shopkeeper's son from Darmstadt, who had worked himself to the point of mental breakdown to achieve success, science was no mere recreation.[60] To remain silent was to give tacit approval—to personal injustice, scientific error, and actions he believed threatened disciplinary development. His colleagues might well have condemned Mitscherlich in private, as a family might handle a relative's bad behavior. But Liebig considered it essential that chemists now establish the kind of public morality suited to a disciplinary community. Passionate as ever, there was, he declared, "only one way . . . only one conviction for which I am prepared to sacrifice myself."[61]

Liebig saw Mitscherlich as a plagiarist whose behavior threatened to undermine disciplinary progress as well as Liebig's research program. Mitscherlich's paper failed to acknowledge prior work and encompass established practice. As a result, Liebig feared it would tempt others into "time consuming and pointless searches" for nonexistent theoretical entities—in this case, the radicals then believed to constitute organic compounds.[62] An incandescent Liebig wrote to Berzelius that Mitscherlich "betrays me like a Judas." After teaching Mitscherlich the skills to enter organic analysis,

Liebig now condemned him as a well-situated "leech" who fed on "everything in his vicinity" and, having "sucked it dry," spat out the residue "like a squeezed lemon."[63]

Matters came to a head following the publication of a new volume of Berzelius's textbook of chemistry in 1837.[64] Writing during 1835, Berzelius based his account of organic analysis on Mitscherlich's recently published textbook.[65] We saw in chapter 2 how Liebig's *Instructions for Organic Analysis* took issue with some procedures that Berzelius endorsed, and how this brought Liebig into conflict with his mentor. We can now appreciate that Mitscherlich—as the originator of these modifications—was the main target of Liebig's criticisms. Not content with using "every means to destroy [Liebig's] reputation and honor," Mitscherlich sought to "undermine" his means of subsistence by using his political connections to seek government restrictions preventing Prussian students from studying in Giessen. Worse still, Mitscherlich had systematically appropriated the prior work of analysts, including Berzelius and Liebig, as his own—hence "*his* method, *his* apparatus for organic analysis." If only, Liebig lamented, Berzelius had taken the trouble to learn about the improvements in analytical practice made in Giessen since 1831, rather than giving Mitscherlich his support and protection.[66]

Mitscherlich meanwhile continued to encroach on organic analysis in ways that presented a material as well as a procedural threat to the standardization at the heart of Liebig's disciplinary vision. Although Mitscherlich's textbook had described the use of Liebig's Kaliapparat, it specified neither the device's dimensions nor how these were to be attained in the process of making the device. Mitscherlich included abbreviated glassblowing instructions that assumed considerable familiarity with lampworking—something Mitscherlich had acquired while studying with Berzelius but which, as we have seen, was not a widely distributed chemical skill in 1834. Even worse, Mitscherlich used his glassblowing skill to modify Liebig's analytical apparatus, including the Kaliapparat (figure 3.10). Increasingly focused on organic analysis, by early 1834, Mitscherlich was convinced that Liebig's Kaliapparat measured too little carbon for compounds containing a high ratio of carbon to hydrogen, such as naphthalene.[67]

In fact, Mitscherlich was not the only chemist in this period to dispute the superiority of Liebig's device and to seek to improve it by the application of glassblowing skill. In 1835, the Swiss chemist Carl Emanuel Brunner published analyses of cornflour and cornstarch made using his own

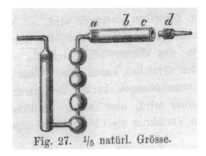

Fig. 27. 1/5 natürl. Grösse.

FIGURE 3.10
Eilhard Mitscherlich used his glassblowing skill to modify Liebig's Kaliapparat. His
device perfectly illustrates the enormous adaptive flexibility chemists gained by working in glass. As the young Berlin chemist Marchand explained, some of Mitscherlich's
modifications made this apparatus easier to make: His device contained only four
bulbs, all of equal size, while other sections were simply constructed from tubing of
larger diameter. Note that in 1876, Eilhard Mitscherlich's son Alexander illustrated
this apparatus at "1/5 natural size." By this time, such an illustration was sufficient
to permit anyone with glassblowing skill to make the device.

Source: Alexander Mitscherlich, "Organische Analyse vermittelst Quecksilberoxyds,"
Fresenius's Zeitschrift für analytische Chemie 15 (1876): 390, Fig. 27. Courtesy of the
University of Wisconsin-Madison Libraries.

modified Kaliapparat (figure 3.11).[68] Although Brunner's Kaliapparat seems
not to have gained much of a following, Mitscherlich's device attracted
some support by 1838.[69] For example, the young Berlin chemist Richard
Felix Marchand concluded that—although Liebig's method was easier
(because it could be performed single handedly)—Mitscherlich's was preferable for analyzing high-molecular-weight compounds with a low hydrogen
content. Not only was Mitscherlich's Kaliapparat easier for a novice glassblower to make, Marchand explained, it also—unlike Liebig's—trapped all
the carbon dioxide produced by burning substances of this kind.[70]

The spread of glassblowing skill during the 1830s equipped European
chemists to make small, cheap glassware for themselves, and they increasingly recognized the benefits of working this way.[71] Yet it remained far
from obvious to chemists beyond Giessen that Liebig's Kaliapparat was the
best device for measuring carbon. As they analyzed a growing diversity of
organic substances, chemists tested—and sometimes exceeded—the limits
of Liebig's device. Many found Liebig's Kaliapparat did not absorb all the
carbon dioxide produced when certain kinds of organic compound burned.

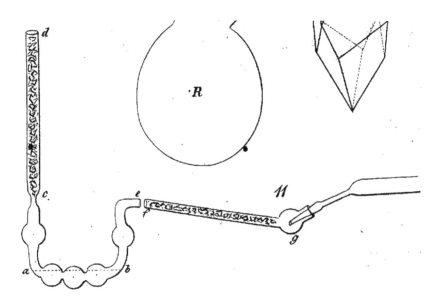

FIGURE 3.11

In 1835, Brunner proposed a new analytical apparatus incorporating a modified Kaliapparat that he used to analyze cornflour and cornstarch. Brunner's major modification—the incorporation of a vertical tube (c-d) filled with small pieces of potassium hydroxide—was intended to prevent loss of water from the Kaliapparat. As a result, Brunner was forced to give the Kaliapparat a more open U-shape. Brunner, who had studied in Berlin, Göttingen, and Paris, could clearly blow glass, but he gave no indication in either words or illustrations of how his apparatus was to be made.

Source: Carl Brunner, "Versuche über Stärkmehl und Stärkmehlzucker," *Annalen der Physik und Chemie* 35 (1835): Tafel III. Courtesy of the Bibliothèque Nationale de France: https://gallica.bnf.fr/ark:/12148/bpt6k151192/f695.item.

From Liebig's perspective, the variants other chemists produced—especially those like Mitscherlich's intended for general use—undermined disciplinary standards. Whatever their merits, such a multiplicity of devices left chemists unable to agree on formulae, effectively obstructing efforts to establish a shared understanding of organic compounds.

GLASSWARE REVOLUTION AND BEYOND

The glassware revolution was a fundamental transformation in chemical practice and material culture. Initiated by personal ambition and professional rivalry, it had far-reaching consequences: Moving chemistry into

lampworked glass fostered and enabled chemists' increasing preoccupa-
tion with the organic world. Lavoisier is typically credited with establishing chemistry as a quantitative science, using expensive, custom-made
apparatus to achieve unprecedented precision.[72] Yet, Lavoisier's brass-and-
glass instruments did not enable quantitatively useful analyses of organic
substances.[73] Overcoming this crucial disciplinary barrier instead relied
on table-top apparatus that chemists made themselves from glass tubing.
Nineteenth-century chemistry owed considerably more to Berzelius and
Liebig than is usually acknowledged.[74]

The Kaliapparat was the key driver of this change. Crucial to Liebig's ability to do organic analysis in Giessen and central to his recognition as one of
Europe's leading analysts, the Kaliapparat was essential in expanding chemistry's geography and demography beyond elite metropolitan centers. Both
the successful scaling up of training in Liebig's laboratory and the spread of
organic analysis away from its Parisian origin required moving chemistry
into apparatus that was cheap to produce in quantity and relatively simple
to use. Apparatus had to get small before science could get big. The Kaliapparat's ease of use and its low cost motivated chemists everywhere to work
in glass, drawing them into organic chemistry. This was how Liebig's innovation had such a profound and long-lasting effect on organic chemistry.

Glassblowing skill freed chemists from instrument makers, ushering in a
period of unprecedented innovation and diversification in chemical apparatus. As the number of known organic substances soared, working in glass
gave chemists much-needed and hitherto unachievable control over their
experiments. From about 1840 on, the quest for improved performance
when handling an increasing diversity of substances drove a rapid expansion in the variety and complexity of chemical glassware, much of it initially
produced by chemists themselves. Will and Varrentrapp's new apparatus for
measuring nitrogen, for example, provided a cheap, accessible solution to an
important chemical problem.

However, the flexibility chemists gained by working in glass created
a tension between standardization and creative innovation. Liebig's personal ambitions were intimately bound up in this tension. While Liebig
strove to standardize analytical apparatus and procedures, glassblowing
chemists, including Mitscherlich, could modify his Kaliapparat according
to individual preference and experimental circumstances. For Liebig, chemists' increasing ability to work in glass was simultaneously necessary to the

distribution of Giessen analysis and—because glassblowing enabled inno-vation—a major threat.

Other chemists' dissatisfaction with Liebig's original Kaliapparat extends this book's previous claims into the material dimension. Liebig had not solved the problem of organic analysis. Nor was his Kaliapparat universally accepted by his contemporaries. According to this study, Liebig's role in prompting chemistry's move into glass constituted a far more remarkable legacy. Later generations of chemists and historians learned to associate Liebig's name with the Kaliapparat and its use in analyzing organic com-pounds from August Wilhelm Hofmann's celebratory account of his teach-er's life and work.[75] For Hofmann, Liebig's transforming effect on chemistry was undeniable. Just as his eulogy helped ensure Liebig's status as "unques-tionably the greatest chemist of his time," so it established the Kaliapparat in the iconography of chemistry.[76]

Liebig's attempts to propagate shared standards and public morality, mean-while, embroiled him in disputes that were largely resolved by later disciplin-ary developments. These resolutions are why Hofmann could justify Liebig's actions and celebrate his foundational disciplinary contributions. Glass and glassblowing were powerful factors in the formation of the discipline, crucial to chemistry's transition from polite pursuit to become an academic disci-pline and an industrial science. Chemists' acquisition of glassblowing skill ini-tially fostered disagreement. But—as chapter 7 explains—the tension between innovation and standardization was ultimately reconciled by establishing standards of practice and apparatus that defined chemistry as a discipline.

By the 1840s, as his students began to make major contributions to organic chemistry, Liebig's role—in the discipline and in this history—began to change. Will and Varrentrapp's work was a notable success. But Liebig was otherwise increasingly disillusioned by the disciplinary stag-nation he attributed to other chemists' failure to recognize the limits of analysis. Unwilling to participate in what he saw as fruitless debate, Liebig henceforth directed his efforts elsewhere—toward physiology and a range of projects intended to fulfil his entrepreneurial ambitions. As chapter 4 explains, Liebig selected his student Hofmann as both his partner in an especially ill-fated commercial venture and his successor in the discipline. In abandoning organic chemistry, Liebig set Hofmann to develop synthesis as a new method of investigating organic compounds—with extraordinary consequences both for Hofmann and the nascent discipline.

4 CAPITAL CHEMIST

August Wilhelm Hofmann entered Justus Liebig's laboratory during the late 1830s.[1] These were tumultuous times for Liebig. As his growing reputation attracted increased institutional support, Liebig at last obtained the improved facilities he had sought for so long. Giessen's new analytical laboratory was designed by Hofmann's father (state architect J. Paul Hofmann) and opened in 1839.[2] From about 1840, students flocked from all over the world to study with Liebig. Despite these outward signs of success, however, his research program in organic analysis struggled to resolve fundamental chemical problems and was consequently approaching deadlock.

It is a major claim of this study that this deadlock prompted Liebig during the early 1840s to make one of the most important innovations in the history of organic chemistry. Recognizing that analysis alone could not enable chemists to understand organic compounds, Liebig set Hofmann—his most capable student and the son of his close friend—to develop synthesis as a new method of investigating organic compounds.[3] Thus, Liebig initiated organic chemistry's transformative mid-century turn to synthesis, providing Hofmann with the basis of a stellar career lasting more than 50 years. No wonder Hofmann would take the relevant lab notebooks with him when he moved to London in 1845—even though these were, by normal convention, Liebig's property.[4]

By the time he received his PhD in 1841, Hofmann was a skilled analyst and a chemist of unusual potential. Remaining in Giessen, he soon became established as one of Liebig's most valued assistants. The intimate personal connection between Liebig and Hofmann is reflected in two complementary strands of their chemistry during the early 1840s. Both concerned natural alkaloids, a group including the important drugs, morphine and quinine.[5] The first made alkaloids the focus of Hofmann's academic chemistry. The second, meanwhile, saw both men seeking to profit from the alkaloids' commercial value.

This chapter deliberately juxtaposes academic and commercial aspects of Hofmann's chemistry, and it does so to highlight their interconnectedness.[6] The quest to understand alkaloids prompted Liebig to initiate Hofmann's groundbreaking "synthetical experiments." This work, as we shall see, was a crucial step in the foundation of synthetic organic chemistry as a field of academic study.[7] Yet Hofmann's synthetic studies of alkaloids sought practical benefits as well as new knowledge. They also depended on the emerging availability of coal tar as a low-value waste product of coal gas manufacture.

By about 1860, aniline derived from coal tar would become the major raw material for manufacturing synthetic dyes. In the 1840s, however, this outcome was far from obvious. Aniline and other substances contained in coal tar were certainly important compounds for academic chemists to investigate—as Hofmann would do. The entrepreneurially inclined Liebig no doubt also anticipated it might be possible to identify useful transformations. Making quinine, for example, would enable large scale commercial production, the availability of vast quantities of cheap coal tar leading to huge profits. Thus, Hofmann simultaneously embarked on a highly innovative research program in academic chemistry and sought to develop coal tar as a potential source of industrial wealth—as indeed would happen later.

This financial and sometimes unprincipled motivation for studying organic compounds is illustrated in this chapter by an exactly contemporaneous scheme involving Liebig and Hofmann. The two chemists risked everything in a highly questionable scheme to profit from supplying "amorphous quinine" as a cheaper alternative to quinine.[8] Ultimately, as we shall see, it was not money men who criticized Liebig and his fellow speculators for what amounted to insider trading but the pharmacists onto whose expert territory they had strayed. Elite pharmacists, it turned out, not only knew more about medicinal quinine than did Liebig, they were also vastly more principled where money and medical treatment were concerned.

Together, these stories shed new light on mid-century organic chemistry, revealing its academic, technical, commercial, and medical aspects as necessarily integrated components of the same whole. We have notable studies of

Hofmann's contributions to chemistry's academic and institutional development. Yet his chemistry has mainly been considered in relation to synthetic aniline dyes' undoubted significance to late-nineteenth-century technology and industry.[9] The few existing histories concerned with organic synthesis in the academy, meanwhile, have largely overlooked Hofmann's contributions in favor of prioritizing the philosophical and intellectual drivers and consequences of synthesis.[10] Such separation obscures the intimate links between academy and industry, and between commercial and abstract chemistry, that this chapter demonstrates. Only in relation to this unified context can we understand Hofmann's career.[11]

His former student and first biographer Ferdinand Tiemann recognized aniline's importance when he called it Hofmann's "lucky star."[12] It seems clear that Liebig and Hofmann hoped aniline might have commercial possibilities. But it did not immediately lead Hofmann to industrially significant discoveries. In the 1840s, aniline's main value to Hofmann was as a model compound for natural alkaloids, studied using the synthetical experiments that provided the organizing core of his research. Despite the enticing prospect of turning coal tar into lucrative commercial products—realized in 1856 by his student William Henry Perkin's preparation of mauve—aniline's primary importance in Hofmann's chemistry at this time was academic.

Hofmann's move to London in 1845 as first director of the Royal College of Chemistry (RCC) similarly spanned commercial and academic chemistry. As we shall see, involvement with amorphous quinine was both central to Hofmann's suitability for this prestigious role and his overriding motive for taking the job. Their common interest in natural alkaloids and coal tar was fundamental to Hofmann's relationship with Liebig. Understanding the extent of Hofmann's debt to Liebig—a research program focused on natural alkaloids, synthetical experiments as his key method, and aniline as his model substance—makes comprehensible much that has previously been puzzling about Hofmann's early career. Discovering how the amorphous quinine scheme came unstuck, and just how narrowly Liebig and Hofmann avoided disgrace, helps explain the future direction of Hofmann's research.

LUCKY STAR

The initial driver of Hofmann's chemistry was indigo, the blue dye extracted from the indigo plant. Formerly originating in many countries, including

Britain's American colonies, by the early nineteenth century indigo was overwhelmingly produced by British-owned plantations in India. All of Europe relied on indigo imported via English ports—a situation brought into sharp focus by the wars that ravaged Europe during this period.[13] Chemists, including Liebig, responded by studying the chemistry of indigo, perhaps hoping to discover more accessible alternatives to the costly Indian product or even to produce artificial indigo in the laboratory. Before his attention became focused on alkaloid analysis in 1830, both Liebig and his first student Heinrich Buff were investigating indigo.[14]

The St Petersburg chemist Julius Fritzsche returned indigo to Liebig's center of attention around 1840. Destructive distillation of indigo produced two new substances: aniline and anthranilic acid.[15] When he read Fritzsche's paper in Otto Linné Erdmann's *Journal for Practical Chemistry*, Liebig found anthranilic acid of such relevance to understanding indigo's constitution (molecular arrangement) that he repeated Fritzsche's work for himself.[16] He was also intrigued by Erdmann's suggestion that Fritzsche's aniline might be identical to "crystallin," the name given to a compound produced from indigo in 1826 by German pharmacist, Otto Unverdorben. This series of events came to a head in 1842, when Nikolai Zinin—a former student of Liebig's—published the laboratory preparation of what he evidently believed was a new compound, called "benzidam," from benzene.[17] In fact, benzidam displayed properties so similar to aniline that Fritzsche immediately suggested the two were identical.[18] Intrigued by the possibility that crystallin, benzidam, and aniline were in fact the same substance, Liebig assigned this problem of chemical identity to Hofmann as the latter's first independent investigation.

Sometime in 1842, Ernst Sell—another Giessen-trained chemist who had recently opened a coal-gas distillery—sent Liebig a sample of coal tar.[19] At this time, coal tar was an unwanted by-product of coal-gas manufacture, and its chemistry was consequently little known. One of the few previous investigations of coal tar was published in 1834 by Friedlieb Ferdinand Runge—a German apothecary and chemist whose work also encompassed indigo. Despite the difficulties of working with the complex, tarry mixture, Runge isolated two compounds he called "kyanol" and "leukol"—also introducing what would prove to be a valuable color test for kyanol. But because he had not obtained either compound in sufficient quantity—or, it would later emerge, purity—for analysis, Runge did not investigate their composition and formulae.[20] Liebig—perhaps suspecting the difficulty of distinguishing and characterizing the many similar constituents of coal tar—now gave this sample to Hofmann, whom he evidently regarded as an analyst of unusual capability.

In 1843, Hofmann published remarkable results. He had solved the original problem of chemical identity by demonstrating that crystallin, aniline, and benzidam were indeed the same compound—which he chose to call "aniline." Hofmann also showed that, when adequately purified, kyanol was identical to aniline.[21] Hofmann's work situated aniline at a crucial nexus linking the commercially important dye indigo and the industrial waste product coal tar to aniline's availability as a product of artificial laboratory preparation. In addition, his study established aniline's close chemical and physiological resemblance to nicotine and coniine, two of the simplest natural alkaloids and consequently the starting point for much chemical investigation of this important family of compounds.[22] It is clear from what Hofmann did next that this similarity was the most significant outcome of this work.

PRIORITY DISPUTE

Before pursuing Hofmann's work on aniline, let us pause to consider what this initial study indicated about leukol. Although it proved far less immediately significant for his future chemistry than would kyanol, leukol thrust Hofmann into a priority dispute with another of Liebig's former students, the Alsatian chemist Charles Gerhardt. Where Hofmann remained close to Liebig, however, Gerhardt—by this time working in Montpellier, following several years in Paris—was becoming increasingly estranged from his former teacher, not least because of his trenchant theoretical opinions.[23]

This dispute confirms the quarrelsome state of chemistry in this period. Liebig's decision to support Hofmann over Gerhardt reveals the extent of his commitment to fostering Hofmann's academic career. It also indicates Liebig's interest in coal tar's commercial potential. Although he privately condemned Gerhardt, Hofmann's public response—guided by Liebig—was decidedly measured. Liebig, meanwhile, defended Hofmann by mounting a very public attack on Gerhardt. As we shall see, this episode begins to explain how Hofmann—unlike his mentor—acquired a reputation as one of the century's most diplomatic chemists.

Hofmann began his study of coal tar convinced that leukol would prove identical to quinoline—a new substance that Gerhardt had recently produced from quinine and which was certainly known to Liebig.[24] If correct, this result would be of extraordinary commercial significance. Whereas aniline was merely similar to nicotine and coniine, the production of quinoline from quinine would indicate the feasibility of the reverse process. In other words, establishing leukol's identity to quinoline would imply the possibility of converting quinoline obtained from coal tar into quinine—an even more remarkable bridge than aniline between the natural alkaloids and coal tar.

At first, experiment persuaded Hofmann that the two compounds were different.[25] Studying leukol's reactivity indicated that—despite its identical source—it belonged to a different family of compounds from kyanol. In 1843, leukol remained for Hofmann a poorly understood substance, ripe for further study. Given its significance, it is no surprise that he would seek to resolve the question of leukol's identity. During the summer of 1844, Hofmann showed that leukol and quinoline were indeed identical. But— perhaps for commercial reasons—Hofmann did not publish this work. Instead, it was left to Liebig—presumably on learning that Gerhardt was once more working in this area—to publish Hofmann's findings as a "preliminary notice" in March 1845. Although Liebig provided no experimental detail, he claimed that Hofmann had definitively established that leukol was identical to quinoline.[26]

Liebig's proclamation apparently passed unnoticed by Gerhardt—even though the latter credited Liebig for his Montpellier appointment having recently been made permanent.[27] In October 1845—around the time Hofmann arrived in London as director of the RCC—Gerhardt published an independent claim to this important discovery. Even worse, he relegated this announcement to a footnote, appended to his highly critical abstract of a recent analytical study of quinoline by another former Giessen student, Johann Conrad Bromeis. Gerhardt merely asserted, without offering any evidence, "that Runge's *leukol* is identical to quinoline."[28]

Gerhardt's behavior angered both Hofmann and Liebig. Writing to Hofmann, Liebig condemned Gerhardt's "banditry" (*Räuberhandwerk*).[29] He also issued a private challenge to Gerhardt, evidently hoping the latter would cede priority over this important result—and the areas of research and possible commercial development it introduced—to Hofmann. In Gerhardt, however, Liebig was confronting someone as obstinate and resistant to social conventions as himself. Gerhardt did not respond.

At the same time, Liebig was adamant that Hofmann should not become embroiled in controversy—an understandable view, given what we know about Liebig's early career. Liebig—whose public battles had undermined many of his key relationships, including with his mentor Berzelius—considered Hofmann too young to become involved in quarrels and enmity. Thus, Hofmann could rely on Liebig to champion his cause. In addition, as explained below, Hofmann's recent appointment at the RCC was by this time extremely precarious, making it doubly important that he should not become involved in public disputes. Thus, Liebig also advised Hofmann to defer publication of his latest work on aniline—a paper intended to ensure his control of a research field encompassing aniline and coal tar.[30]

Around the end of the year, with no admission forthcoming from Gerhardt, Liebig published a blistering attack denouncing Gerhardt's failure to acknowledge Hofmann's priority—even though Hofmann had yet to publish results establishing the identity of leukol and quinoline. His fury at Gerhardt's neglect of Hofmann's work was as partisan as it was predictable. In Liebig's eyes, attack was certainly the best form of defense. Equally incensed by Gerhardt's response to some of his own work, Liebig condemned Gerhardt's arrogance, experimental sleight of hand, and theoretical intolerance in characteristic style.[31]

When Hofmann read Liebig's article, he thought it a "delicious snub" to the man he had privately dubbed a "liar."[32] Yet Hofmann—although he clearly shared his mentor's assessment of Gerhardt's behavior—played no public part in the dispute. And when—again following Liebig's guidance—he published "New Researches on Aniline" in December 1845, Hofmann began by neatly sidestepping Gerhardt's claim. He and Gerhardt, Hofmann explained, had independently prepared some of the same compounds—but Hofmann now declared these of subordinate importance. Instead, as chapter 5 explains, this paper launched a research program focused on using aniline to study the chemistry of natural alkaloids.[33]

Despite his anger, Hofmann's published response to Gerhardt was a model of diplomacy—restraint that was no doubt made easier by the knowledge that his mentor had so roundly and publicly condemned the plagiarist. In fact, Liebig's combined attack on Gerhardt and defense of Hofmann attracted considerable censure—even though many shared his negative assessment of Gerhardt's chemistry.[34] Meanwhile, Hofmann was learning the virtues of adopting a measured tone in his public dealings with colleagues—no matter what he expressed in private. While Liebig's actions confirmed his tendency to disputatious behavior unbecoming to an academic chemist, Hofmann was being shaped as the century's paradigmatic gentleman-chemist—and the dominant researcher in alkaloid chemistry.

ALKALOID ANALOG

Now that we understand the extent of Liebig's commitment to Hofmann's career, let us return to Giessen—where Hofmann remained as one of Liebig's assistants until early 1845. Following publication of his 1843 paper on coal tar, Hofmann began work to establish aniline as a model compound for natural alkaloids—a step that would be crucial in enabling his future work in this area. Three things made this possible. The first was aniline's chemical similarity to nicotine and coniine. Second, it was significantly easier to obtain aniline—whether from indigo or from coal tar distillation—than to isolate nicotine and coniine.[35] The third crucial factor was aniline's far greater experimental tractability.

Where coniine and nicotine and their derivatives were hard to crystallize and consequently almost impossible to purify, aniline's tendency to form crystalline compounds was a highly significant advantage.[36] At a time when recrystallization was chemists' only reliable method of producing individual, pure compounds, this made aniline very much easier to work with—thereby greatly increasing the likelihood of producing valid results. For example, only pure compounds produced useful analytical results, making purity of sample an essential prerequisite for accurate formula determination.[37] These qualities suggested that aniline might be a practical as well as plausible model for the natural alkaloids. There was just one problem. In the 1840s, it was not obvious to chemists that compounds derived from natural and artificial sources could be identical.

Establishing that two otherwise identical compounds—one derived from nature, the other produced in the laboratory—were the same implied a significant change in how chemists characterized substances (examined in chapter 7). Origin—unlike almost every other property—was certain, traditionally playing a prominent role in approaches to differentiating and identifying substances. Amid the uncertainties of composition and chemical identity, chemists could at least be sure how they had sourced the substances they studied.

By the 1830s, however, Liebig's development of analysis as a reliable method of determining composition and formula began to transform this situation. Chapter 2 described how his students Rodolphe Blanchet and Ernst Sell showed—contrary to previous studies—that camphor derived from distinct sources shared the same composition and formula.[38] Analysis taught Liebig that exactly the same compound could be obtained from multiple sources. This conviction—as we shall see—was fundamental to the validity of using model compounds in studying organic nature. It also underpinned mid-century chemists' resilient faith in their eventual ability to produce natural products in the laboratory—no matter how unattainable this goal was at that time.[39]

The idea of using laboratory products to study natural substances was not new to Liebig. In fact, he proposed using artificial organic bases to unravel the chemistry of natural alkaloids during his landmark 1838 study of organic acids. Undertaken with the explicit intention of unifying chemistry's various branches into "one harmonious whole," this work led Liebig to link the acidity of both organic and inorganic acids to a single element, hydrogen. And—since not all hydrogen in organic acids was acidic—Liebig inferred the existence of hydrogen in multiple constitutional forms.[40]

This molecular explanation of acidity encouraged Liebig to take two crucial yet speculative steps. First, he used a series of closely related artificial compounds that spanned the acid-base spectrum to claim that a similar relationship linked the organic acids that chemists frequently isolated alongside natural alkaloids. Quinic acid, for example, was found together with quinine in cinchona bark. This common origin led Liebig (erroneously) to speculate that quinic acid and quinine were similar substances, whose acidic or basic properties were dictated merely by the substitution of one atomic grouping for another.[41] Thus, Liebig extended his explanation of acid-base phenomena by relating the alkaloids' basicity to the presence of an atomic grouping containing nitrogen and hydrogen that he initially called "M" and later reformulated as "amidogen." In a second move of enormous significance for Hofmann's subsequent work, he also proposed that artificial bases—by virtue of their "simpler constitution"—might shed useful light on the essential nature of organic bases (and, by extension, their related acids).[42]

Liebig's proposal clarifies how Hofmann recognized aniline's plausibility as a model compound. But—despite Liebig's reiteration of the same notion in 1840—other chemists still required convincing of its validity.[43] As I now show, this was exactly what Hofmann did in his 1845 publication "On the Metamorphoses of Indigo."[44] This paper built on his earlier review of indigo chemistry—a comprehensive survey of existing work that left Hofmann well versed in the subject.[45] The title of Hofmann's new experimental study, however, was misleading: It was more concerned with aniline than with indigo. The most noted outcome of Hofmann's paper—which Hofmann, almost certainly guided by Liebig, claimed as its cause—was to provide corroborating evidence for a novel constitutional theory that was rapidly gaining traction: substitutionist type theory (type theory).

Type theory—introduced by Dumas and developed by Gerhardt and Laurent—was a challenger to the dominant electrochemical dualistic theory (dualistic theory), championed by Berzelius. Briefly, dualistic theory explained reactivity in terms of the attraction between oppositely charged atomic groupings, stronger attractions overcoming weaker ones in reactions that changed molecular constitution and, therefore, chemical properties. In contrast, type theory interpreted many reactions as the substitution of one atomic grouping for another in an underlying framework, or type. This type, which had yet to be given distinct molecular form, was believed to determine chemical properties. Since the type was unchanged by substitution, both starting material and product ought to display almost identical properties.[46] Type and dualistic theories thus made fundamentally different predictions about the consequences of substituting one atom in an organic compound for another. In the present state of organic chemistry, however, chemists struggled to evaluate these competing claims—leading to persistent and sometimes highly acrimonious disagreements (discussed in later chapters).

According to Tiemann, entering this controversy was what "first drew the attention of the entire chemical world to the young experimenter," Hofmann.[47] The focus of Hofmann's paper was his preparation of chloraniline, a novel artificial derivative of aniline in which one hydrogen atom was substituted by an atom of chlorine. Unable to make chloraniline from aniline directly, Hofmann proceeded instead via isatin, another decomposition product of indigo isolated by Auguste Laurent in 1840.[48] Despite the introduction of electronegative chlorine, Hofmann found that chloraniline—like aniline—was basic. This similarity was much more easily explained in type theory than in dualistic theory—leading Liebig to highlight this result as "definite, irrefragable proof" of type theory's correctness. By contrast, Hofmann—with what we shall later recognize as characteristic interpretational circumspection—merely concluded "that in certain circumstances chlorine or bromine can perform the part of hydrogen in organic compounds."[49]

In fact, Hofmann's paper indicates that his primary goal had been to demonstrate the validity of using artificial laboratory products to ascertain chemical principles that also applied to organic nature. For Hofmann, the key result was that chloraniline, despite its artificial origin, was an organic base. Where most chemists defined organic bases as the group of natural alkaloids, Hofmann—again following Liebig—now asserted the equivalence of artificial and natural compounds.[50] Thus, organic bases became "azotized [i.e., nitrogenous] bodies, which possess all the properties of basic metallic oxides"—a definition that included aniline and chloraniline as well as coniine and nicotine.[51] As we shall see, this work's paramount importance for Hofmann was to establish studying aniline as a method of acquiring knowledge about natural alkaloids, thereby making aniline into a viable alkaloid analog.

THE TURN TO SYNTHESIS

Having established this important principle, Hofmann now focused his attention on aniline: its purification; its production from indigo and coal tar; and—above all—its reactivity. Where Liebig's research program had been exclusively analytical, Hofmann now pioneered synthesis as a complementary route to new understanding of organic compounds. The turn to synthesis did not constitute an absolute discontinuity in the science. Analysis continued to be vital as a way of establishing a substance's composition and formula. Nor was studying reactions novel. But—whereas Giessen analysis sought to understand a starting material by decomposing it into simpler constituents—Hofmann's synthetic approach explicitly shifted focus from starting material to products. For Hofmann, synthesis meant using reactions that transformed the starting material into products of equal or greater complexity to shed light on the nature of the product.[52] As we shall see in this and two subsequent chapters, Hofmann developed this approach into a powerful investigative method, acquiring unmatched skill in inferring reliable constitutional information from its experimental results. Thus, synthesis became Hofmann's primary mode of laboratory reasoning.

The term "synthesis" was first used in the context of organic chemistry in April 1845 by Hofmann—writing with his Giessen colleague, the Scot John Blyth.[53] While investigating the plant-derived resin "styrole" (styrene), Blyth and Hofmann found that the action of heat produced a transparent, vitreous substance they named "metastyrole"—a transformation that was reversed by distillation.[54] Since metastyrole was formed from styrole (and vice versa) "without loss of or addition to any one of its elements," Blyth and Hofmann inferred that the two substances differed only in their "molecular structure." "Analysis as well as synthesis," they concluded, "has equally proved that styrole and . . . metastyrole . . . possess the same constitution [sic] per cent."[55] Thus, Blyth and Hofmann's use of the term synthesis was directly analogous to its established meaning in inorganic chemistry.[56] Synthesis here implied the production of a more complex product from a simpler starting material. But its primary usefulness lay in the new knowledge—rather than the novel substance—that it produced.[57]

Shortly afterward, while still under Liebig's direction, Hofmann embarked on a more extended collaboration in synthetic organic chemistry with his fellow student, the Liverpudlian James Sheridan Muspratt. Unsurprisingly, given Hofmann's research interests, the core of this project was Zinin's recently published procedure for converting benzene into aniline. Muspratt and Hofmann set out to demonstrate the wider applicability of this transformation, work that resulted in three coauthored publications. The earliest of these, an April 1845 article "On Toluidine," reported their first "synthetical experiment"—and it launched organic synthesis.[58]

The status of "synthetical experiments" as the systematic investigative complement to analysis was crucial to Muspratt and Hofmann's neology.[59] Indeed, they did not claim absolute novelty for their attempt to use a known reaction to produce "certain compounds which suggested themselves to the theoretical inquirer." Instead, highlighting synthetical experiments as both enabled by previous studies of the transformations of organic compounds and unlikely—at least in the short term—to be generally successful pointed to their most significant innovation. The immediate purpose of synthesis was as a means of overcoming those "suppositions contrary to nature" that Muspratt and Hofmann believed obstructed intentional laboratory preparation. In other words, as I now show, synthetical experiments' most important outcome was to shed light on the constitution of the substances they produced.

Synthetical experiments worked by applying a known chemical transformation to related starting materials. Where Zinin began with benzene, the simplest member of the family of aromatic compounds, Muspratt and Hofmann selected toluene (methylbenzene, which contained one additional carbon) as their starting material.[60] Because they started with a similar compound, Muspratt and Hofmann anticipated a similar outcome—as proved to be the case: Toluidine indeed related to aniline just as toluene related to benzene. This result indicated that Zinin's transformation worked reliably on other starting materials—a result that suggested it might be widely applicable.

Crucially, knowing the process by which toluidine had been formed enabled Muspratt and Hofmann to infer its constitution. Because they believed this transformation involved replacing one atomic grouping by another, Muspratt and Hofmann could use analogy to reason from the benzene-aniline pair to toluene-toluidine. Representing aniline ($C_{12}H_7N$) as $C_{12}\{H_5/Ad\}$ (Ad = Liebig's amidogen) was the cutting edge of notational representation, an essentially linear view of constitution reflecting Muspratt and Hofmann's current understanding of how groups of atoms were ordered and linked in the molecule.[61]

In making toluidine, therefore, Muspratt and Hofmann did not add a new artificial substance at random to the growing pool of known organic compounds. On the contrary, because of how they made toluidine, they were quickly able to establish its relationship to aniline and its identity as a new organic base. Thus, synthetical experiments—because they operated on families of related starting materials—produced similarly related series of products, in this case artificial organic bases. As a result, they systematically expanded the taxonomy of known organic substances while simultaneously guiding chemists' attempts to elucidate constitution and structure.[62]

Synthetical experiments' ability to produce novel substances such as toluidine makes it worth emphasizing that early organic synthesis did not permit chemists to make defined molecular targets at will—whether these were natural products or solely the results of chemical artifice. Like others before them, Muspratt and Hofmann certainly aspired to the laboratory preparation of molecular targets, including such natural products as quinine and coniine.[63] Yet they recognized this ambition as both beyond their present abilities and distinct from the more immediate investigative purpose of synthesis.[64] Indeed, as Hofmann soon learned, it would be a long time before chemists were able reliably to make "a given compound in one or the other way."[65] This realization perhaps explains why, by 1846, Hofmann's associate John Gardner contended that realizing the promise of "constructive chemistry" would require the development of a "new branch of chemistry, which is neither analytical nor synthetical."[66]

In the first instance, synthetical experiments were a means of expanding chemical knowledge of substance and transformation rather than a reliable method of making anything. In 1849, while elucidating the significance of his new, synthetic approach, Hofmann explained that: "it is not the host of new substances, which we are continually discovering, that interest us so much, but the new methods of operation, by which we may imitate, for our own purposes, the special forces of nature. Every new reaction . . . is a step nearer to the solution of this grand problem."[67]

Synthetical experiments—and the reactions that underpinned them—produced a multitude of new compounds whose origin meant they were easier to classify. Some of these coincidentally proved identical to known natural products.[68] Many more were novel compounds not previously obtained from any other source—with the result that early organic synthesis rapidly and systematically expanded the number and diversity of artificial organic substances. This was how synthesis could become a powerful tool for the manufacture of industrially useful compounds, beginning with dyes, while remaining almost entirely incapable of making natural products, such as quinine and the sugars—or even the much-simpler coniine.

Despite the limitation of synthetical experiments as an approach to target synthesis, whether the target was natural or artificial, the laboratory preparations of aniline from benzene, and of toluidine from toluene, were of great significance to Hofmann. By providing an insight into the constitution of alkaloids, they suggested to Hofmann how these might plausibly be made in the laboratory. Having established that aniline and toluidine, as organic bases, were good models for natural alkaloids, Hofmann now reasoned that alkaloids consisted of Liebig's amidogen combined with a hydrocarbon moiety—that is, an atomic grouping containing only carbon and hydrogen. He therefore anticipated that the correct hydrocarbon might be converted into the corresponding alkaloid by the same reactions that transformed benzene into aniline, and toluene into toluidine.[69] In fact, Hofmann's own work contributed to chemists' recognition that the alkaloids were a much more complex and diverse family than Hofmann anticipated at this time. Yet, as we shall see in subsequent chapters, Hofmann would remain faithful to essentially this view of alkaloid synthesis throughout his long career.

COMMERCE AND CONTROVERSY

Hofmann interrupted his experiments on aniline in 1845 to pursue goals that transcended academic chemistry. Relations between Liebig's family and Hofmann's were closer than ever. Hofmann had recently become engaged to Liebig's niece by marriage, Helene Moldenhauer. Dearly wishing to marry but prevented by his financial circumstances, Hofmann had to act. Fearing his advancement in Giessen was blocked by others with a prior claim to Liebig's preferment, Hofmann arranged—during Liebig's absence, and apparently without consulting him—to move to Bonn as Privatdozent in the spring of 1845. Hofmann's move—though it caused Liebig some embarrassment—did not lead to any break between the two.[70] Instead, Hofmann's personal, financial, and professional ambitions were about to catapult him into business with Liebig, and so to London.

AMORPHOUS QUININE
Wishing to marry but lacking the money to do so, Hofmann entered a commercial partnership with Liebig from which he anticipated significant profits. Liebig—almost certainly aided in the laboratory by Hofmann—had recently developed a process for purifying the active constituent of quinoidine, the residue of commercial quinine production.[71] Quinoidine remained after quinine, then the most efficacious remedy for fevers, including malaria, had been extracted from cinchona bark. So named because it shared many of quinine's medicinal properties, including the ability to combat fever, quinoidine was widely used as a cheaper substitute for the more effective—and vastly more expensive—quinine. This practice, however, was not without its problems. The composition of quinoidine varied hugely, depending on which type of cinchona bark was used, and who performed the extraction of quinine. As a result, it was difficult to predict quinoidine's effects—both therapeutic and damaging—a situation that seriously limited its market value.

Quinoidine's physiological activity, however, suggested that it might contain either residual quinine or some similar alkaloid. This possibility led Liebig to develop a new assay for quinine based on Gerhardt's 1842 production of quinoline from quinine by destructive distillation. In fact, Gerhardt had used conversion of cinchonine and strychnine (but not codeine) into quinoline as a way of classifying alkaloids into two distinct groups.[72] Isolating an identical decomposition product here defined a family of compounds, production of quinoline identifying quinine and quinine-like alkaloids.[73] This test indicated quinoidine contained a lot more quinine-like alkaloid than Liebig had expected.[74] Liebig now isolated this alkaloid from quinoidine—albeit in a form that could not be crystallized and therefore remained less than completely pure.

Liebig was quick to see the commercial potential of this result. Just as coking produced unwanted coal tar, so quinine production resulted in quinoidine waste. After more than 20 years of medical use of quinine, vast stocks of quinoidine had been amassed by European pharmacists. In outline, Liebig's plan was to buy a large quantity of quinoidine, while taking care not to alert the market. He would then publish an account of how quinoidine could be converted into an effective drug. Liebig anticipated his authority would drive up the market price of quinoidine—which, established as the source of a safe and effective remedy for fever, could then be sold at a profit. This was how quinoidine joined fertilizers in Liebig's portfolio of commercial speculations.[75]

To execute this plan, however, Liebig needed an agent—preferably one with access to the world's largest quinine market: London. John Lloyd Bullock must have seemed the ideal candidate. Bullock had spent time with Liebig in Giessen in 1839 and was now established as a pharmaceutical chemist in the British capital.[76] He knew enough chemistry to carry out Liebig's purification procedure, and his role as a pharmacist gave him access to patients on whom to test the new drug. He was also a practiced entrepreneur, capable of discreetly acquiring raw quinoidine on the European markets and overseeing its eventual sale for profit. It is a mark of his entrepreneurial skill that it was ostensibly Bullock, not Liebig, who dubbed the new drug "amorphous quinine"—a name that prepared customers for its noncrystalline appearance while highlighting its relationship to quinine.[77]

By the early summer of 1845, all seemed to be set fair. Liebig and Hofmann were in partnership with Bullock and his associate John Gardner in

London. Liebig's article on amorphous quinine—approved by both Hofmann and Bullock—was ready for publication, and Bullock had already purchased a significant quantity of quinoidine for import to Britain. Hofmann—no doubt eager to see how the speculation would improve his matrimonial prospects—was already calculating his share of the proceeds.[78] Liebig and Hofmann, however, grossly underestimated the risks and complexities of the venture. Indeed, Bullock—situated in London at the scheme's capitalist center, and with far more relevant commercial experience—was to exert vastly more power over its development than they had anticipated. Moreover, both Liebig and Hofmann were about to take steps that hugely increased their exposure.

In Liebig's case, naïve conviction that his command of the science entitled him to ownership of—and ought therefore to secure his control over—what was rapidly becoming a commercial venture now led him to make an almost catastrophic error. In June 1845, he invited the chancellor of Giessen University, Justin von Linde, to participate in the scheme. Liebig's stated purpose was to thank Linde for securing Heinrich Will's recent professorial appointment.[79] But it is surely more than coincidence that the chancellor was about to become a leading advocate for Liebig's ennoblement. By involving Linde, Liebig enormously increased the scheme's chances of failure. For Linde, quite naturally, soon sought to include his brother-in-law in such a profitable venture—a proposal Liebig, who was increasingly beholden to Linde, was powerless to block.[80] As the number of speculators increased, so their interest in quinoidine became more noticeable—a situation to which both the quinoidine market and the London excise soon responded.[81]

Hofmann's involvement with amorphous quinine, meanwhile, was about to be transformed. Not only was the chemistry involved squarely in his area of chemical expertise, Hofmann was also emerging as Liebig's business representative where amorphous quinine was concerned. Toward the end of June, he was persuaded by none other than Bullock to accept a position as first director of a new college of chemistry in London. This appointment instantly elevated Hofmann to the elite levels of British science. It also transported him to amorphous quinine's commercial center. This combination, as I now explain, soon subjected Hofmann to a bruising and potentially disastrous introduction to chemistry, capital, empire, and trade in mid-nineteenth-century Britain.[82]

ROYAL COLLEGE

Gardner and Bullock had long advocated bringing Liebig's instructional approach to London—but no government funding was forthcoming. Nor was any existing institution willing to house the necessary teaching laboratory. By the summer of 1845, they at last secured adequate sponsorship from among the scientific elite and the landowning aristocracy to found a private institution for this purpose. Queen Victoria's personal physician, Sir James Clark, took a leading role in these developments, while the Queen's German husband Prince Albert emerged as a committed patron of the new college—which thus became the Royal College of Chemistry (RCC).[83]

The RCC was intended not merely as a *"national Institution,"* it was also created with explicitly commercial goals that were overtly connected to chemistry's importance to the effective exploitation of Britain's imperial dominions.[84] The RCC's founders were convinced that only Liebig or one of his former assistants could succeed as its director. But it was not merely Liebig's skills as an academic chemist that they sought to bring to London. His commercial and entrepreneurial instincts were equally desirable. This was especially true for Gardner and Bullock, who—though their search for personal gain would later bring them to grief—at this point still exercised considerable power over the new college.

Liebig soon declined—though not without drawing the offer to Linde's attention. Such a prestigious invitation might help secure his baronetcy.[85] In early June—after his seniors Carl Remigius Fresenius and Heinrich Will had declined the appointment as too risky—Hofmann was offered the job. Indeed Hofmann, because of his involvement with amorphous quinine, was surely Gardner and Bullock's preferred candidate (figure 4.1). Yet even Hofmann, whose situation as Privatdozent in Bonn was far less secure than those of either Fresenius or Will, was not at first inclined to accept. In fact, Hofmann at this time was far more interested in developing plans for launching amorphous quinine on the British market than he was in moving to London.[86]

Soon after hearing of Hofmann's intention, Bullock deviated from touring Germany and buying quinoidine to pay a surprise visit to Bonn. Accosting Hofmann at the railway station, he spent "a few hours" with Hofmann—with remarkable effects. While in Bonn, Bullock delivered both good and bad news about amorphous quinine, which Hofmann relayed to Liebig. On the upside, Bullock's London trials indicated that amorphous

Fresenius   Will   Bullock   Gardener   Hofmann
im Liebigschen Laboratorium in Gießen nach einer Daguerrotypie
anfangs der vierziger Jahre.

FIGURE 4.1

Carl Remigius Fresenius, Heinrich Will, John Lloyd Bullock, John Gardner, and August Wilhelm Hofmann (left to right) in Liebig's laboratory ca. 1843.

Source: Courtesy of the Edgar Fahs Smith Memorial Collection, Kislak Center for Special Collections, Rare Books and Manuscripts, University of Pennsylvania.

quinine was "completely effective" and could therefore be sold for "only a slightly lower price than quinine itself." Less welcome, his ongoing acquisitions of quinoidine—together with those of an unidentified third party—were already attracting attention. There was, according to Bullock, "a vague feeling on the Quinin [sic] market," and the price of quinoidine was rising, threatening the venture's profitability.[87]

Bullock's visit also produced a dramatic reversal in Hofmann's attitude toward the London job. Determined, until this point, that his prospects were better in Bonn, Hofmann now believed he was just "the man to make something out of the conditions there." Moving to England was a "wonderful opportunity," such as one might be offered once in a lifetime. And, although Hofmann initially linked the prospect of advancement to science, he quickly continued to a calculation of the financial benefits he anticipated—money that "would bring [him] significantly closer to [his] goal" of marrying Helene. Only Bullock could have been the source of these figures—and they surely included some estimate of the expected profits from amorphous quinine.[88] Liebig, moreover, was clearly in favor—since he now persuaded Helene of the move's benefits.[89]

Even now, however, Hofmann's acceptance did not come without strings attached. In fact, his requirements led to a significant increase in the stakes attached to both the new college and the plan for amorphous quinine. Unwilling to leave Bonn without the promise that he could return at the end of 2 years, should the London position fail, Hofmann sought Liebig's help to secure this option. Expressing somewhat implausible skepticism that his patronage would be effective in England, Liebig was initially reluctant to comply. This time it was Gardner's turn to act. Within a month, he was in Giessen, leaving with a letter from Liebig to Sir James Clark extolling Hofmann's virtues.[90]

Liebig's letter had the desired effect, and toward the end of August, Sir James Clark confirmed that arrangements were in place for Hofmann's arrival in London. The possibility that Hofmann might return to Bonn had been guaranteed by none other than Prince Albert's brother Ernest, Duke of Saxe-Coburg-Gotha. Liebig and Hofmann consequently became indebted to royalty in both England and Germany, circumstances that—when combined with the RCC's aristocratic backers and high visibility among London's chemists and pharmacists—placed Hofmann's appointment in the spotlight. In October 1845, Hofmann arrived in London determined to establish what he called the "London branch" of Liebig's Giessen laboratory, and eagerly anticipating the profits he would accrue from amorphous quinine.[91]

UNEXPECTED DEVELOPMENTS

Hofmann's expectations proved false on both counts. Establishing a school for chemistry in London was a good deal more difficult than he had supposed. Many resources taken for granted in Germany were much harder to obtain in Britain. For example, Hofmann not only imported glassware from Germany, he also relied heavily during the RCC's early years on German assistants with glassblowing skill. Compared with Liebig's analytical research program, Hofmann's synthetical experiments greatly expanded the necessary range of skills, apparatus, and material provisions—with important consequences for the laboratory environment (examined in chapter 8).

When planning the RCC's new laboratory, Hofmann soon encountered unfamiliar British approaches to science and science funding. The new college might have royal endorsement, but it remained an essentially private venture, reliant on individual donations. Despite Gardner's tireless efforts, it was difficult to persuade mid-Victorian landowners and capitalists that a college of chemistry could really produce anything to benefit them. As director, Hofmann would need to master diplomacy, compromise, budgeting, and shameless self-promotion. Meanwhile, his business involvement with Liebig, Bullock, and Gardner was about to turn sour.

During the summer of 1845, Liebig increasingly wondered why his paper on amorphous quinine had not appeared in print. In fact, it seems likely Bullock already had possession of—and thus control over—this crucial component of the scheme.[92] Even Liebig's visit to Britain in September did not resolve the growing tensions between Liebig and Bullock.[93] Liebig's ongoing advice concerning a pricing strategy for amorphous quinine—supposing that Hofmann communicated this to Bullock—no doubt did not help.[94]

As Bullock bought ever more quinoidine, its market price was rising quickly, and British import tolls had also increased—leading Bullock to avoid the Port of London and to conceal his purchases as asphalt or resin before storing them in the Muspratts' Liverpool warehouse.[95] In fact, unbeknownst to Bullock, Linde also continued acquiring significant quantities of quinoidine during this period—for himself and his brother-in-law—which Liebig was underwriting by providing Linde with unpaid bills.[96] Unsurprisingly, the state of the quinoidine market concerned Bullock, and in February 1846, he arrived in Giessen proposing a radical change of plan.[97]

Bullock wanted to patent amorphous quinine in Britain, so as to gain control over its London production and sale. This news must have been unwelcome for Liebig—who, thanks to Linde's backing, had recently received his baronetcy.[98] It meant that—rather than simply disposing of their stockpiled raw quinoidine as the market peaked following Liebig's announcement—profits would only be made from Bullock's sales of amorphous quinine. Such a move made apparent the extent to which Bullock had the whip hand. For one thing, Liebig now felt obliged to admit that he had invited Linde to join the scheme. This explained the market's behavior, legitimating Bullock's earlier suspicions. By exposing Liebig's involvement with Linde, it also revealed his vulnerability—further increasing Bullock's power over amorphous quinine.

At this point, Liebig was caught among numerous evils. On one hand, he could see that a patent for its production from quinoidine would secure amorphous quinine's long-term profitability by establishing a "monopoly" on its supply throughout the patent's 14-year life.[99] On the other hand, Bullock, as sole patentee, would in principle dictate exactly when and how those profits were made. Especially where Linde was concerned, this arrangement was unacceptable. Liebig therefore secured Bullock's agreement to process 10 pounds of Linde's quinoidine for every 100 of his own—on the understanding that Linde would buy no more quinoidine. But this crucial protection of Linde's interests came at quite a cost.[100] When Liebig proposed Hofmann's name should also appear on the patent, Bullock persuaded him that this was not possible in English patent law. His hand fatally weakened, Liebig was forced to concede.[101]

Liebig now began to consider the consequences of patenting amorphous quinine. His first step was to instruct Hofmann to consult a London patent agent using opium (rather than cinchona) residues as an example.[102] By mid-March, he wanted Hofmann to make a contract with Bullock, stipulating how the patent's profits would be shared. And by the end of the month, having received the terms of Bullock's proposed contract, he was clearly panicking.[103] At this point, Liebig claimed his trust in the "respectable" Bullock remained "rock solid." He nevertheless considered legal advice on the patent as literally a matter of "life and death." It may well be true—as Liebig explained—that he feared Bullock might die unexpectedly. But Liebig was equally concerned that Bullock's patent might be challenged.[104]

In fact, as we shall see, Liebig had very good reasons to be worried. His immediate priority, however, was to ensure his patron Linde's anticipated profits. Liebig therefore decided Bullock should buy Linde's quinoidine, thereby releasing Linde from the scheme. Bullock, meanwhile, was convinced Linde had bought more quinoidine since February—possibly with Liebig's connivance—and was therefore unwilling to go along with Liebig's proposal. On this occasion, even Hofmann was suspicious—until a wounded Liebig produced Linde's written assurance that this was not the case.[105] In fact, Liebig was apparently not satisfied even by this, asking his brother-in-law Friedrich Ludwig Knapp—at that time in Britain—to investigate in person.[106] Following an inspection of Muspratt's Liverpool warehouse, Knapp confirmed Linde's account of the situation.[107]

In the midst of all this, Liebig was furious to learn that Bullock had applied for the patent without first agreeing to his demands. Committed to protecting Linde, Liebig now threatened drastic measures. It was imperative that Hofmann prevent publication of Liebig's paper on amorphous quinine. If the paper was still in Hofmann's possession, it must not leave his hands. If not, Hofmann must take steps to get it back. And if, in the very worst case, his paper should appear, Liebig declared that he would disown it—thereby dealing what he anticipated would be a fatal blow to the undertaking.[108]

Having delivered this ultimatum, Liebig finally reached a satisfactory agreement with Bullock. Hofmann even offered to share the financial burden that Liebig had shouldered in freeing Linde.[109] Had the standoff not been resolved, of course, all the parties involved would have lost out—and Liebig admitted as much to Linde. Relief, however, quickly gave way to a sense of injustice. "The idea to develop a speculation based on my investigation [of amorphous quinine]," Liebig lamented, "is my property, in which I may allow anyone I choose to participate."[110] Undoubtedly stung by the harsh commercial realities required to profit from his discovery, Liebig nevertheless seems to have anticipated the worst was over. He was wrong.

PATENT MANEUVERS

Bullock's English patent application was granted on May 13, 1846. Ten days later, Liebig's article "On Amorphous Quinine"—in Gardner's translation—was printed in *The Lancet*.[111] Thomas Wakley, the periodical's founder and editor, admired Liebig's work and therefore highlighted this "remarkable paper." Wakley's endorsement no doubt helped attract the attention of chemists, pharmacists, and surgeons throughout the capital.[112] Liebig's name, moreover, was again prominent in mid-June, when Prince Albert laid the foundation stone of the RCC's new laboratory in Mayfair's Hannover Square. Liebig had a poor opinion of English chemists and pharmacists.[113] It turned out they were even less impressed by amorphous quinine and the patent for its preparation.

The London response to Liebig's paper and Bullock's patent centered on the Pharmaceutical Society of Great Britain, whose founder Jacob Bell and professor of pharmacy Theophilus Redwood initially regarded the RCC as unwelcome competition for the society's pharmaceutical training laboratory.[114] When Redwood prepared amorphous quinine using Bullock's procedure, he found that the product was far from pure. Not only was Bullock's chemistry flawed, Redwood considered it to be quite improper that such a medicinally useful discovery should have been patented. Even worse, Bullock had taken out the patent in secret, just days before Liebig's article appeared. Throughout the summer, Redwood used the pages of *The Lancet*, the *London Medical Gazette,* and the society's own *Pharmaceutical Journal and Transactions* to lambast Bullock.[115]

Meanwhile in Germany, trouble was brewing for Liebig in the person of Ferdinand Ludwig Winckler. Winckler was a respected Darmstadt pharmacist with a PhD in pharmacy from Giessen university—and the co-founder of Hessen's pharmaceutical society (*Apothekenvereins im Großherzogtum Hessen*). He also had an established research interest in cinchona alkaloids including quinine and quinoidine stretching back some fifteen years. When Liebig, still jockeying with Bullock for control over amorphous quinine, published an anonymous account of his discovery in the *Augsburger Allgemeine Zeitung* in early June 1846, this soon came to Winckler's attention—with potentially disastrous consequences.[116]

For the most part, Liebig's report confirmed results that Winckler knew only too well—since he had obtained and published them in 1843. In fact, Winckler had not only coined the term "amorphous quinine" in German, he had also proposed this substance was chemically identical to quinine—albeit without supporting his claim with the results of elemental analysis.[117] But when Winckler approached the Augsburg newspaper, hoping to challenge the "very cleverly calculated justification" published by their "Giessen correspondent," he found the editors unwilling to publish his response—which he instead placed in the Frankfurt *Oberpostamtszeitung*.[118]

The situation quickly took a turn for the worse when Liebig published his article "On amorphous quinine"—this time bearing his name—in the June 12 issue of the local Darmstadt paper (*Großherzoglich Hessische Zeitung*). This time, the editors proved slightly more resistant to Liebig's status and, a week later, Winckler's response appeared.[119] Given the extent to which he believed that Liebig had plagiarized his work, Winckler's commentary was remarkably measured. No doubt aware of Liebig's superior social standing and professional status, he limited himself to expressing "some modest doubts." Though Liebig's findings in general agreed with Winckler's own, some of Liebig's results "contradicted established facts." In

Winckler's opinion, Liebig significantly overestimated quinoidine's medicinal efficacy—especially while it remained colored, and therefore impure.

Winckler now referred explicitly to his 1843 preparation of "pure, completely white" quinoidine. Even in this state, however, quinoidine was neither as effective nor as reliable as quinine, which—according to Winckler—remained the only dependable remedy for intermittent fevers such as malaria. Winckler also challenged Liebig's chemistry—very much as Redwood was doing in London. Indeed, when Winckler—referring to a "safe source"—expressed concern that a speculation based on Liebig's name was in progress in England, it seems probable he and Redwood were in communication and sharing their findings.[120] Liebig's fears were being realized—in Germany as well as in London.[121]

Liebig again used his status, persuading the Darmstadt paper to reprint Winckler's 1843 article so that Liebig might undermine its originality. Liebig had three main strategies. First, he claimed Winckler's proposal that amorphous quinine and quinine were essentially the same substance was already widely known in 1843. It followed that the crucial novelty was to determine amorphous quinine's composition—as Liebig, the master analyst, had done. His second step was thus to condemn Winckler's failure to analyze amorphous quinine—thereby bringing the controversy onto his own expert territory. Finally—and somewhat implausibly, given the circumstantial evidence discussed below—Liebig declared he had been unaware of Winckler's work.[122]

This was too much for Winckler. By October 1846, following a further 3 months of intensive experimental work on amorphous quinine, he had important new results and even more damning information from England. Winckler now turned one of Liebig's favorite techniques against him. As coeditor of the Hessen pharmaceutical society's journal (*Jahrbuch für praktische Pharmacie*), Winckler had ready access to publication. Reproducing his public exchange with Liebig in full, Winckler accompanied this with a highly critical commentary, rejecting each of Liebig's strategic claims in turn.

Winckler began by claiming the term "amorphous quinine" as his own invention and restating his originality in identifying amorphous quinine as a noncrystalline form of quinine. He challenged Liebig to substantiate his refutation of this claim, adducing significant evidence to the contrary. Published shortly before Winckler's article, the 1843 edition of Geiger's *Handbook of Pharmacy* (*Handbuch der Pharmacie*)—revised by none other than

Liebig—made no reference to the possibility that quinoidine might contain residual quinine.[123] Nobody other than Liebig, Winckler argued, would doubt his claim to originality—as indeed seems largely to have been the case.[124]

Winckler was not afraid to engage Liebig on his preferred analytical territory. While Liebig had reported the outcome of elemental analysis, Winckler believed that the apparent impurity of Liebig's sample rendered these results dubious. In any case, Winckler argued, numerous cases made clear that composition alone was a far-from-adequate means of distinguishing alkaloids. Liebig's failure to describe a qualitative investigation of amorphous quinine's behavior and properties, moreover, made his contribution just as deficient in Winckler's eyes as Winckler's was in Liebig's.

Finally, Winckler found it highly unlikely that someone as widely read and well informed as Liebig should have missed his article. Indeed, given Liebig's interests and geographical proximity to Winckler, this does seem extraordinary. In fact, Winckler—though he did not say so—had already presented suggestive evidence in this regard. For although, according to Hofmann, it had been Bullock's suggestion that Liebig call their new product "amorphous quinine," there seems little doubt that either Bullock or Liebig had come up with this name only after reading Winckler's article.[125]

Having tackled Liebig's arguments against him, Winckler proceeded to extend his expertise in alkaloid chemistry, reporting new experiments that further developed chemical understanding of the complex alkaloid mixture found in cinchona bark.[126] Everyone—except Liebig—had abandoned Friedrich Sertürner's "adventurous opinion" that quinoidine was a "true fever killer." Indeed, Liebig's quinoidine was far from consisting almost entirely of an alkaloid identical to quinine. On the contrary, Winckler's study of 10 quinoidine samples, each from a different source, confirmed the widely varying amounts of quinine they contained—thereby explaining their varying efficacy as treatments for fever. "[P]ure amorphous quinine," Winckler concluded, was present in Liebig's quinoidine, but this was only because Liebig's quinoidine was the residue of incomplete quinine extraction. In other words, therapeutically useful quinoidine from which amorphous quinine could be produced resulted from wasteful, inferior quinine manufacture.[127]

Winckler's final blow was decisive. A friend in England—presumably Redwood—had provided a sample of Bullock's patented amorphous quinine.[128] This, Winckler declared, was a poor product, worth very much less than its sale price. Moreover, Bullock's accompanying leaflet justified his patent on precisely the opposite grounds: Only by protecting his process with a patent, Bullock claimed, could he afford to supply such a valuable product so cheaply. In exposing this "misuse of Liebig's name," Winckler gave no indication that he suspected Liebig's involvement. His expectation that Liebig would subject Bullock to "public censure," however, was surely intended as an implicit challenge. For if Liebig chose not to act—as indeed he could not—this would clearly implicate him in the affair.[129]

Bullock's truculent responses to Redwood, meanwhile, merely served to fan the flames of the English scandal, especially when it emerged that Gardner—whose misappropriation of the lease on the RCC's new laboratory led to his forced removal from the premises and resignation from the College's governing Council in August—was involved.[130] His critics were shocked that Gardner should use public lectures held at the RCC to promote amorphous quinine. Yet Gardner was guilty of no more than the contract with Bullock, Liebig, and Hofmann for amorphous quinine required of him.[131]

As the summer wore on, Liebig's "pecuniary interest" in the affair became the subject of public speculation.[132] His discovery of amorphous quinine had been "(apparently) so disinterestedly given to the public," yet "a mystery . . . remains to be unravelled." How, Redwood and Bell wondered, had Liebig's paper come to be published "ten days after a patent was actually granted for the same discovery?" Although some continued to defend Liebig, Redwood and Bell clearly suspected his collusion. And Wakley—who knew better than most the close business ties binding Liebig, Bullock, and Gardner—now publicly distanced himself from the affair, expressing his disapproval of the patent as the product of "collusion with quacks and quackery."[133]

We can only imagine Liebig and Hofmann's reaction when, in October 1846, the surgeon John Groves began lobbying "the [medical] profession to consider whether it would be advisable to appeal to Her Majesty . . . to . . . declare [Bullock's] patent as void."[134] Coming in the wake of Gardner's scandalous behavior, amorphous quinine placed the RCC in the eye of a mounting storm that threatened to destroy both Liebig and Hofmann's profits and—a yet more catastrophic prospect—Liebig's reputation and Hofmann's career.

In the end, for reasons that remain buried, the storm did not break—at least, not in public. Had the full extent of Liebig and Hofmann's involvement with amorphous quinine come to light, the affair would surely have led to the RCC's closure, Liebig's censure in Britain, and Hofmann's disgrace. Liebig no doubt learned painful lessons about doing business in England. Buying out Linde certainly left him carrying a substantial loss.[135] He may even have had cause to revise his estimation of English chemists and pharmacists. Liebig avoided explicit criticism and, crucially, Hofmann's name was not publicly connected with this affair—so that Queen Victoria and Prince Albert were not embarrassed. The Prince continued in the role of patron to both Hofmann and the new College, while Sir James Clark retained his intimate involvement as a member of the College's newly instituted managing committee.[136]

It is impossible to know how badly rumors surrounding amorphous quinine affected the RCC's early subscribers, many of whom were medical men.[137] Entirely dependent on private support, the RCC's finances were certainly fragile from the outset. In 1846, ostensibly in response to the financial crisis precipitated by building the new laboratory, Hofmann took a 50% pay cut to secure the College's continuation. Presented as an act of loyalty to Sir James Clark, it is also the case that Hofmann's role in the amorphous quinine scandal would have made it hard for him to resist such a proposal.[138] Despite the twin financial blows of amorphous quinine's failure and his reduced RCC income, Hofmann was nevertheless able to marry Helene in August 1846—an outcome that perhaps indicates he was receiving financial support from elsewhere.

From this point on, Hofmann was understandably far more cautious in his business dealings with Liebig.[139] Somewhat ironically—having made the transition to the world center of trade and capital—Hofmann now devoted himself to an overwhelmingly academic research program. He certainly recognized and made frequent rhetorical use of chemistry's role in isolating natural alkaloids such as quinine, and the desirability of producing such medicinally valuable substances in the laboratory.[140] Yet, as we shall see in chapters 5 and 6, Hofmann henceforth eschewed the alkaloids' commercial possibilities in favor of an almost entirely academic study of their chemistry.

ABSTRACT CHEMISTRY

Although he put his relationship with his mentor on a new footing in the wake of the amorphous quinine debacle, Hofmann never forgot his debt to Liebig. To some extent, the debt was shared across the discipline. According to Hofmann, Liebig had provided "the instruments and the means of prosecuting the researches by which the domain of chemistry must be enlarged, . . . show[ing] us how to keep up the supply of intellectual agents to carry on the work."[141] In other respects, however, Hofmann's debt was both more personal and of a more specific disciplinary significance. By the time he left Giessen, synthetical experiments were the organizing core of Hofmann's research. This was surely what Hofmann had in mind when he later credited Liebig with teaching him the "art of carrying out experimental enquiries."[142] Organic synthesis was the foundation both of Hofmann's career and of organic chemistry's rise to dominance—and it began in 1840s Giessen.

This turning point originated in the very particular context provided by industrializing Germany and, brought to London by Hofmann, it took root at the heart of imperial capitalism. Coal tar was a rich source of organic compounds. But its low cost and plentiful availability were crucial. Without such starting materials, chemists' search for commercially viable products would be in vain. Hofmann's research into coal tar thus had more in common with the amorphous quinine affair than might at first appear. Both promised profits, because quinoidine—like coal tar—was an industrial waste product. But, whereas amorphous quinine vanished amid barely suppressed scandal, aniline derived from coal tar would ultimately transform late-nineteenth-century science and industry.

Despite their similarity, these two ventures required distinct knowledge and different resources. Amorphous quinine provided a valuable lesson in the limitations of "abstract chemistry"—showing clearly what could go wrong when chemists abandoned their area of expertise.[143] Liebig's scheme quickly took him out of his depth—commercially as well as chemically. His conviction that command of the science ensured his control over amorphous quinine as a market speculation reflected a naïve lack of appreciation of what was entailed in bringing the matter to a successful conclusion. In launching a scheme that relied so heavily on his name, moreover, Liebig took an enormous risk—particularly when we consider the likelihood that his findings were in fact plagiarized. Especially in Britain, Liebig's reputation suffered considerably when his scientific claims were refuted by experienced pharmacists.

Liebig's laboratory ca. 1845 failed to make significant progress in resolving the complex chemistry of cinchona alkaloids. Despite the power of Giessen analysis, apothecaries such as Winckler and pharmaceutical chemists including Redwood were far more capable of producing relatively pure alkaloids from cinchona bark. Notwithstanding a growing body of scholarship devoted to practical science, research in academic laboratories such as Liebig's still tends to dominate accounts of cutting-edge chemical capability, reflecting deeply held assumptions about how nineteenth-century science worked.[144] By further undermining the "pure science" rhetoric so successfully deployed by academic chemists, including Liebig and Hofmann, this chapter calls for an even more extreme revision of the established mid-century relationship between academic and commercial chemistry.

In a similar vein, while several accounts of the RCC's early years correctly note its practical goals and connections to trade, industry, and empire, this study indicates that they have not gone far enough in emphasizing the commercial preoccupations of its founders and first director.[145] Revealing the extent of Hofmann's involvement with amorphous quinine and how this speculation threatened not only his career but also the RCC's survival show that the new college was anything but academic. As Sir James Clark explained to Liebig in 1845, the chemical exploitation of Britain's industrial and colonial resources was in fact the RCC's primary purpose.[146] Quite separate from any existing university or academic institution, the RCC's foundation as a private venture left it without precedent in Britain—circumstances that help explain why one of its critics nicknamed it "the amorphous college."[147]

Where amorphous quinine foundered on Liebig's limited pharmaceutical and entrepreneurial expertise, Hofmann's synthetical experiments on aniline tested the limits of cutting-edge academic chemistry, revealing questions of chemical identity as fundamental to chemists' ability to make sense of both the natural world and the artificial products of synthetical experiments. The rapid expansion of the artificial realm following the turn to synthesis, coupled with chemists' growing desire to produce natural products in the laboratory, made it increasingly necessary to know whether any two substances were the same or merely similar. As we shall see in chapter 7, this problem emerged around 1850 as one of the most serious obstacles to organic chemistry's development.

Transplanted to London, Hofmann's developing research program in synthesis depended on considerable resources. Access to aniline was facilitated by the RCC's location at the heart of Britain's imperial, industrial capital. But, as future chapters will show, glassware and glassblowing skill, the assistance of junior chemists, and a suitable laboratory environment were equally necessary and proved hard to acquire in London. In the first instance, Hofmann's work at the RCC focused on overcoming these difficulties.

These circumstances help account for the academic and institutional focus of Hofmann's early career. But they are less capable of explaining why, throughout the almost 20 years he spent in London, Hofmann's research produced largely abstract knowledge.[148] Indeed, this focus modifies the widely accepted view that Hofmann was mainly active in industrially applicable chemistry while in London.[149] Although he seemed in 1845 to be on the brink of entrepreneurial success, Hofmann largely avoided direct commercial involvement during this period. He certainly took on nonacademic work—for example, as an expert witness and on government committees. Yet, despite the close association between the RCC and William Henry Perkin's serendipitous and much-celebrated discovery of the purple dye mauve in 1856, Hofmann took little part in the emerging synthetic dye industry until about 1860—by which time his academic credentials were firmly established, and he was being courted for professorial appointments in Germany.

The context provided by his brush with scandal over amorphous quinine—doubtless a salutary exposure to the dangers awaiting an academic chemist choosing to operate in chemistry's related commercial fields—offers a plausible explanation of Hofmann's decision to step back from his early entrepreneurial aspirations. The conceptual riches that he anticipated would derive from studying aniline were surely an effective and enduring inducement to pursue academic goals. Practical and commercial aspects of alkaloid chemistry thus emerge as powerful, linked forces shaping Hofmann's early career and the direction of his research.

Artificial and natural compounds were similarly enmeshed in Hofmann's chemistry. Whereas amorphous quinine concerned natural alkaloids, Hofmann's work with aniline has typically been viewed in relation to the artificial realm. This outlook, however, neglects aniline's significance as a model for natural alkaloids. Establishing aniline in this role promoted the equivalence of nature and artifice, and thus the feasibility of making nature-identical compounds in the laboratory. Aniline's status as a model compound also became fundamental to Hofmann's ability to tackle alkaloid chemistry. As we shall see in chapters 5 and 6, laboratory reasoning from synthetical experiments on aniline enabled Hofmann to produce the mid-century's most important development in the theoretical understanding of natural alkaloids.

5 LABORATORY REASONING

This book promised an explanation of the relationship between experiment and theory in nineteenth-century organic chemistry. This chapter and the next fulfil that promise by showing how what I have called "laboratory reasoning" led Hofmann to the ammonia type, his major contribution to chemical theory. The ammonia type was a significant advance in how chemists understood nitrogen-containing substances, such as the alkaloids. But it would also change how they conceptualized the constitution of organic compounds more generally. These two chapters therefore address the question: What is a chemical theory?

The present chapter focuses on the process of laboratory reasoning, showing how this enabled Hofmann to make progress in the absence of a stable master theory. We shall see how he ordered the results of synthetical experiments on aniline, building a taxonomy of organic bases linked to dualistic theory, his chosen working framework. Initially successful, this taxonomy ultimately proved incompatible with experiment. At this point, laboratory reasoning led Hofmann to abandon his provisional taxonomy. But it also identified a plausible way forward, directing Hofmann to a fresh series of experiments.

As chapter 6 will show, this new avenue enabled Hofmann's research to continue without significant interruption, guided by the same laboratory reasoning. This time, he would succeed in creating a taxonomy that was compatible with experiment, based on substituting organic groups for hydrogen in ammonia. In doing so, Hofmann established ammonia as the type molecule for a certain group of organic bases, providing a powerful impetus to the further development of type theory and introducing what would prove to be a durable classification of simple organic bases.

Hofmann's chosen problem was one of mid-nineteenth-century organic chemistry's most pressing challenges: the constitution of natural alkaloids. To understand how he tackled this problem, and why this work was of such great significance in shaping both his career and the development of organic chemistry, we must follow Hofmann into his new laboratory at the Royal College of Chemistry (RCC). This account of how Hofmann used synthetical experiments to investigate alkaloids develops many of this book's established themes, including aniline's unique position at the nexus of industry, nature and artifice, and its role as alkaloid analog. It also explains how Hofmann's chemistry began to diverge from Liebig's as he developed the investigative possibilities of synthesis.

Reconstructing Hofmann's laboratory reasoning—and what he accomplished using it—requires entering his world and engaging the technicalities he wrestled with. Because they concern what Hofmann did and how he worked, this chapter and the next are intrinsically technical. This signals a necessary departure from what has gone before. Only such technicality can make visible the role of laboratory reasoning in bringing order to organic chemistry and re-directing chemists' efforts toward synthesis—as well as in establishing Hofmann's mastery of this field.

These two chapters take us to the heart of the relationship between experiment and theory in the new science of organic synthesis. This chapter shows how Hofmann's laboratory reasoning elevated synthetical experiments to become a method capable of producing conceptual knowledge about organic compounds—rather than merely serving as a way of making novel artificial substances of practical value. Although, by the time he left London some 20 years later, Hofmann was responsible for several commercially important aniline dyes, this was an unanticipated outcome of—not the motivation for—his synthetical experiments on aniline.

This chapter explains how laboratory reasoning functioned in the work of one chemist. Unlike so many stories in the history of science, it does not lead up to a great discovery or triumphal moment of theoretical insight. Instead, it shows how laboratory reasoning was constitutive of Hofmann's daily work, crucial in managing the uncertainties he faced in his struggle to develop a useful theory of the complex constitution of alkaloids. Where other chemists operated entirely in the framework of a preferred theory, laboratory reasoning enabled Hofmann to avoid such allegiances, making it possible for him to transcend the breakdown of his working hypothesis. As we shall see in chapter 6, the same laboratory reasoning would ultimately lead him to establish a new theory that came to belong to all chemists. In seeing how Hofmann approached and overcame failure, we shall see his brilliance as a chemist in action as nowhere else.

THEORIZING ALKALOIDS

Before examining Hofmann's work, it will be helpful to summarize the state of alkaloid chemistry ca. 1845. As we shall see, chemists' ability to interpret the constitution of natural alkaloids was at something of an impasse. Despite improvements in analytical technique, many of them developed in Giessen, chemists still struggled to assign stable formulae to this diverse and complex group of natural products. Such persistent uncertainty regarding composition made it difficult to investigate constitution (the arrangement of elements in the molecule) and how it shaped the alkaloids' properties and reactivity. But their medical and commercial importance meant that understanding alkaloids was one of mid-century organic chemistry's most significant goals.

Chemists' quest to unravel the relationship between the alkaloids' chemical nature and their physiological activity—like their desire to make alkaloids such as quinine in the laboratory—had two motives. As we saw in chapter 4, quinine was big business. Its artificial production therefore offered the prospect of personal advancement as well as academic distinction. At the same time, improving chemical understanding of the alkaloids was a sure way to establish organic chemistry's relevance to medicine and physiology as well as to commerce. Thus, the alkaloids became a focus of acrimonious yet largely irresolvable mid-century disagreements between adherents of two competing constitutional theories (introduced in chapter 4): electrochemical dualistic theory (dualistic theory) and substitutionist type theory (type theory).

Dualistic theory—championed by Jöns Jacob Berzelius—explained molecular constitution and reactivity in terms of additive processes linking two atomic groupings or radicals that were held together by forces of electrostatic attraction. Thus, dualistic theory traced the behavior of organic substances to the nature of the constituent elements in the molecule. Around 1840, Dumas introduced type theory as a challenger to Berzelian dualistic theory. Investigating the action of chlorine on organic compounds—a process he modeled using Berzelian formulae—led Dumas to interpret these reactions as the substitution of one atom or group of atoms for another.[1] As he became convinced of its wider applicability, Dumas extended this concept into a law of substitution, associated with a new theory of chemical or molecular types. For Dumas, compounds belonged to the same type when they contained the same number of equivalents (atoms) combined in the same manner and displayed the same fundamental chemical properties.[2] For example, he regarded acetic acid and chloroacetic acid, two organic acids that differed by the substitution of chlorine for hydrogen, as belonging to the same type. In Dumas's system, therefore, it was the number and arrangement of the atoms (rather than their nature) that determined reactivity.[3]

This neglect of the elements was anathema to those in the Berzelian camp, with the result that type theory intensified disciplinary discord. Focusing on molecular taxonomy rather than on constituent elements also meant that type and dualistic theories made significantly different predictions. For example, whereas dualistic theory anticipated that the introduction of the electronegative element chlorine would impact a substance's chemical properties, type theory indicated no such alteration.[4] This was why, as we saw in chapter 4, Liebig could interpret Hofmann's demonstration that chloraniline retained aniline's basicity as favoring type theory.[5]

Despite Dumas's repeated assertion that his theory was experimentally based, others—notably Berzelius—found it disturbingly arbitrary. Indeed Liebig, like many German chemists, had initially been strongly opposed to Dumas's proposals—even though he was by no means convinced that established dualistic theory offered an adequate explanation of the behavior of organic compounds. Despite some individuals' strong endorsement of one or the other theory, the constitution of many organic compounds was a puzzle that remained to be solved.

This was especially true for the alkaloids. In this case, basicity was the heart of the matter, and nitrogen—by now the established origin of the alkaloids' ability to neutralize acids—was the crucial element. The dualistic view was that preformed ammonia in the molecule was responsible for the alkaloids' basicity. Dualistic theory therefore interpreted alkaloids as composed of ammonia added to the remainder of the molecule. Liebig had begun to question this orthodoxy as early as 1830, during his study of the relationship between how much nitrogen the alkaloids contained and their "capacity to neutralize acids."[6] Indeed, the failure of alkaloid analysis to solve this problem prompted Liebig to tackle it by other means.

One of these approaches—proposed during his 1838 study of organic acids—was the suggestion that artificial bases could shed useful light on the essential nature of organic bases and their related acids.[7] In fact, Liebig had already begun putting this plan into action—using the artificial organic base melamine and its sequential decomposition products (ammeline, ammelide, and cyanuric acid) to model the spectrum from basic to acidic behavior.[8] Quantitative analysis led Liebig to infer the existence in all these bases of an atomic grouping "M" containing nitrogen and hydrogen that reacted with acids.[9]

By 1840, Liebig had developed a variant of Berzelius's dualistic theory in which the alkaloids' nitrogen existed not as ammonia (NH_3) but as what he called "amidogen" (NH_2), simultaneously redefining the "organic bases" as organic compounds containing nitrogen—a move that drew on his attribution of acidity to the element hydrogen.[10] Just as organic acids contained hydrogen that did not contribute to their acidity, moreover, Liebig's new definition implied that organic bases might contain nonbasic nitrogen. Therefore, in addition to reformulating the alkaloids' basic nitrogen as amidogen, Liebig's constitutional interpretation suggested that nitrogen could exist in more than one form in organic bases—a possibility that would be crucial to Hofmann's future research.[11]

Liebig's work contributed to chemists' growing recognition of the complexity of organic compounds—a situation reflected in developing constitutional theory, notably Dumas's type theory and his law of substitution. Dumas's theory found favor with the young Alsatian chemist and former Liebig pupil Charles Gerhardt (whom we briefly encountered in chapter 4). A crusading reformer intent on placing organic chemistry on what he regarded as a sound theoretical footing, Gerhardt in 1842 sought to promote type theory by unifying atomic weights—and hence molecular formulae—on the two-volume system common in France, to the exclusion of the four-volume system developed by Berzelius and widely accepted in Germany.[12]

Chemists working with two-volume atomic weights assigned different molecular formulae from those using the four-volume system. This had important consequences for attempts to establish taxonomic relationships between molecules, especially those that underpinned Dumas's type theory.

Gerhardt's attempt to enforce the two-volume system without an accepted empirical justification was therefore perceived as seeking unwarranted control over chemical theory and was condemned by his critics, eventually leading to his exclusion from elite Parisian chemistry.[13]

Despite such opposition, Gerhardt built from this atomic weight reform, and in 1842, his heavily taxonomic approach led him to introduce what would prove to be a powerful concept: homologous series. By sequencing compounds of the same type containing an increasing number of carbon atoms, homologous series highlighted the similar reactivity of compounds that differed only in their hydrocarbon portion. Where Liebig's series of melamine bases plus cyanic acid contained compounds with the same hydrocarbon component but distinct reactive atomic groupings, members of the same homologous series consisted of identical reactive groups combined with a steadily growing hydrocarbon component (e.g., methyl and ethyl alcohol, the first two members of the alcohol series). Homologous series therefore introduced an additional dimension into attempts to bring taxonomic order to organic chemistry.[14]

Although these developments enabled chemists to rationalize the behavior of relatively simple organic compounds, it remained unclear how they might clarify the chemistry of more complex substances, such as natural alkaloids. For example, identifying taxonomic patterns in the alkaloid family was difficult—partly because alkaloid formulae remained unstable, but also because many natural alkaloids contained complex hydrocarbon portions. As a result, it was all but impossible to assign alkaloids to homologous series with confidence.[15]

In a situation where decisive progress was so hard to achieve, disputes between adherents of dualistic and type theories continued to rage. As long as chemists saw dualistic and type theories as mutually exclusive alternatives, they remained locked in fruitless disagreements and almost whimsical changes of allegiance. Dumas, for example, adopted, defended, and then abandoned trenchant opinions regarding both theories.[16] Indeed, the inability of analysis to support well-founded conclusions regarding how to conceptualize the constitution of organic compounds contributed to Liebig's disillusionment with the field and the breakdown of his relationship with his former mentor Berzelius.[17]

Synthesis was Liebig's response to this state of affairs, a new investigative approach that he offered to his protégé Hofmann. Analysis relied on destructive processes whose vigor frequently destroyed constitutional information. Hofmann's synthetical experiments, as chapter 4 explained, used well-understood reactions to convert related starting materials into novel substances whose underlying constitution either remained intact or had been altered in predictable ways. Thus, the turn to synthesis highlighted standard reactions as reliable means of effecting chemical change and thereby elucidating constitution.

This focus on reliable reactions was crucial to Hofmann's ability to bypass the stagnation produced by other chemists' theoretical convictions. As the remainder of this chapter will show, studying reactions whose outcome he could predict provided a stable basis for Hofmann's laboratory reasoning, so that he was able to make reliable inferences without committing to any overarching constitutional theory. Synthetical experiments and laboratory reasoning—rather than allegiance to either dualistic or type theory—gave stability to Hofmann's research.

Some early results that were easier to rationalize in type theory did not lead Hofmann to abandon dualistic theory. Indeed, his first attempt to taxonomize the artificial organic bases produced from aniline by synthesis would be formulated in dualistic theory—at that time, the better established of the two theories. Yet the more he learned about the alkaloids, the more complex and diverse Hofmann recognized this family of natural products to be. As we shall see, usefully theorizing the constitution of artificial bases and natural alkaloids would take Hofmann beyond the disputes that preoccupied so many of his contemporaries.

RESEARCH RESOURCES

Hofmann's research built on foundations established by his earlier work, as well as work by Liebig and others in Giessen. For example, it was vital to Hofmann's synthetic approach that two compounds with identical composition and constitution were recognized to be the same, regardless of how they had been obtained.[18] As chapter 4 explained, this equivalence was crucial in establishing aniline as a plausible model compound for natural alkaloids.

By the mid-1840s, Hofmann was poised to exploit the possibilities of synthetical experiments on aniline. Indeed, his reliance on aniline made London the ideal place to be. Located at the heart of Britain's imperial trade routes, both coal tar and Indian indigo were far more readily available there than in Germany. One of Hofmann's first goals was to secure a supply of aniline from indigo—his source of aniline until at least 1849.[19] But, as I explain below, he also began to investigate an alternative possibility—one that would ultimately transform aniline into a cost-effective industrial raw material.

His London location gave Hofmann access to other resources that synthesis required. The RCC's manufacturing connections significantly improved the employment prospects of its trainees compared with their Giessen counterparts.[20] As a result—and somewhat to his surprise—Hofmann quickly attracted competent chemists to work with him. Two of his first assistants, Edward Chambers Nicholson and Frederick Augustus Abel, arrived at the RCC with significant prior chemical experience, and both went on to have highly successful careers encompassing chemistry's industrial and military applications.[21]

Hofmann was quick to exploit this skilled manpower. Producing aniline from indigo by distillation was a difficult manipulative task—one he immediately delegated to his assistants.[22] Although this secured his supply of aniline from indigo, Hofmann recognized that indigo was not an ideal raw material for research on the scale he anticipated. Despite the greater availability of indigo in London, obtaining aniline this way was expensive as well as time consuming. While in Bonn earlier that year, however, Hofmann had uncovered an exciting alternative. Although aniline itself was a relatively minor component of coal tar, he showed—using a color test introduced during his Giessen collaboration with John Blyth—that coal tar contained benzene in significantly greater quantity.[23]

Identifying benzene in coal tar suggested to Hofmann that aniline might be obtained by converting benzene to aniline using Nikolai Zinin's procedure—provided pure benzene could be isolated in sufficient quantity. He therefore directed another of his assistants, Charles Blachford Mansfield, to study the purification of benzene from coal tar by fractional distillation—a technique for separating mixtures by boiling point that developed rapidly during the mid-century, thanks to chemists' increasing access to glassblowing.

Although coal tar contained a good deal of benzene, the separation process turned out to be difficult and dangerous. It was not until 1848 that Mansfield patented a reliable procedure for isolating benzene from coal tar. And, tragically, he perished in 1855, nine days after suffering severe burns at his north London lodgings when a glass still he was using to isolate benzene in large quantity caught fire. Mansfield was 35 years old.[24] In fact, aniline did not become available in large quantities from coal tar until Perkin—after leaving the RCC to found a business for the manufacture of the synthetic dye mauve—successfully scaled up Mansfield's process in 1856.[25] Thus, British industry and commerce were both crucial in providing Hofmann access to essential resources, and ultimately they were transformed by the chemistry he brought to London.

EXPERIMENT AND THEORY

Following his arrival at the RCC, Hofmann began a series of synthetical experiments on aniline, hoping to produce compounds containing more nitrogen than was present in aniline and so to ascertain new information about the constitution of such molecules. This work illustrates how synthetical experiments enabled Hofmann to address key constitutional questions. The new substances that synthetical experiments made were neither a routine consequence of theory nor produced by unsystematic trial and error. On the contrary, the experimental system that guided their production was fundamentally necessary to Hofmann's reasoning and was crucial in achieving the theoretical advances he sought.

Hofmann began by reacting aniline with cyanic acid—a simple acid containing nitrogen. He believed aniline, an organic base, contained nitrogen in some form related to ammonia—whether as ammonia itself or as Liebig's amidogen. His experiment hinged on whether aniline's basic behavior would dominate (in which case, the product would be a salt formed by neutralizing cyanic acid) or whether aniline would mimic ammonia, which produced urea under these conditions. The result was clear. Rather than making "the ordinary cyanate of aniline" (i.e., a salt of cyanic acid and basic aniline), Hofmann isolated a substance whose composition and properties indicated it was "a kind of urea in which the ammonia is replaced by aniline." Showing that aniline reacted like ammonia allowed Hofmann to infer the product's constitution from the relationship between aniline and ammonia.[26]

Ever since Giessen, Hofmann explained, he had interpreted "urea of aniline" (i.e., urea in which aniline replaced ammonia) as demonstrating the "remarkable analogy" between aniline and ammonia. This analogy opened "an important new field of research," because "it was to be expected that all the relations exhibited by ammonia, and the metamorphoses it undergoes, might be realised with aniline." In other words, the success of Hofmann's first synthetical experiment on aniline implied that any of ammonia's known reactions might reasonably be attempted on aniline and interpreted using that analogy—thereby providing a rational basis for future experiments.[27]

It was also highly significant in light of Liebig's 1840 definition of the organic bases that Hofmann's experiment produced a novel organic substance containing more nitrogen than aniline—as his subsequent work makes clear. By the time he published this work in December 1845, Hofmann had already begun reacting aniline with cyanogen and chloride of cyanogen. Like cyanic acid, these substances also contained nitrogen, and Hofmann anticipated that they would introduce additional nitrogen into aniline. From the outset, Hofmann's synthetical experiments on aniline were intended to produce information about the relationship between the alkaloids' nitrogen content and their basicity, and ultimately to elucidate the constitution of nitrogen in organic bases.

Hofmann's next publication—whose appearance in January 1846 marked the start of a 2-year hiatus, during which he focused on his duties as the RCC's director—also reported the outcome of a synthetical experiment that introduced additional nitrogen into aniline. Coauthored with James Sheridan Muspratt—who briefly joined Hofmann in London before returning to the Muspratt family's Liverpool chemical business—this completed the project begun in Giessen when they used Zinin's procedure to make "toluidine, a new organic base."[28] Muspratt and Hofmann now detailed the preparation and behavior of another novel artificial compound they called "nitraniline." Produced by applying known reactions to aniline, nitraniline contained a second equivalent of nitrogen. Whereas aniline contained only the basic nitrogen responsible for its analogy to ammonia, nitraniline also contained nitrogen in the form of its acidic "peroxide."[29]

Despite its production by "the exchange of hydrogen for the elements of the hyponitric acid," Muspratt and Hofmann found nitraniline neutralized the same amount of acid as did aniline. Nitraniline's basic action was therefore more compatible with type than with dualistic theory—according to which, this property should be destroyed by the presence of acidic nitrogen (i.e., as peroxide). However, nitraniline's basicity was "exceedingly weak"—suggesting, contrary to type theory, that substitution did affect the molecule's properties. Even though their results were insufficient to establish the correctness of (or to falsify) either theory, Muspratt and Hofmann believed that they demonstrated the inadequacy of dualistic theory. "If we reject [type] theory," Muspratt and Hofmann reasoned, "we can scarcely understand the constitution of this base [i.e., nitraniline] and its relations to aniline."[30]

Muspratt and Hofmann's report on nitraniline prompted another Giessen-trained chemist, Carl Remegius Fresenius, to publish a highly speculative hypothesis concerning alkaloid constitution. In 1847, Fresenius claimed that nitrogen existed in alkaloids in just two distinct forms. The first, related to ammonia, was responsible for their basicity, while some form of nitrogen oxide was the second. Although it was almost exclusively built on the limited evidence supplied by nitraniline, Fresenius offered this suggestion as a general explanation of the fact that the alkaloids' "saturating capacity" (i.e., how much acid they neutralized) was not proportional to their nitrogen content.[31]

Fresenius's theoretical speculation was both threat and opportunity for Hofmann. Just as Liebig had previously defended Hofmann's dominion over research on the constitution of natural alkaloids, so Hofmann now took steps to dissuade Fresenius from further incursions into this field.[32] At the same time, Fresenius's publication provided the perfect foil to Hofmann's next move. Where his former Giessen colleague claimed a generality that lacked a solid empirical basis, Hofmann responded with characteristically meticulous research that used extensive evidence and logical reasoning to build on established foundations.

CONSTITUTIONAL DIVERSITY

As the demands of his directorial duties eased in the late 1840s, Hofmann resumed studying the reaction between aniline and cyanogen—a reaction that he had first performed but not fully interpreted in Giessen—and in April 1848, he published his findings.[33] Cyanogen's high toxicity made this investigation hazardous in the RCC's limited laboratory facilities.[34] Caution, as we shall see, was Hofmann's watchword in reasoning as well as in manipulation. Building on an exhaustive knowledge of others' work as well as his own results, Hofmann's experiments enabled him to refute Fresenius's proposal. They also led him to define an entirely new class of alkaloids—thereby beginning the expansion of alkaloid taxonomy that became the foundation of his future work.

Hofmann was interested in aniline's reaction with cyanogen, because cyanogen, like the oxide of nitrogen present in nitraniline, could be considered

an "electronegative radical"—that is, an atomic grouping that dualistic theory predicted would severely limit or even destroy aniline's ability to act
as a base.[35] Containing only carbon and nitrogen, cyanogen was known
to behave rather like chlorine and bromine, whose reactions with aniline
Hofmann had already studied. Just as he had previously observed, aniline's
reaction with cyanogen introduced an additional equivalent of nitrogen
into aniline in the form of cyanogen, producing a compound Hofmann
called "cyaniline."[36]

Cyaniline, like nitraniline, retained aniline's "saturating capacity" but was
a much weaker base than aniline.[37] Hofmann inferred that cyaniline was an
organic base containing a second equivalent of nonbasic nitrogen. Because
this was in the form of cyanogen rather than oxide of nitrogen, however, it
was powerful evidence against the general validity of Fresenius's suggestion.
"In a class of compounds so rich as that of the organic bases," Hofmann
concluded, "a greater variety is certainly to be expected, than is admissible on
the theory of my friend, Dr. Fresenius."[38] Thus, Hofmann began to argue for
unprecedented diversity in alkaloid constitution—an expansive move that
shaped his ability to design and interpret future experiments.

Up to this point, Hofmann had focused on cyaniline's similarities to
nitraniline—evidence most easily interpreted using type theory. Yet there
were also puzzling differences between the two compounds' composition
and reactivity that now led him to invoke dualistic theory. Nitraniline's
formation was consistent with substitution, and so was easily reconciled
with type theory. In contrast, cyaniline's formula ($C_{14}H_7N_2$)—which Hofmann determined by multiple analyses, paying close attention to tricky
nitrogen[39]—clearly indicated the addition of cyanogen to aniline ($C_{12}H_7N$).
Hofmann therefore described cyaniline as a "conjugated organic base"—
terminology that referenced additive dualistic theory.[40]

It is plausible—given his previous results—that product formation by
addition rather than substitution surprised Hofmann. Yet he soon reconciled this result within a framework that—while not explicitly encompassing both type and dualistic theories—was certainly grounded in diversity.
This step built on Liebig's earlier study of acids and bases—work that led
Liebig to a dualistic view of acid constitution, and to identify a close constitutional relationship between the artificial melamine bases and cyanuric
acid. Hofmann now suggested there were conjugated organic bases to complement the established class of conjugated organic acids. This proposal,

however, did not rule out the existence of other bases better explained within type theory.[41] Though previously it has been construed as a major reversal in his theoretical beliefs, this step was in fact key in Hofmann's growing appreciation that the alkaloids were a varied as well as complex family.[42]

Hofmann's next move was to subject new starting materials to the same reaction—a classic synthetical experiment—with even more curious results. On treating toluidine and cumidine—two artificial organic bases closely related to aniline—with cyanogen, he found that they reacted like aniline, producing "a new series of splendid alkaloids" that bolstered the evidence against Fresenius's proposal. The natural alkaloids nicotine and quinolone, however, did not react this way. Neither did ammonia. Yet these findings did not cause Hofmann to question aniline's crucial similarity to the natural alkaloids and to ammonia. Quite the reverse. Instead, Hofmann assigned nicotine and quinoline to "a perfectly different class of alkaloids" from aniline.[43] Thus, synthetical experiments enabled Hofmann to refine and extend the taxonomy of organic bases.

Hofmann's approach reveals the stability of empirically established analogies in his reasoning. Drawing on either dualistic or type theory, whichever seemed more helpful, Hofmann displayed much greater commitment to experimentally derived principles. Recognizing the primacy of experiment over theory in Hofmann's work, however, should not be taken to imply his disinterest in theory. Instead, these synthetical experiments on aniline illustrate how he made use of theory without becoming locked in restrictive dogma. For Hofmann, dualistic and type theories were important, because they indicated there should be systematic relationships between the substances he worked with, and that these relationships might be elucidated by experiment and reason. In producing these constitutional and taxonomic insights, Hofmann's focus remained the natural alkaloids—whose study was otherwise "greatly hindered by the costliness of the materials."[44]

COMPREHENSIVE KNOWLEDGE

Toward the end of 1849, Hofmann completed his study of how the introduction of electronegative atomic groups affected aniline. This work exemplifies Hofmann's thoroughness—a quality that was crucial to his acquisition of comprehensive knowledge of the substances and reactions he studied, and his consequent ability to spot results that did not fit expected trends. Having previously reacted aniline with cyanogen as well as the "general agents" chlorine and bromine, Hofmann now turned to the similar element, iodine. Where chlorine mainly decomposed aniline, bromine substituted three equivalents of hydrogen, and cyanogen formed only an addition compound, iodine replaced just one hydrogen atom.[45]

This trend from decomposition (chlorine) through substitution (bromine, iodine) to addition (cyanogen) exhibited "the decreasing affinities which these radicals manifest for hydrogen."[46] Thus, Hofmann brought order to his results—much as Liebig had previously arranged the melamine bases in order of decreasing basicity. But there was one crucial difference between their approaches. Liebig had inferred the relationship between basicity and constitution solely from decomposition reactions—which, by definition, destroyed some aspects of the starting molecule. In contrast, Hofmann's constitutional inferences were primarily based on synthetical experiments—evidence that was more reliable, because it was produced by less destructive means.[47]

Hofmann's next move was also a continuation of previous work. Reacting aniline with chloride, bromide, and iodide of cyanogen was a logical development from earlier synthetical experiments performed with chlorine, bromine, iodine, and cyanogen. Hofmann found these reactions produced "new confirmations of a general view of the organic bases which is daily gaining ground"—that is, dualistic theory. These experiments therefore prompted Hofmann to reinterpret some of his previous results, ultimately leading him to extend the fundamental analogy between aniline and ammonia in what would—somewhat ironically, since it involved dualistic theory—prove to be a decisive step toward the ammonia type.[48]

Hofmann began by surveying the existing literature concerning chloride of cyanogen and its action on organic compounds. This mode of presentation—though now standard—was then relatively unusual, and it highlights an important change in research and publication practice following the turn to synthesis. As chemists identified new reactions, they produced many novel substances—with the result that there was much more chemistry that a researcher at the cutting edge had to know. Analysis required minimal familiarity with prior work and a relatively limited set of techniques. Planning and executing useful synthetical experiments, by contrast, depended on wide reading as well as a greater diversity of laboratory methods. Hofmann's inclinations and Giessen training had prepared him to work in this way. An accomplished linguist with a good memory, Hofmann had long understood the importance of knowing the literature—as his 1843 review of indigo chemistry illustrates.[49]

Hofmann hoped that reacting aniline with cyanogen chloride would substitute cyanogen for hydrogen, producing a compound analogous to chloraniline. But when he performed the reaction—which was extremely hazardous, because cyanogen chloride, like cyanogen itself, was highly toxic—both "the mode in which [the product was] formed" and the results of analysis suggested that addition rather than substitution had occurred. Using a comma to indicate the additive conjunction between its two parts, Hofmann therefore designated the product by "cyanilide,aniline"—a dualistic interpretation that conceptualized it as additively composed of cyanilide (explained below) and aniline. In other words, Hofmann believed that the new compound—like cyaniline—was a conjugated organic base.

This was not the only curious feature of Hofmann's unexpected product. Although it was formed by the coming together of multiple molecules of basic, nitrogen-containing starting material, this compound—which Hofmann, by analogy with Liebig's melanine, called "melaniline"—saturated "only *one* equivalent of acid."[50] Having dismissed Fresenius's unsatisfactory proposal concerning the alkaloids' nonbasic nitrogen, Hofmann continued to seek an alternative constitutional explanation. Melaniline—because it represented a previously unknown kind of conjugated organic base—might shed light on this important issue.

Hofmann therefore embarked on a systematic study of melaniline, exposing it to the "various agents" previously applied to aniline: chlorine, bromine, iodine, nitric acid, and cyanogen. In all, Hofmann isolated, purified, and analyzed more than 20 novel compounds, which he arranged in a "synoptical table" (figure 5.1). Many of these compounds were the salts whose analysis confirmed melaniline's composition, and whose formation established its saturating capacity. However, by introducing "reagents" such as cyanogen, Hofmann also made compounds containing more carbon than the starting material—thereby expanding melaniline's organic (i.e., carbon-containing) basis, as well as organic taxonomy.[51]

It is important not to allow the later significance of carbon-adding reactions in target synthesis to distort their contemporary importance. Hofmann's primary focus was not on the substances he made but on the acquisition of comprehensive knowledge of melaniline's properties and

Melaniline . . . . . . $C_{12} H_7 N, C_{12} H_6 N \, Cy$
Sulphate of melaniline . . $C_{12} H_7 N, C_{12} H_6 N \, Cy, H \, SO_4$
Nitrate of melaniline . . $C_{12} H_7 N, C_{12} H_6 N \, Cy, H \, NO_6$
Binoxalate of melaniline . $C_{12} H_7 N, C_{12} H_6 N \, Cy, 2\,(H \, C_2 O_4)$
Hydrochlorate of melaniline $C_{12} H_7 N, C_{12} H_6 N \, Cy, H \, Cl$
Hydrobromate of melaniline $C_{12} H_7 N, C_{12} H_6 N \, Cy, H \, Br$
Hydriodate of melaniline . $C_{12} H_7 N, C_{12} H_6 N \, Cy, H \, I$
Platinum-salt . . . . . $C_{12} H_7 N, C_{12} H_6 N \, Cy, H \, Cl, Pt \, Cl_2$
Gold-salt . . . . . . $C_{12} H_7 N, C_{12} H_6 N \, Cy, H \, Cl, Au \, Cl_3$
Silver-salt . . . . . $2(C_{12} H_7 N, C_{12} H_6 N \, Cy), Ag \, NO_6$

Dichloromelaniline . $C_{12} \left\{ {H_6 \atop Cl} \right\} N, C_{12} \left\{ {H_5 \atop Cl} \right\} \quad N, Cy$

Platinum-salt . . $C_{12} \left\{ {H_6 \atop Cl} \right\} N, C_{12} \left\{ {H_5 \atop Cl} \right\} \quad N, Cy, H \, Cl, Pt \, Cl_2$

Dibromomelaniline . $C_{12} \left\{ {H_6 \atop Br} \right\} N, C_{12} \left\{ {H_5 \atop Br} \right\} \quad N, Cy$

Hydrochlorate of dibromomelaniline } $C_{12} \left\{ {H_6 \atop Br} \right\} N, C_{12} \left\{ {H_5 \atop Br} \right\} \quad N, Cy, H \, Cl$

Platinum-salt . . $C_{12} \left\{ {H_6 \atop Br} \right\} N, C_{12} \left\{ {H_5 \atop Br} \right\} \quad N, Cy, H \, Cl, Pt \, Cl_2$

Diiodomelaniline . $C_{12} \left\{ {H_6 \atop I} \right\} N, C_{12} \left\{ {H_5 \atop I} \right\} \quad N \, Cy$

Platinum-salt . . $C_{12} \left\{ {H_6 \atop I} \right\} N, C_{12} \left\{ {H_5 \atop I} \right\} \quad N \, Cy, H \, Cl, Pt \, Cl_2$

Dinitromelaniline . $C_{12} \left\{ {H_6 \atop NO_4} \right\} N, C_{12} \left\{ {H_5 \atop NO_4} \right\} N, Cy$

Hydrochlorate of dinitromelaniline } $C_{12} \left\{ {H_6 \atop NO_4} \right\} N, C_{12} \left\{ {H_5 \atop NO_4} \right\} N, Cy \, H \, Cl$

Platinum-salt . . $C_{12} \left\{ {H_6 \atop NO_4} \right\} N, C_{12} \left\{ {H_5 \atop NO_4} \right\} N \, Cy, H \, Cl, Pt \, Cl_2$

Dicyanomelaniline . $C_{12} \quad H_7 \quad N, C_{12} \quad H_6 \quad N, Cy \left\{ {Cy \atop Cy} \right.$

FIGURE 5.1
Hofmann's synoptical table displayed the formula of every melaniline derivative he had made, organized in order of increasing complexity. In some cases, Hofmann's formulae provided some information about molecular constitution. These formulae nevertheless make clear that numerical relationships and linear connectivity, rather than visual thinking, were fundamental to Hofmann's view of the molecular world. It is also worth noting that Hofmann, in true Giessen style, used quantitative analyses of multiple different salts to increase the reliability of the formulae he assigned. At this time, Hofmann's formulae were based on $C_{12}H_7N$ for aniline, reflecting his use of an atomic weight for carbon that was roughly half of the present-day value. He used Cy as an abbreviation for cyanogen, C_2N.

Source: August Wilhelm Hofmann, "Researches on the Volatile Organic Bases. III: Action of Chloride, Bromide, and Iodide of Cyanogen on Aniline. Melaniline, a New Conjugated Alkaloid," *Quarterly Journal of the Chemical Society* 1 (1849), 311–312. Courtesy of the Royal Society of Chemistry Journals Archive: https://pubs.rsc.org/en/content/articlelanding/1849/qj/qj8490100285.

reactivity. This information was essential to his taxonomic overview. It enabled Hofmann to recognize compounds he had previously encountered and to identify where no compound fitting taxonomic predictions yet existed—gaps he frequently attempted to fill. This was why—much as Liebig had previously tabulated the melamine bases—the "synoptical table" became one of Hofmann's key research tools.[52]

TAXONOMIC EXPECTATIONS

Melaniline was not merely "a new conjugated base" that expanded the series that began with cyaniline. Its complexity also confirmed what Hofmann by now recognized as a fundamental feature of the alkaloids: a diversity that was far easier to reconcile with dualistic than with type theory. Surveying chemists' knowledge of artificial and natural bases, Hofmann now declared it was "exceedingly probable" that ammonia "impressed its character [i.e., its basicity] on the whole compound." As a result, it was "scarcely to be doubted that the natural alkaloids . . . are formed in a similar manner"—that is, by combining an organic portion with ammonia or, in the case of nonbasic nitrogen, some other form of ammonia. Thus, "the generation of so numerous a class of bodies like the organic alkaloids, may be explained without difficulty."[53]

Dualistic theory increasingly appealed to Hofmann because of its ability to explain diversity. But this did not mean that he was simply reverting to the theory Berzelius had championed for two decades. Nor, as we shall see, was he abandoning the systematizing utility of type theory. In fact, Hofmann now proposed a significant development of dualistic theory, in which ammonia could be present in alkaloids without exhibiting its basic characteristics. In arguing his case, Hofmann reverted to the product of his previous synthetical experiment: "the urea of the aniline series" produced by reacting aniline with cyanic acid. And now, for the first time, he specified its analogy to urea in terms of constitution as well as composition.

Expressing urea as "NH_3, C_2HNO_2" and aniline-urea as "$NH_3, C_2HNO_2, C_{12}H_4$" indicated that both substances contained "one [central] equivalent of hydrogen" (bold added) in "a peculiar state of mobility." Identifying this reactive hydrogen allowed Hofmann to explain the conversion of urea into nonbasic amides in terms of its substitution by alkali metals, including sodium and potassium. "Exactly the same" kind of substitution, he then reasoned, occurred when an organic base displaced aniline-urea's analogous hydrogen—producing compounds Hofmann called "anilides." Although Hofmann believed anilides—like amides—contained ammonia, these compounds were not basic. They nevertheless formed addition compounds with organic bases such as aniline—as in "cyanilide,aniline" (melaniline). Hofmann therefore attributed both basic and nonbasic nitrogen in the alkaloids to the presence of ammonia, albeit in different forms. Formulating melaniline as a conjugated compound of basic aniline with a nonbasic anilide thus explained its ability to neutralize just one equivalent of acid.[54]

Hofmann now drew on his knowledge of the literature to propose an extended four-part taxonomy for the organic bases. Just as ammonia salts were known to lose either two or four equivalents of water to produce "amides," "nitryles," "amidogen acids," and "imidogen compounds," so Hofmann believed that "this series . . . can likewise be reproduced with the organic bases." Among a vast body of evidence, Hofmann referred to carbamide (produced by the action of ammonia on phosgene) and its aniline analog, carbanilide (which he had previously prepared).[55] In other words, just as aniline gave rise to anilides—the conjugated analogs of amides—so Hofmann now anticipated the existence of a series of similarly conjugated nitriles, amidogen acids, and imidogen compounds derived from aniline.

At this point, the new taxonomy contained numerous gaps—which Hofmann was working hard to fill. "The imidogen compounds are as yet the least known," while in no single case "have all the four classes as yet been obtained." Yet Hofmann was confident that "[a]ll the bases analogous to aniline, will most likely, show the same comportment." At first, the work proceeded according to plan. Decomposing aniline-urea into carbanilide and urea prompted Hofmann to reconceptualize this compound as "carbamide, carbanilide"—a move that confirmed its conjugated nature, while also making sense of the most striking difference between it and urea: its failure to act as a base. It was "exceedingly probable," Hofmann concluded "that the organic bases are indeed conjugated ammonia-compounds."[56]

Hofmann's proposal accounted for every known organic base, providing the first adequate explanation of nonbasic organic nitrogen. Of his four categories, only "conjugated imidogen compounds" and conjugated nitriles had yet to be observed. But Hofmann confidently predicted that experiments "just now in progress in my laboratory" on dicyanomelaniline (produced by reacting melaniline with cyanogen) would remedy the situation. As we shall see below, the failure of this prediction—which indicated a fundamental flaw in Hofmann's taxonomic proposal—proved crucial in prompting Hofmann's decisive next step.[57]

A NOSE FOR THE UNEXPECTED

Hofmann's study of dicyanomelaniline produced "a result as remarkable as any . . . elicited in the course of these researches." Hofmann successfully isolated a compound he called "melanoximide" and identified as the imidogen derived from melaniline by hydrochloric acid dehydration. Melanoximide's composition and properties provided "unequivocal" support for this classification. The reverse reaction—that is, the conversion of melanoximide back to melaniline in the form of its oxalate salt—was successful. But Hofmann also noticed "the peculiar somewhat cyanic odour, which I had frequent opportunities of noticing during these investigations, being evolved in a remarkable manner."[58]

Things soon took an even more surprising turn. When Hofmann tried to dehydrate melaniline's oxalate salt—a process he anticipated would also produce melanoximide—the products were not as expected. On heating, the salt melted and began to boil, producing "torrents" of carbon monoxide and carbon dioxide, together with sublimed crystals of carbanilide—all of which were explicable as decomposition products. But the gas evolved also "possessed in a very powerful degree, the peculiar odour to which . . . I have repeatedly alluded." Smell told Hofmann that this reaction, rather than making melanoximide, in fact produced a large quantity of "this peculiar compound . . . this enigmatical body, which I had vainly chased on so many occasions."[59]

At first, Hofmann failed to isolate this mysterious, malodorous substance. Evidently reasoning that it was formed via the expected product, melanoximide, he heated melanoximide instead. Now Hofmann identified the source of the smell: another new compound he called "anilocyanic acid." This was Hofmann's "remarkable" result. Examining how he studied anilocyanic acid, and how knowledge of its properties directed his laboratory reasoning, explains how synthetical experiments on aniline and its derivatives brought Hofmann's dualistic interpretation of the organic bases to a crisis point.

PREDICTIVE FAILURE

Hofmann now set out to produce enough anilocyanic acid for investigation—an onerous task that illustrates the experimental difficulties of this work, and how these were overcome by careful manipulations in glass. Dry distillation of melanoximide formed crude anilocyanic acid as a yellow liquid, along with a large volume of gaseous products resulting from decomposition on exposure to moisture. Hofmann minimized decomposition by carefully drying the starting material and apparatus, but—since the reaction itself evolved water—he nevertheless obtained anilocyanic acid in only about 10% yield.

This distillation was made even more difficult by the yellow liquid's "most powerful odour, recalling at once the odour of aniline, of cyanogen, and hydrocyanic acid [hydrogen cyanide], provoking lachrymation in a most fearful manner, and exciting too in the throat, the suffocating sensation produced by the latter." Hofmann was clearly familiar with the physiological effects of highly toxic substances, including hydrogen cyanide. The large volume of gas evolved during distillation also evaporated anilocyanic acid, with the result that it was "necessary to collect the gas [responsible for the smell], in order to escape its irritating action on the nose and eyes." Hofmann did not describe or illustrate how its collection was accomplished—but it almost certainly relied on modifying the glass distillation apparatus to incorporate a gas trap.

Moreover, the crude product was significantly contaminated by another compound with a very similar boiling point—which made it impossible to purify by further distillation. Hofmann therefore cooled the mixture instead, crystallizing a pure product he collected by filtration—a simple-sounding process that was complicated by the mixture's instability when exposed to atmospheric moisture. Finally, anilocyanic acid was distilled from a glass flask "in order to obtain it in a state fit for analysis." It "entered into regular ebullition [boiling]" at 178 °C, while "during the latter part of the distillation, the thermometer rose very gradually to 180 °C."[60]

Only skillful use of chemical glassware and minute attention to laboratory technique—combined with what chapter 7 will show were newly established criteria of purity—enabled Hofmann to isolate, purify, and begin to characterize anilocyanic acid. In addition to reporting its boiling point, Hofmann noted properties that were strongly reminiscent of cyanic acid, a compound he knew well. Anilocyanic acid was "a colourless, very mobile liquid, heavier than water, strongly refracting light, and possessing the powerful odour in undiminished intensity."[61]

The difficulties of this dangerous work meant that Hofmann struggled to obtain suitable samples for analysis, publishing only two carbon determinations, together with one analysis in which nitrogen was "estimated" according to a method developed during his earlier study of cyaniline. In fact, Hofmann based anilocyanic acid's formula ($C_{14}H_5NO_2$) on the second carbon determination alone, since the analysis had been performed on a purer sample. Synthetical experiments, however, "corroborated" this formula, "presenting perhaps, when taken as a whole, a higher order of experimental evidence than can be obtained from any elementary analysis, however accurate it might be." Thus, Hofmann argued for synthesis as a powerful, independent investigative method, more than capable of making good the limitations of analysis.[62]

Hofmann's reliance on synthesis was vital to his laboratory reasoning. As he explained, "The formula derived from analysis, although correctly enumerating the elements grouped in the yellow compound [anilocyanic acid], leaves us quite in the dark respecting their actual arrangement, and consequently of the [taxonomic] position which has to be assigned to the substance."[63] Having assigned a molecular formula, synthetical experiments enabled Hofmann to interpret the constitutional relationship between his product and cyanic acid. Experimental evidence, rather than "theoretical speculations," lay behind Hofmann's identification of this product as "represent[ing] in the aniline-series the term cyanic acid, . . . in other words the odorous liquid [anilocyanic acid] would stand in the same relation to aniline, as hydrated cyanic acid stands to ammonia."[64]

This analogy now led Hofmann to incorporate a new reaction into his research program. His good friend and former Giessen associate, Adolphe Wurtz, had recently established that cyanogen chloride reacted with alcohols to produce urethanes—compounds first prepared by Dumas.[65] Wurtz showed that the reaction between cyanic acid and alcohol—previously studied by Liebig and Wöhler—also produced urethane.[66] This result—the outcome of work performed at Dumas's instigation and in his laboratory—spanned the boundary between inorganic and organic chemistry.[67] In fact, as we shall see in chapter 6, it began Wurtz's transition from the inorganic borderlands to organic chemistry proper.[68]

Hofmann's recent work with anilocyanic acid meant that he perceived more significance in this result than did Wurtz. Analogy suggested that a similar reaction might take place with anilocyanic acid and—when he tried it—this proved to be the case. Despite having only "approximative combustions, performed with substances not absolutely pure," Hofmann confidently identified the products of the reaction between anilocyanic acid and various alcohols as urethanes. If, he reasoned, urethanes could be viewed as "compound ethers, in which a peculiar acid, carbamic acid, pre-exists, it would follow that the preceding series would have to be regarded as the ethers of carbanilic or anthranilic acid, with whose composition they coincide."[69]

At this crucial juncture, anilocyanic acid's "long and complicated" preparation left Hofmann "in want of material"—bringing this aspect of his experimental work to a less-than-satisfactory conclusion. Unable to complete his investigation of anilocyanic acid's "beautiful reactions" with alcohols, Hofmann lamented the "expenditure of time and labour" required "to follow out the manifold ramifications of the aniline-family in its numerous and often so intricately related derivatives." Even aided by "the unremitting zeal, and the remarkable experimental skill" of his assistants, Abel and Nicholson, Hofmann lacked "sufficient leisure" to complete this investigation—presumably a reference to his onerous directorial duties. He therefore concluded by "enumerat[ing] . . . the various facts concerning anilo-cyanic acid," offering an unusually speculative "probable theory" concerning its formation from melaniline and melanoximide.[70]

FACING CRISIS

Hofmann's study of anilocyanic acid failed to satisfy his own exacting requirement to produce a "strict interpretation of unequivocal facts."[71] His extensive knowledge of this curious new substance nevertheless had decisive consequences. As explained above, Hofmann's four-part dualistic taxonomy of organic nitrogen compounds included conjugated nitriles—compounds not yet observed, but which Hofmann expected would be produced by dehydrating oxalate salts of aniline and melaniline. He already knew that dehydrating either melaniline's oxalate salt or melanoximide produced anilocyanic acid. Equipped with this knowledge, Hofmann now recognized that the product of dehydrating oxanilide was also anilocyanic acid—rather than "anilo-nitrile," as his dualistic taxonomy required. "The odour," he explained, "could not leave the slightest doubt upon this point." Even in the absence of "the conclusive test of a combustion [analysis]," Hofmann had "not the slightest hesitation in asserting [the] identity" of his new reaction product with anilocyanic acid.[72]

This predictive failure brought Hofmann to a fundamental paradox. Unable—despite his mastery of aniline chemistry—to complete the four-part taxonomy he anticipated would rationalize aniline's dualistic analogy with ammonia, Hofmann's laboratory reasoning led him to wonder, "Why aniline, which so faithfully imitates all the habits of ammonia, refuses to follow its example with respect to the formation of the nitriles?" Although the dualistic interpretation of conjugated bases was "in accordance with all the observations which had [previously] been made," this view of aniline "gives no satisfactory answer to the above question. There is no comprehensible reason why the oxalate of ammonia in the aniline-salt should not be deprived of its four equivalents of water."[73] In other words, dualistic theory satisfactorily accounted for the dehydration of oxamide (NH_2,C_2O_2) to produce oxalonitrile (cyanogen, C_2N), but it could not explain why dehydrating oxanilide ($NH_2,C_2O_2,C_{12}H_4$) produced anilocyanic acid ($C_{12}H_5N,2CO$) rather than an analogous nitrile in the aniline series.[74]

Toward the end of 1849, Hofmann knew that his dualistic taxonomy of organic bases could not be reconciled with the results of his experiments. Absolute confidence in his results and their interpretation in relation to dualistic theory enabled Hofmann to identify definite flaws in his provisional taxonomy, so that he was forced to abandon his working hypothesis. At this point, Hofmann faced a crisis. What should he do next?

6   AMMONIA TYPE

We left August Wilhelm Hofmann confronting the failure of his dualistic taxonomy of organic bases. This chapter picks up the story, explaining how laboratory reasoning enabled him both to transcend this crisis and to develop a new classification of organic bases linked to type theory. The same confidence in experiment and reason that forced Hofmann to abandon his dualistic taxonomy meant that he could also pinpoint a useful direction for further work. Thus, Hofmann identified the central importance of anilocyanic acid—the malodorous, highly toxic substance that had dogged his recent experiments.

Focusing on anilocyanic acid highlighted two results recently published by Adolphe Wurtz. One, encountered in chapter 5, was the reaction between cyanic acid and alcohols to produce urethanes. Preliminary indications were that anilocyanic acid behaved similarly to cyanic acid, suggesting that the two substances were analogous, but practical difficulties meant these experiments remained unfinished. The other result—which Wurtz came upon by chance during further investigations into how cyanic and cyanuric acids reacted with alcohols—was his recent preparation of a series of new organic bases called "amines."[1]

When Wurtz published this result in early 1849, it caused something of a sensation. But this was not because other chemists immediately accepted either his experimental findings or his assertion that ammonia "should be for all chemists the type of this numerous class of compounds."[2] Indeed, Wurtz's claim—which reflected a prior commitment to Gerhardt's type theory—was based on rather limited empirical evidence. His results nevertheless attracted considerable interest, so that the Parisian Academy of Sciences appointed Jean Baptiste Dumas to review them. Dumas—who also stood in the role of mentor to Wurtz—was understandably positive but focused on confirming Wurtz's experimental results and explaining how they supported the unification of organic and mineral (inorganic) chemistry. It is especially worth noting that Dumas's report made no reference to ammonia as type, or indeed to type theory.[3]

For Hofmann, Wurtz's theoretical speculations offered nothing new.[4] As we saw in chapter 5, by this time he had concluded that dualistic theory did not provide a viable framework for classifying organic bases, leaving type theory as the only rational alternative. It is also implicit in what Hofmann did next that he did "not regard [Wurtz's] production of organic ammonias . . . as proof of their constitution."[5] In other words, Hofmann did not recognize a valid theory in Wurtz's speculations. What Hofmann did see, however, was how laboratory reasoning from synthetical experiments based on Wurtz's new reactions might help build such a theory. Where Wurtz had merely stumbled on a curious result and used it to buttress his preferred constitutional theory, Hofmann understood what would be required to produce a useful theory of the constitution of the organic bases.

In explaining how Hofmann established the ammonia type, this chapter extends our consideration of the relationship between laboratory reasoning and theory. We saw in chapter 5 that laboratory reasoning was constitutive of Hofmann's daily work, confined in his personal experimental world. For theory to be useful, however, it had to be accepted by, and available to, the disciplinary community. Thus, we may wonder how what Hofmann identified as the ammonia type moved from being the localized product of his laboratory reasoning to becoming a durable component of chemists' understanding of organic compounds. In answering this question, we shall learn what distinguished Hofmann's mode of theorizing from the more speculative approaches of other chemists, including Wurtz and Gerhardt.

HOFMANN'S FIRST TYPE

Hofmann's ability to move beyond the failure of his dualistic taxonomy depended on a fresh approach to clarifying the relationship between ammonia and aniline—one that was consistent with his synthetic method and could be applied to anilocyanic acid. It was therefore the two reactions that

Wurtz had published—rather than his interpretation of their outcome—that most interested Hofmann. Thus, Hofmann set out to overcome the problems of working with anilocyanic acid and to develop Wurtz's new reactions as the basis of synthetical experiments. I now explain how this work prompted Hofmann to produce a revised taxonomy of organic bases linked to type theory.

Before proceeding it will be helpful to step back in time, summarizing Wurtz's findings during the late 1840s to explain how—despite the limited experimental detail his publications provided—these suggested a way forward to Hofmann. By 1848, Wurtz had extended his original study of urethanes to examine other reactions of cyanic and cyanuric acids. During this period, he prepared cyanate esters, including methyl and ethyl cyanate.[6] Both esters reacted with ammonia to form addition products. And when treated with water, they behaved just like cyanic acid, losing carbon dioxide to produce substances that Wurtz suggested—using unpublished analytical data—were "homologue[s]" of urea.[7]

This use of the term "homologue" explicitly referenced the additional classificatory dimension introduced by Gerhardt's homologous series. By adding the elements of methylene (CH_2) to his simplest product, Wurtz asserted, "we climb a rung of the ladder" linking members of a "series." He therefore predicted the existence of other ureas containing more carbon. This proposal, however, was based purely on arithmetical relationships between assigned formulae. Crucially, in the absence of supporting evidence, other chemists could not assess its veracity. Wurtz also inferred from formulae and differences in reactivity that his products included isomers, offering a hypothetical explanation of how these might have been formed. This was highly speculative theorizing well beyond empirically established facts and quite unlike the careful reasoning that we have seen characterized Hofmann's work.[8]

Continuing to study the esters of cyanic and cyanuric acids, and their related ureas, Wurtz found that they reacted with potassium hydroxide to produce gaseous products. This tricky experimental work tested Wurtz, reflected in the fact that theory-driven expectation led him for quite some time to misidentify his products as ammonia. As Hofmann later recalled, a chance observation—rather than systematic enquiry—prompted Wurtz to recognize his mistake: When one of his gaseous products was ignited by a nearby flame, Wurtz realized that it could not be ammonia.[9]

Wurtz now identified his gaseous products as methyl- and ethylamine, compounds in which ammonia's basic properties were barely affected by the introduction of the elements of organic matter. These results, communicated in preliminary form to the Parisian Academy of Science in February 1849, were startling, and they disconcerted Hofmann. Despite his conviction that Wurtz's identification of ammonia as the "type of this numerous class of compounds" was nothing more than a restatement of an idea implicit in Liebig's earlier work on the organic bases, Hofmann was taken aback to see Wurtz encroach on his research area.[10] He would now take steps to establish his unique command of this field.

As chapter 5 explained, Hofmann already believed there was an analogy between anilocyanic and cyanic acids. Just as cyanic acid reacted with alcohols to produce "compound ethers, in which a peculiar acid, carbamic acid, pre-exists" (i.e., urethanes), so anilocyanic acid produced a similar "ether of carbanilic, or anthranilic, acid." He had therefore named the product of reacting anilocyanic acid with ethanol "anilo-urethane."[11]

Hofmann now built on this analogy, reconceptualizing Wurtz's production of "compounds of the alcohol radicals with cyanic acid" (the cyanate esters) as a series of synthetical experiments. Focusing initially on their common method of preparation, Hofmann reasoned that these esters were "a series of compounds of which cyanic acid is, as it were, the type."[12] This classification enabled him to make sense of all the reactions Wurtz had reported—including the formation of the amines. Thus, Hofmann's first use of the word "type" was coupled with cyanic acid, not ammonia and—because it ordered what had previously seemed chaotic—it was reminiscent of Liebig's earlier linking of the melamine bases to cyanuric acid.[13]

As evidence of the insufficiency of dualistic theory mounted, Hofmann was beginning to countenance type theory as an alternative. Yet his new terminology signified neither wholesale abandonment of dualistic theory nor sudden commitment to type theory in its entirety. On the contrary, Hofmann continued according to his established pattern—building on previous work by using carefully designed experiments whose results provided a secure basis for further reasoning.

Hofmann's reinterpretation of the behavior of cyanate and cyanurate esters "when treated with the alkalis" would have particularly significant consequences. In this reaction, as Hofmann then understood it, such esters lost carbon dioxide, becoming "conjugated ammonias," which he incorporated in a reorganized taxonomy. Thus, Hofmann renamed Wurtz's "valeramine" as "amylamine" to clarify its position in his new taxonomy.[14]

Aniline similarly took its place beside methyl-, ethyl-, and amylamine, from which Hofmann inferred it might be produced by converting phenol into its corresponding cyanate ester and then decomposing this ester according to Wurtz's procedure. This was a logically sound proposition. But when Hofmann attempted the first reaction (between cyanic acid and phenol), it did not proceed as expected.

This outcome did not cause Hofmann to doubt Wurtz's chemistry, which suggests that he had already successfully mastered Wurtz's methods for converting simple alcohols into their respective cyanate esters and amines. Instead, Hofmann interpreted this failure as confirming the distinct behavior and unusual stability of the phenyl group in aromatic substances, such as phenol and aniline. Phenol, he concluded, "is no homologue but only an analogue of common alcohol."[15] This was a highly important result that significantly improved understanding of the distinction between aromatic and aliphatic (nonaromatic) compounds, helping refine Gerhardt's notion of homology. As we shall see, it also prompted Hofmann to interpret the constitution of organic bases in a fundamentally new way.

INTERPRETATION IN TRANSITION

As he proceeded, Hofmann simultaneously acknowledged Wurtz's contribution, asserted Liebig's primacy, and cleared the path for his own future work. Where Berzelius's dualistic theory stipulated that organic bases contain preformed ammonia (NH_3), Hofmann now recalled that Liebig's amidogen theory indicated aniline's constitution was "($C_{12}H_5$),H_2N." This formulation readily accounted for the observed impossibility of removing four equivalents of water from aniline's neutral oxalate salt, which Hofmann now represented as "($C_{12}H_5$),H_2N,HC_2O_4" in place of "H_3N,($C_{12}H_4$)HC_2O_4." Hofmann's new constitutional interpretation, unlike its predecessor, clarified why his previous efforts to dehydrate oxanilide had failed to produce the "anilo-nitrile" his dualistic taxonomy required, instead forming anilocyanic acid. As Hofmann explained, this would entail "destruction of the term $C_{12}H_5$ (phenyl)"—a group whose unusual stability had been demonstrated by his recent attempt to react cyanic acid with phenol.[16]

In arguing for amidogen as the key to "more rational views respecting the constitution of the organic bases," however, Hofmann did not merely adopt Liebig's theory.[17] Where Liebig's original amidogen was formulated in dualistic theory, Hofmann now reasoned that: "It is probably more in conformity with truth to consider aniline as a substitution product, as ammonia, in which part of the hydrogen is replaced by phenyl." This shift implicitly incorporated the kernel of Wurtz's speculative proposal into Liebig's amidogen concept, naturalizing the notion that organic bases were substituted ammonias in Hofmann's chemistry. It also, as I show below, allowed him to reconceptualize his existing dualistic taxonomy of organic bases in type theory.[18]

Hofmann's next step was to react aniline and ammonia with "the bromides of the alcohol-radicals" (modern alkyl bromides)—reactions that enabled him "actually to replace the basic hydrogen of these substances [i.e., aniline and ammonia], equivalent for equivalent, by the alcohol-radicals, and to produce in this manner a numerous series of new alkaloids which appear to admit of no other mode of interpretation." By early November 1849, Hofmann was convinced that organic bases were indeed formed by substituting organic radicals for hydrogen in ammonia. This was Hofmann's ammonia type. Before the end of the year, he had assembled evidence for presentation to London's prestigious Royal Society.[19]

HOFMANN'S AMMONIA TYPE

Sir James Clark—one of his foremost London backers—read Hofmann's paper "Researches Regarding the Molecular Constitution of the Organic Bases" to the Royal Society on January 17, 1850. Writing for this august audience, Hofmann sought both to justify his new theory to nonchemist members of the Royal Society and to persuade his fellow chemists of the power of the synthetic method. These goals led Hofmann occasionally to diverge from the most plausible discovery narrative that I have been able to reconstruct.[20] The account presented here was built on an intensive study of Hofmann's prior publications, which—in the absence of surviving manuscript materials—are the most relevant primary sources. It offers further insight into how laboratory reasoning from synthetical experiments produced theory, explaining why this approach relied so completely on new organic reagents and on using innovative glassware to manage persistent problems of purity and chemical identity.

Building on his earlier response to Wurtz—and on his mentor's status—Hofmann formulated his conclusions regarding the organic bases' constitution as a natural extension of Liebig's theory, in which the amidogen bases (primary amines) "now present[ed] themselves . . . only as particular instances of the permutations possible among the elements of the primary type ammonia." In light of this theory, "It seemed but logical to look among the bases for analogues too of the imidogen-compounds and the nitriles."[21]

Finally abandoning the dualistic taxonomy, Hofmann now identified "three classes of organic bases, derived from ammonia by the replacement respectively of 1, 2 or 3 equivalents of hydrogen." Plenty of amidogen-bases (present-day primary amines) were already known—including aniline, as well as Wurtz's methylamine, ethylamine, and amylamine—but examples of imidogen- and nitrile-bases (secondary and tertiary amines, respectively) were scarce. These latter were what Hofmann had spent much of 1849 attempting to make, building a new substitution-based taxonomy.[22] Thus, Hofmann's laboratory preparation of compounds that his revised taxonomy predicted should exist—but did not tell him how to make—served to demonstrate type theory's ability correctly to account for the chemistry of amines.

Recognizing that theory did not tell Hofmann how to make these novel compounds undermines the assumption that the predictive power of chemical theory encompasses exactly this laboratory capability. In fact, as Hofmann explained, his initial attempts to substitute basic hydrogen in aniline by a phenyl group utterly failed.[23] According to Hofmann, the crucial factor in bringing this work to a successful conclusion was not his reconceptualization of organic bases in type theory as substituted ammonia compounds. Instead, it was the availability of a new family of reactive compounds, the alkyl halides, which Hofmann—by offering a comprehensive account of their reactivity—now began transforming into some of organic chemistry's most valuable reagents.

Although Hofmann did not explain how he came to experiment with the alkyl halides, we may reasonably infer that his choice was driven by Edward Frankland's chemistry. Hofmann had recently learned how Frankland used alkyl halides to prepare alkyl zinc compounds—organometallic substances that have since become established as powerful synthetic reagents. When Hofmann declared the alkyl halides "the most appropriate substances" for his intended purpose, his choice reflected an expectation that the alkyl halides' ability to react with zinc (a basic metal) indicated that a similar reaction would occur with an organic base, such as aniline.[24]

Hofmann's reasoning proved justified—but this did not mean that the reactions he attempted proceeded without difficulty. Using the alkyl halides in a new context, he had to overcome numerous practical difficulties. He soon discovered, for example, that alkyl chlorides were too volatile to be easily handled, while alkyl iodides often produced complex mixtures of products that he was unable to separate and purify. In general, therefore, Hofmann worked with alkyl bromides (intermediate between chlorides and iodides). Yet he still found it difficult to know what his reactions made. In fact, this issue became so significant following the turn to synthesis that chapter 7 is devoted to explaining how chemists tackled and ultimately resolved it.

Both innovative glassware and new ways of using existing items were crucial to Hofmann's work, frequently warranting explicit mention. The reaction between aniline and dry ethyl bromide, for example, took place either by leaving the unheated mixture for several hours or by "gently heating the mixture in an apparatus which will allow the volatilized bromide to return to the aniline."[25] Because Hofmann provided no illustrations, it is impossible to know exactly what apparatus he used. Hofmann, however, was certainly familiar with the condenser that Giessen chemists used when performing distillations—the device later named after Liebig. Originally consisting of a long glass tube encased in a metal sleeve, Liebig's condenser used flowing water to cool distilling vapor.[26] It therefore seems plausible that Hofmann adjusted this device's usual orientation so that it could perform the same function while standing vertically above the reaction vessel (figures 6.1 and 6.2).

When heated, the mixture of reactants and any solvent boiled and vaporized, only to condense and return to the reaction vessel. This method was useful, because it made it possible to maintain the reaction mixture at its boiling point for extended periods without increasing pressure—thereby encouraging reaction without introducing the dangers associated with high pressures that were a routine consequence of sealed tube reactions. By 1870, the term "to heat/boil under reflux" (*am Rückflusskühler hitzen/kochen*) had entered common usage, and within two decades, this was one of synthetic organic chemistry's most widely applied methods.[27]

It is interesting to consider why Hofmann in 1850 provided such limited information concerning what seems to have been an innovative technique.

FIGURE 6.1

This photograph shows Hofmann (seated front row, center) with students at London's Royal College of Chemistry. The presence of a young William Perkin (back row, 5th from right) fixes the date to ca. 1855. In addition to showing a rack of test tubes (center right) and Will and Varrentrapp's apparatus (bottom left), one of Hofmann's students holds a Liebig condenser in the vertical orientation required for reflux (bottom right). Although the inner condensing tube was made of glass, the outer casing of Liebig's condenser was at this time made of a metal, such as zinc. By the latter decades of the nineteenth century, reflux was an established, standard method for maintaining reactants at elevated temperatures for extended periods without increasing pressure.

Source: Courtesy of the University of St Andrews Libraries and Museums, ID: JR-N-2344.

The available evidence suggests that he wanted others to be able to repeat his work and use his methods, as indeed would happen. His description made it possible in principle for other chemists to attempt the procedures he described and for those who had trained with him—or perhaps with Liebig in Giessen—it may have been sufficient. For others, however, Hofmann's reticence surely constituted a real barrier to replication. Thus, it appears that the community of chemists working at the cutting edge of synthesis was limited in this period to those linked—either by personal connection or through lineages of training—to Hofmann and Liebig.[28]

Hofmann judged the reaction complete when the mixture solidified. This sometimes happened relatively quickly. But on other occasions—such as when preparing diamyl aniline—it took several days.[29] Although the basic procedure was always similar, success in any individual instance

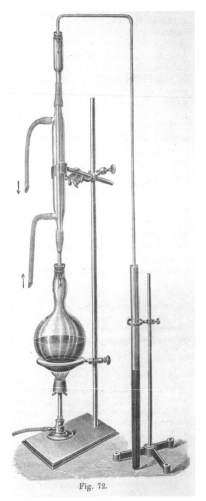

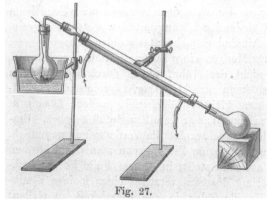

Fig. 72.

Fig. 27.

FIGURE 6.2

Gattermann illustrated (left, labeled "figure 72") how a reaction mixture could be continuously maintained at its boiling point by the vertical arrangement of a condenser above the round-bottomed flask in which the reaction occurred. This illustrates two key features of chemists' use of glass apparatus: innovative combination and variable purposing. Elsewhere, he showed (right, labeled "figure 27") the condenser being used in distillation, as originally intended. The condenser had been introduced to condense distilling vapors by cooling them as they passed from the still head to the receiver. The reader may notice that this image shows cooling water flowing in the opposite direction from that now recommended. Reoriented to a vertical position and combined with other items of apparatus, the same device fulfilled the same function to distinct ends.

Source: Ludwig Gattermann, *Die Praxis des organischen Chemikers,* 4th edition (Leipzig: Veit, 1900), 239, figure 72 (above); and 32, figure 27 (below). Courtesy of the University of Wisconsin-Madison Libraries.

required constant attention to—and often variation of—practical details of time and temperature, purification of starting materials, and the proportions in which they were mixed. In the reaction between ethyl bromide and aniline, for example, Hofmann found that a large excess of ethyl bromide was required to produce the desired product.

These experiments invariably produced mixtures, from which pure products were separated in multiple steps—each requiring considerable manipulative skill and relying on specialized techniques and glassware. First, the crude crystalline product mixture—hydrobromide salts of various aniline bases—was washed with potassium hydroxide solution to release the free bases as oils immiscible with the aqueous solution. In preparing ethyl aniline, his first and key synthetical experiment, Hofmann explained how a "brown basic oil [rose] at once to the top of the liquid" (because organic compounds tend to be less dense than water). Once separated, this oil could be purified by "rectification" (distillation), but its initial separation from the aqueous layer was accomplished using "a pipette or a tap-funnel."[30]

Although, as before, Hofmann did not illustrate either device, his decision to specify the equipment used indicates that he viewed this as significant for other chemists' ability to repeat his work and successfully perform similar experiments. Investigating contemporary methods of separating immiscible liquids makes it possible to infer the likely apparatus Hofmann used for this purpose ca. 1850. As we shall see, understanding how nineteenth-century chemists met this functional requirement confirms their mounting dependence on lampworked glassware.

Separating immiscible liquids, such as ether and water, was an established part of pharmaceutical practice, required in the purification of essential oils isolated by steam distillation. From the early nineteenth century, manufacturing pharmacists used a device called a "separatory funnel" for this purpose (figure 6.3). This glass vessel, fitted with a glass stopper and handle for convenient manipulation, held liquid volumes of a liter or more. Such an item was typically made by furnace glassblowing and was therefore relatively costly. Yet some chemists no doubt adopted the separatory funnel despite its size, cost, and limited availability, thereby perpetuating established material links between chemistry and pharmacy.[31]

Instrument makers, including the firm of Wenzel Batka (briefly encountered in chapter 3), used glassblowing skill to modify the pharmacists' separatory funnel by incorporating a glass tap—a change that improved

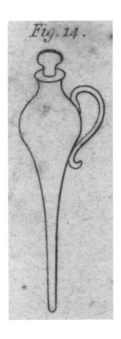

FIGURE 6.3

This early separatory funnel combined a bulbous portion capable of holding a considerable volume of liquid with a finely tapered, open-ended tip to aid the separation of two immiscible liquid layers. In pharmaceutical practice, one of the liquids was aqueous, while the other was usually an ethereal solution of essential oil that formed the upper layer. Separating the two layers required skillful control of the liquid flow exiting the funnel, accomplished either using a finger to cover the opening or by means of the stopper.

Source: Anthony Todd Thomson, *The London Dispensatory* (London: Longman, Hurst, Rees, Orme, and Brown,1811), Plate I, fig. 14. Courtesy of the University of Wisconsin-Madison Libraries.

functionality and ease of use but more than doubled the cost (figure 6.4, items V59 and V60). However, such expensive apparatus was beyond the means of many chemists. Nor did the separatory funnel with tap—of whatever size—always meet the needs of academic chemists.[32]

In 1827, the legendary experimental chemist and pioneer of lamp-worked glassware Michael Faraday had addressed the problem of separating immiscible liquids. Faraday recognized the availability of "glass vessels, furnished with a stop-cock beneath"—most likely the kind of device that Batka and others were selling. But Faraday's preferred alternative was "a

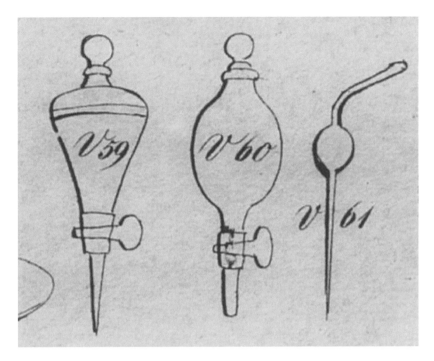

FIGURE 6.4

Batka's (1832) catalog offered at least three pieces of glassware for separating immiscible liquids: V59 Filtrirtrichter (filter funnel); V60 Scheidetrichter (separatory funnel); and V61 Heberpipette (lifting pipette), where "V" indicates verre (glass). The filter funnel was available with or without tap—a difference in construction that was reflected in a significant price differential (2fl.30 kr. with tap; 1fl. without). The pipette was far more affordable, priced at just 12kr. (60 Kreutzers =1 florin.)

Source: Wenzel Batka, *Verzeichniss der neuesten chemischen und pharmaceutischen Gerätschaften mit Abbildungen: Herausgegeben bei Gelegenheit der Versammlung deutscher Naturforscher in Wien* (Leipzig: Barth, 1832). Courtesy of the National Library of the Czech Republic.

glass tube with a bulb an inch in diameter blown in it, and drawn out below to a moderately fine aperture" (figure 6.5). As with Liebig's Kaliapparat, lampworking enabled chemists to produce home-made apparatus whose functionality matched or exceeded that of the more expensive devices sold by instrument makers.

Pipettes like the one Faraday recommended were smaller than separatory funnels, making it easier to work with limited amounts of material—as when purifying the products of synthesis. Increasingly in demand following the

FIGURE 6.5

Michael Faraday recommended this simple device—later called a "pipette"—for separating immiscible liquid layers. As he explained: "The aperture being immersed in either the upper or under liquid, the mouth is to be applied above, and the air withdrawn, when the liquid consequently will enter. The finger being then placed on the upper end of the tube, so as to close it, the instrument may be removed, and the fluid within transferred to any convenient vessel."

Source: Michael Faraday, *Chemical Manipulation: Being Instructions to Students in Chemistry, on the Methods of Performing Experiments of Demonstration or of Research, with Accuracy and Success* (London: Phillips, 1827), 248. Courtesy of the Cole Collection, University of Wisconsin-Madison Libraries.

turn to synthesis, by mid-century such pipettes were available in a variety of forms, including a bent version that made it easier to see the liquid level, thereby helping chemists avoid sucking potentially toxic or corrosive liquids into their mouths (figure 6.4, item V61; figure 6.7).[33] Here, as elsewhere, lampworking innovations produced worthwhile improvements in functionality that were subsequently translated into standard commodity items.

Glass pipettes were easier to buy in Hofmann's native Germany than in London ca. 1850.[34] Because there were skilled glassblowers among his assistants at the RCC, however, Hofmann nevertheless had ready access to home-made pipettes similar to those for sale in Germany. Indeed, the RCC's glassblowing pattern book (*Musterbuch*) included a life-sized drawing of just such an item (figure 6.6)—along with several other similar variants. The *Musterbuch* makes it clear that Hofmann's London research school relied

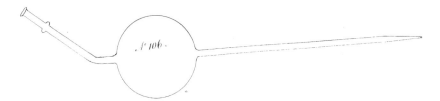

FIGURE 6.6

This diagram appeared in the *Musterbuch* ("pattern book") kept in Hofmann's laboratory at the RCC. Its pages were filled with lampworked glassware, mainly illustrated at life-size or, occasionally, to a specified scale. Item No. 106, drawn life-sized, was about 11 inches long, its bulb just under 2½ inches in diameter. Thus, every item could be made to consistent form and dimensions, helping standardize the apparatus used in Hofmann's laboratory.

Source: Courtesy of the Archives Imperial College London, item 1005, RCC *Musterbuch* (undated).

on apparatus produced by in-house lampworking, while its specification of size and shape confirms the importance of standard apparatus—regardless of who made it.[35]

Having established the most likely form of Hofmann's pipette, let us now consider what he meant by "tap-funnel." Separatory funnels like those Batka offered for sale were not among the items in Hofmann's *Musterbuch*—consistent with their likely furnace origins. Nor did London suppliers, such as John Joseph Griffin, then advertise separatory funnels of this kind—perhaps reflecting their limited English market. Yet Hofmann continued to import glassware from Germany until at least January 1850—making it possible that he sourced his tap-funnel this way and that it resembled those illustrated by Batka (figure 6.4, items V59 and V60).[36]

However, the *Musterbuch* presents a persuasive alternative. It illustrated a glass vessel fitted with a tap that might be called a "tap-funnel," also conforming to Faraday's description of "glass vessels, furnished with a stopcock beneath" (figure 6.8). Such an item could be made by lampworking, although incorporating the tap would require considerable skill.[37] Open at the top, this device did not permit thorough mixing of the liquid layers prior to their separation. Thus, it was similar to the pipette in functionality and, like the pipette, was suitable for small-scale operations. At the same time, it promoted the superior separation achievable with the separatory funnel. Whereas effective separation using a pipette required outstanding

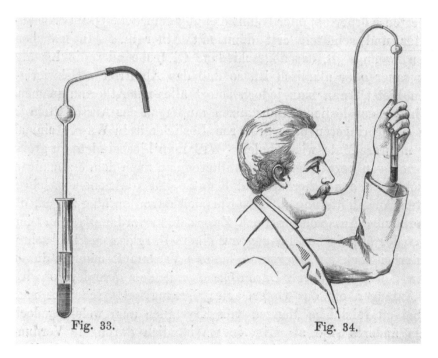

FIGURE 6.7

When using a pipette to separate immiscible liquids, chemists held the apparatus at eye level to optimize separation of the two layers. Note that, although the chemist pictured here is wearing a lab coat, he is not wearing any eye protection.

Source: Ludwig Gattermann, *Die Praxis des organischen Chemikers,* 4th edition (Leipzig: Veit, 1900), 39, Figs. 33 and 34. Courtesy of the University of Wisconsin-Madison Libraries.

hand-eye-mouth coordination (figure 6.7), these manipulative demands were reduced by an apparatus that incorporated a tap and could be supported in a stand. This, it seems reasonable to suppose, was Hofmann's tap-funnel.

Despite their availability from such makers as Batka and their later status as ubiquitous components of chemists' toolkits, separating funnels in something resembling their modern form were not in widespread use by chemists ca. 1850. In fact, chemists had many methods of separating immiscible liquids in addition to those discussed here.[38] It therefore seems likely that Hofmann, having found "a pipette or a tap-funnel" particularly effective in purifying his synthetic products, sought to alert others to their use.[39] One plausible reason for his choice concerns scale. At this point, Hofmann did not routinely

FIGURE 6.8

Hofmann's *Musterbuch* included this funnel fitted with a tap and long, pipette-like tube beneath. Drawn to actual size, the funnel was approximately 2 inches across. Griffin, *Chemical Handicraft*, 173, listed for sale an item of precisely this form: "Separatory funnel, with stopcock, form of Fig. 1641 [on 174] small size, funnel 2 inches wide, tube 12 inches long, 2s." Griffin's description supports the inference that Hofmann used this piece of apparatus to achieve effective separation when handling quantities that were too small for separatory funnels like those Batka sold.

Source: Courtesy of the Archives Imperial College London, item 1005, RCC *Musterbuch* (undated).

report the quantitative yield of his reactions but, as we saw in chapter 5, he frequently worked with small quantities of product. Here, as elsewhere, lamp-worked glassware was essential to his ability to obtain pure substances.

The problem of chemical identity came to the fore during purification. Hofmann's products were generally very similar to one another and the starting material, aniline. As a result, it was hard to identify each new substance, differentiating it from known compounds. This was where Hofmann's experience and extraordinary attention to detail paid dividends. Before establishing composition and formula by analysis, Hofmann applied the qualitative test for aniline to each base. Ethyl aniline, for example, "was distinguished [from aniline] by a slight difference in the odour, perhaps imperceptible to an inexperienced nose, by a higher boiling-point and a lower specific gravity." Nor did it "exhibit the violet coloration with chloride of lime which characterizes aniline."[40] Thus, distinguishing features only accessible to someone with vast practical experience of this area of chemistry were essential to Hofmann's success.

In addition to these qualitative tests, Hofmann provided quantitative characteristics for each compound he identified. Ethyl aniline's boiling point (204 °C, measured as the constant temperature at which the pure compound boiled from platinum) was far higher than pure aniline's (182 °C). As well as analyzing ethyl aniline, Hofmann also prepared and analyzed its hydrobromide, oxalate, and platinum salts—a multiplication of analyses that served, in true Giessen style, to corroborate the formula he had assigned. Once its identity, composition, and formula were established with certainty, ethyl aniline became a known compound that Hofmann could use as the basis for further synthetical experiments. He therefore reacted it with established reagents including cyanogen, using it as the starting point for further reactions with alkyl halides.

Hofmann's meticulous, systematic approach entailed a vast amount of work. Despite overall similarities in the reactions and products he studied, each case presented its own challenges. Making each step as secure as possible, moreover, was essential in enabling Hofmann to summarize what he had learned in a "synoptical view" (figure 6.9). This table contained the bases he had produced by substituting the ethyl group for aniline's basic hydrogen. It also included bases similarly produced from chloraniline, bromaniline, nitraniline, and ethyl amine.[41]

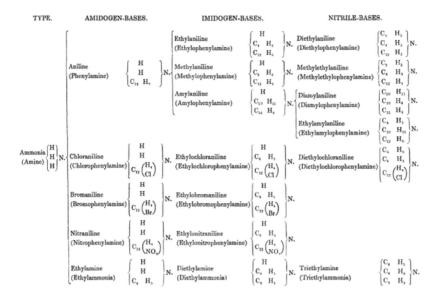

FIGURE 6.9

Hofmann's 1850 paper displayed the constitutional relationship between the artificial bases that he had prepared and their underlying type molecule, ammonia.

Source: August Wilhelm Hofmann, "Researches Regarding the Molecular Constitution of the Volatile Organic Bases," *Philosophical Transactions of the Royal Society* 140 (1850): 131. Courtesy of the Royal Society of London under a Creative Commons License: Attribution 4.0 International (CC BY 4.0).

In total, Hofmann presented more than a dozen fully characterized new compounds—each classified as amidogen, imidogen, or nitrile base. Relations among them were established by their method of preparation—exactly as synthetical experiments were intended to operate. By demonstrating the successive substitution of each of the basic hydrogen atoms in aniline— and all the hydrogen atoms in ammonia—by ethyl, Hofmann showed that organic bases were formed when hydrocarbon groups replaced the hydrogen in "the primary type ammonia."[42] This evidence, Hofmann believed, "established in a sufficiently satisfactory manner, the point of theory which is here in question." In other words, simple organic bases were best understood as "derived from ammonia," the underlying inorganic type molecule, by successive "replacement of its three hydrogen-equivalents."[43] Other chemists agreed.

THE AMMONIA TYPE'S SIGNIFICANCE FOR HOFMANN

The ammonia type has previously been explained as merely the resolution of Hofmann's alternating commitment to dualistic and type theories. Thus, it reflected—and in turn cemented—type theory's increasing dominance, establishing the classification of amines that chemists use today. Seen from this perspective, the ammonia type appeared to be a comprehensive, theory-led endpoint to the problem of how nitrogen existed in organic bases. In fact, as this book makes clear, the ammonia type provided only a provisional, limited answer to that question. As Hofmann explained, "I am far from supposing that the above formulae include the constitution of all the complex bases with which we are acquainted. It is chiefly the volatile alkaloids whose constitution may be represented by this view, but even here exceptions may occur."[44]

By exposing the crucial link between Hofmann's ammonia type and the experimental methods that produced it, the present account offers insight into the nature of his theorizing. Wurtz's theoretical speculations were reminiscent of Gerhardt's tendency to theorize in advance of evidence.[45] In contrast, Hofmann's reshaping of type theory to encompass organic nitrogen was tightly connected to its empirical basis, inseparable from synthetical experiments; glassware; and his new reagents, the alkyl halides.

This embedding in a systematic research program that supported laboratory reasoning made the ammonia type productive, despite its limited scope. Crucially, it indicated fresh directions for future research. One was the investigation of organic compounds containing arsenic and phosphorus, elements whose hydrides were already known to behave analogously to ammonia. A second was that Hofmann's taxonomic overview allowed him to suggest how existing series of organic bases might be extended.[46]

Hofmann's key starting point for this reasoning was the existence of numerous isomeric pairs among the bases he had studied. Isomers—because they are distinct compounds with the same molecular composition—were especially useful to Hofmann's constitutional investigations. In his "most striking" case, methyl aniline and toluidine (both $C_{14}H_9N$) displayed quite different properties. Toluidine, Hofmann reported, "is a beautiful crystalline compound, boiling at 198° [C], yielding difficultly soluble, perfectly stable salts with almost all acids. . . . Methylaniline, on the other hand, is an oily liquid, boiling at 192° [C], whose salts are distinguished by their solubility and by the facility with which they are decomposed, aniline being reproduced."[47]

Such isomeric pairs indicated that "basic compounds [such as methyl aniline and methyl ethyl aniline] derived from aniline [an amidogen base] must be either imidogen- or nitrile- bases." They therefore implied the existence of entirely new series of bases derived from other amidogen bases including toluidine and cumidine. Thus, the ammonia type enabled Hofmann to predict the existence of new compounds whose production was already being attempted in his laboratory. And although Hofmann had thus far focused on producing aniline derivatives in the ethyl series, he now anticipated that "phenyl-, toluyl-, xylyl- and cumyl-series" could also be made, once suitable new reagents had been established.[48]

Hofmann's ammonia type therefore drove an expanding research program aimed at producing novel artificial organic bases and, by means of laboratory reasoning from this experimental basis, shedding further light on the constitution of all alkaloids—whether natural or artificial. For example, he reasoned that di- and tri-phenylamine—compounds whose preparation so decisively eluded him at the start of this work—would be made without difficulty, given a suitable phenyl halide. Hofmann had already produced "three alkaloids of exactly the same composition, namely, ethyltoluidine, methylethylaniline, and cumidine," and he confidently anticipated that the addition of a unit of methylene would lead to "six alkaloids, having all the same numerical formulae, but widely differing in their construction."[49]

This multiplication of isomers as the number of atoms of carbon increased gave Hofmann unprecedented insight into the inadequacy of relying on composition and molecular formula in understanding the alkaloids. The largest aniline derivative he had so far made was amylamylaniline, whose "numerical formula" ($C_{32}H_{27}N$) Hofmann now calculated "represent[ed] not less than twenty different alkaloids which the progress of science cannot fail to call into existence—a striking illustration of the simplicity in variety that characterizes the creations of organic chemistry."[50]

Confidence in synthetical experiments and laboratory reasoning as a way of learning more about the alkaloids' constitution, coupled with a belief in the underlying simplicity of nature, featured in Hofmann's explanation of how the ammonia type—and its associated methods—applied to his area of fundamental interest: the natural alkaloids. The ammonia type remained useful here precisely because it could "in a variety of ways assimilate several other groups of elements without forfeiting its original character." Indeed, this quality was essential in accommodating the "complexity" Hofmann now reasoned was likely to exist in natural alkaloids such as quinine and morphine.[51]

At the same time, the "powerful and defined action" of his new reagent, ethyl bromide, offered a method for "ascertaining the state of substitution in which the ammonia exists in these compounds, or in other words, whether the alkaloid in question is an amidogen-, imidogen-, or nitrile-base." For Hofmann, the value of the ammonia type was not that it endorsed any specific constitutional theory. Instead, its associated methods made it possible to infer how nitrogen existed in organic bases. It therefore strengthened Hofmann's conviction that laboratory reasoning from synthetical experiments would help establish the natural alkaloids' constitution—and so lead to their laboratory construction.[52]

In fact, Hofmann had already begun to follow this line of inquiry. Quite reasonably, he started with two of the simplest natural alkaloids, nicotine and coniine—compounds whose similarity to aniline had legitimated Hofmann's use of aniline as an experimentally more tractable model compound. Now, expertise in aniline chemistry led Hofmann to expect "a harvest of interesting results" concerning nicotine and coniine. Such optimism—perhaps coupled with his august audience—also resulted in some uncharacteristic speculation concerning medicinal alkaloids, including quinine and morphine. Although their "constitution is at present still very enigmatical," Hofmann hoped they would "perhaps be found of a surprising simplicity when subjected to a closer examination."[53]

This was an important rhetorical move—for two reasons. First, Hofmann reminded his audience that naturally occurring medicinal alkaloids, not artificial aniline derivatives, were his primary objects of interest. Second, he refocused attention on organic chemistry's practical and commercial importance, and its power to solve complex problems of material and medical as well as theoretical significance. Already in 1850, Hofmann was a powerful advocate for the discipline of organic chemistry—a role he would develop in future decades.

In fact, only one major research area was absent from Hofmann's prospective. Though his work on aniline has often been depicted as directed toward the chemistry of aniline dyes, Hofmann in 1850 did not anticipate this outcome for his work.[54] This realization prompts reinterpretation of Perkin's 1856 preparation of the first synthetic dye, mauve. Previously viewed as both a brilliant insight and the serendipitous outcome of a ludicrously misguided attempt to make quinine, Perkin's attempts to oxidize first allyl toluidine and then aniline are surely better construed as logically designed synthetical experiments reflecting then current knowledge of constitution and reactivity. Perkin chose to exploit the unexpected outcome of these experiments for immediate commercial advantage, while neglecting their scientific implications, and this choice—rather than envy of his student's success—may well have attracted Hofmann's disapproval.[55] At this stage, Hofmann was a profoundly academic chemist.

HOW THE AMMONIA TYPE CHANGED CHEMISTRY

Hofmann's ammonia type proved an important step toward greater disciplinary consensus in how chemists conceptualized organic compounds. While type and dualistic theories competed to explain and predict how organic compounds behaved, mid-century organic chemistry remained riven by chemists' rapidly changing theoretical allegiances and the acrimonious disputes these prompted. For his obituarist, Ferdinand Tiemann, Hofmann's investigations were decisive in overcoming this situation, paving the way for a reconciliation between the two camps.[56] August Kekulé went further, suggesting that once the ammonia and water types had been identified, the development of Gerhardt's types might be considered "obvious."[57]

Yet the ammonia type was not the result of changing theoretical allegiance. Indeed, Hofmann's reinterpretation of compounds whose position in his dualistic taxonomy had seemed secure confirmed the primacy of experiment in his reasoning. Nor did Hofmann, having formulated the ammonia type, entirely abandon dualistic theory. Instead, by recognizing the continuing but bounded relevance of both theories, he helped establish the possibility of their coexistence, providing a fruitful basis for later developments.

Revising our understanding of Hofmann's mid-century contributions to theory and experiment suggests that we need a new explanation of type theory's increasing dominance. In its mature form, type theory is usually attributed to Gerhardt, who in 1853 proposed water, ammonia, methane, and hydrogen as the four defining molecular types.[58] Until now, Alexander

Williamson's 1850 studies of ether formation—the basis of the celebrated "Williamson ether synthesis"—have been prioritized as the decisive factor supporting Gerhardt's proposition. Yet the greater significance accorded to Williamson's work is called into question by this account of Hofmann's introduction of the ammonia type almost a year before.[59]

Trained in Giessen and Paris, Williamson had recently been appointed professor of chemistry at University College London when he showed that ethers formed when two alcohol molecules combined with the loss of water. It is relevant that ethers were much simpler to work with than organic bases. Precisely because he selected an unusually straightforward class of organic compounds, Williamson was able to produce decisive results much faster than Hofmann could—and so was able to develop a comprehensive constitutional rationalization of ethers and ether formation.

These differences suggest why Williamson's experimental work permitted, and even encouraged, him to engage in speculations concerning a realistic, mechanistic, and proto-visual interpretation of atomic motion and chemical reactions, while Hofmann remained committed to experiment. The prefiguring in Williamson's work of what later became fundamental chemical notions has understandably led historians to track the development of type theory through that work. Yet Hofmann's struggles with the complexity of organic bases and natural alkaloids were, I contend, far more typical and are therefore more illuminating concerning the state of organic chemistry ca. 1850.[60]

This brings us to the question of how other chemists heard about and responded to Hofmann's ammonia type. Many would have learned of it from Hofmann's own publications, which soon appeared in German and French as well as in English.[61] Liebig immediately promoted the ammonia type through his editorship of Poggendorff's *Handbook*.[62] By 1853, its inclusion—albeit without attribution—in Gehardt's *Treatise* would certainly have attracted the attention, if not the approval, of others who were following mid-century chemistry's theoretical debates.[63]

The ammonia type's rapid appearance in the leading textbooks of the period is even more telling, indicating both its ready acceptance by some of those best placed to judge its value, and its speedy transition from research result to stable component of the curriculum. For Kekulé, Hofmann's ammonia type constituted "a significant extension of the original type theory."[64] Even Hermann Kolbe—whose reactionary theoretical views are well established—promoted the classificatory value of Hofmann's innovation.[65] In 1856, William Gregory considered type theory "no longer purely controversial," describing the ammonia type as one of the "great" types, on a par with water.[66] And by 1874, when Hofmann's former student Henry Edward Armstrong published an elementary textbook of organic chemistry for schools, his presentation of the amines began with Hofmann's classification, rendered in the now-familiar terminology of primary, secondary, and tertiary amines.[67] Although by this time, the ammonia type had been subsumed within structural theory—type theory's successor—Hofmann's underlying taxonomy endured.

From mid-century on, an increasing array of textbooks and manuals of practice introduced students to the reagents, reactions, and glassware that made the ammonia type powerful and gave durability to Hofmann's approach. Both Kolbe and Kekulé explained how Hofmann's exhaustive methylation established the number of hydrogens attached to nitrogen in an organic base.[68] From its earliest editions, George Fownes's *Manual of Elementary Chemistry: Practical and Theoretical*—a favorite in Hofmann's London laboratory—explained glassware-based manipulations, including the use of Liebig's condenser for distillation. By 1863, Fownes's *Manual* also described how reaction with alkyl halides could be used to produce a multitude of artificial organic bases.[69] Thus, glassware and reagents assumed a central position in chemical training.

By about 1890, as Hofmann's career drew to a close, Ludwig Gattermann's canonical text *The Practice of the Organic Chemist* integrated glassware and reagents with stable theory, including Hofmann's established classification of amines as primary, secondary, and tertiary. As we have seen (figures 6.2 and 6.7), Gattermann provided detailed instructions for using a wide variety of glassware. In addition to techniques for distillation and reflux and the separation of immiscible liquids, he explained how to prepare, manipulate, and—most hazardous of all—open the sealed glass tubes that were a staple of Hofmann's chemistry.[70]

Glassware was implicit throughout Gattermann's text, including his account of Hofmann's approach to amine chemistry. In addition to presenting "Hofmann's reaction"—later the Hofmann degradation, a method for converting a primary amide into a primary amine "of the next highest series," that is, containing one fewer carbon atom—Gattermann described the use of alkyl halides to produce amines of all kinds. The simpler volatile amines were to be distinguished from ammonia by heating them gently with strong alkali in a small test tube and igniting the liberated gas—a simple test in glass to help young chemists avoid the error that had threatened to undermine Wurtz's original preparation of primary amines.[71]

Thus, Hofmann's taxonomy, his reactions and reagents, and his reliance on chemical glassware outlasted the ammonia type—a durability of experiment and reasoning that helps explain Hofmann's status as one of the nineteenth century's greatest chemists. It also suggests why Hofmann's contributions to theory have come to be perceived as limited. The ammonia type's success in promoting mid-century theoretical advance was such that it was soon subsumed within developing structural theory. This apparent transience is misleading. Even as chemists' interpretational framework developed, the practical methods and laboratory reasoning that produced and sustained the ammonia type retained their productive power.

ADVANCED ART

Wurtz's terminology has led some scholars to see his contribution as the crucial impetus driving Hofmann's introduction of the ammonia type.[72] It has even been suggested that credit for this new theory should be shared by Hofmann and Wurtz—although, as far as I can discover, Wurtz never made such a claim.[73] By explaining Hofmann's route to the ammonia type, this chapter has clarified the relationship between Wurtz's work and Hofmann's, calling for a revised estimate of Hofmann as theorist.

Where previously Hofmann's theoretical contributions have been characterized as limited, this study suggests otherwise.[74] It explains why other chemists accepted that his "observations"—rather than Wurtz's work—established the nature of organic bases "in an unambiguous manner," recognizing Hofmann's ammonia type as "significant" in "shaping the development of chemical theory."[75] Although the ammonia type was far from providing a complete explanation of the constitution of natural alkaloids, or the chemistry of organic nitrogen, Hofmann's inability fully to decipher this chemistry reflects the problem's extreme difficulty rather than any indifference to theory.

Recognizing this difficulty is vital if we are to understand Hofmann's world—and the world of organic chemistry ca. 1850. Hofmann countered difficulty with work on an unprecedented scale. Yet, despite his comprehensive grasp of existing knowledge and meticulous analysis of how its remaining gaps and puzzles might be examined, Hofmann's synthetical experiments frequently did not produce their expected outcomes. It was hard to make the wished-for reaction happen, hard to analyze what it produced, and even harder to theorize what took place in the transformation.

For Hofmann, theory was not a partisan issue. As Tiemann observed, Hofmann "was never prejudiced by the narrow horizon of a currently dominant theory." Nor was it a matter of speculation. Instead, he saw theory as a powerful tool that must "emerge from the facts with compelling necessity." This was why Hofmann advised young colleagues that, when theoretical stability eluded them, they "must . . . proceed according to the rules of the advanced art of experimentation."[76]

Hofmann's ammonia type perfectly exemplifies this approach: grounded in experiment and stabilized by reason. This reasoning required a systematic framework to give experimental results meaning—a need that was particularly acute in a period when chemistry was so minimally visual. As Hofmann's synoptic tables and formulae make clear, his approach relied on taxonomic order rather than visual thinking. To the extent that Hofmann's thinking was visual, this was reflected in his tabulation of family resemblance based on reactivity rather than constitution or structure.

This observation highlights the intrinsic relationship between the ammonia type and Hofmann's new reagents, the alkyl halides. Just as the ammonia type conceptualized organic bases as formed from ammonia by substitution, so the alkyl halides functioned as the "most valuable agents of substitution in the hands of chemists."[77] Because they replaced hydrogen associated with nitrogen, the alkyl halides enabled chemists to count the number of such hydrogens in a molecule. In the context of the ammonia type, they became powerful tools of constitutional investigation, providing a partial answer to Hofmann's original question about the constitution of natural alkaloids.

Such was the power of Hofmann's new reagents that his successors named the relevant reactions after him. Although chemists could also use the alkyl halides to prepare substituted anilines—a capability that acquired great commercial importance following the introduction of aniline dyes—it was this conceptual advance that they most admired. Hofmann's exhaustive methylation and the Hofmann elimination, like the Hofmann degradation, became enduring staples of analytical organic chemistry—displaced only by the mid-twentieth-century advent of instrumental analysis. Thus, Hofmann's methods long outlasted type theory and chemists' use of the ammonia type—illustrating the durable utility of laboratory reasoning.

Hofmann's papers provided a meticulous record of experiment and interpretation, highlighting his assistants' contributions as well as the crucial roles played by glassware and reagents. These characteristics distinguished his publications from those of contemporaries, including Wurtz, and—thanks to Hofmann's increasing eminence and commitment to teaching—they helped establish new norms. They also point to what was distinct and novel in Hofmann's theorizing. Whereas type theory functioned as a speculative umbrella hypothesis that was loosely tied to experiment, Hofmann's ammonia type was intrinsically linked to how it was made, embedded in the methods that produced it.

Integration with experiment was fundamental to the ammonia type's success as a theory. We have seen how laboratory reasoning enabled Hofmann to produce a revised taxonomy of organic bases. But this reasoning was tied to its particular empirical basis and rooted in his personal practice. Hofmann's development of organic reagents and his use of novel glassware were essential to the ammonia type, because they made it possible to perform the reactions and purify the compounds this new taxonomy required.

To become a viable theory, the ammonia type had to become the property of the community of chemists. What made this possible, however, was not abstraction from experiment. Instead, an embedded theory of this kind required a shared foundation of technique and instrumentation to move beyond the individual experimental world of its creator. The ammonia type's ability to transcend the disputatious nature of mid-century theorizing in organic chemistry was therefore inseparable from the reagents and glassware that produced it. This was why Hofmann's publications specified the conditions under which reagents worked best and the most effective glassware for purifying products.

In Hofmann's hands, synthesis supported a kind of theorizing that was unlike the speculations engaged in by many of his contemporaries. This account of how Hofmann produced the ammonia type therefore helps explain why and how chemical theory became so different from its counterpart in physics. Incapable of the mathematical abstraction that characterizes modern theories of matter and motion, chemical theory remained embedded in experiment and encumbered by the behavior of the material world it sought to rationalize and control.

"Synthesis" was certainly "the key to Hofmann's thinking."[78] But this was not because of the compounds he made. Indeed, Hofmann—despite a career extending over a further four decades—would never produce any natural alkaloid in the laboratory. By explaining how he arrived at the ammonia type and how his innovation changed chemical understanding of the alkaloids, this chapter and the one before it have begun to explain what organic synthesis meant—and, crucially, did not mean—to Hofmann and his generation.

7 CHEMICAL IDENTITY CRISIS

We saw in chapter 6 that August Wilhelm Hofmann struggled to differentiate and identify the products of synthesis. In fact, what I have called the
"chemical identity crisis" was a major problem associated with the synthetic method.[1] Following the turn to synthesis, it became a critical barrier
to disciplinary progress. This chapter explains how chemists—working with
glassblowers—tackled this problem, ultimately making melting and boiling
points measured in glass serve as decisive indicators of identity and purity.

This accomplishment implied more fundamental changes than might
initially be apparent. It required improved instruments and techniques.
But, as this chapter shows, it also entailed a new definition of individual,
pure substance. In their struggle to stabilize melting and boiling points
as "Constants of Nature," chemists established new connections between
purity and identity, experiment and theory that are continuous themes
throughout the remainder of this book.[2]

This study has situated the origins of synthetic organic chemistry in two
laboratories: Justus Liebig's in Giessen and Hofmann's in London. That
pioneering work's broader disciplinary context now comes to the fore.
Although Liebig and Hofmann were prime movers in the turn to synthesis,
many people contributed to this vast project. Such expansion introduced
its own problems. Liebig's former students frequently did not adhere to his
analytical methods after leaving Giessen. Similar variations in experiment
and interpretation threatened organic synthesis, highlighting standardization as a necessary component of chemists' solution to the chemical identity crisis.

By 1850, glass was *the* material medium for chemical innovation. This chapter shows how collaborations between chemists and glassblowers transformed glass apparatus and instruments into an equally significant resource for mid-century standardization—in chemistry and beyond. It explains why the quest for shared standards furthered the emergence of a new professional: the scientific glassblower. Scientific glassblowers—exemplified here by Heinrich Geissler and his successor Franz Müller—enabled chemists to realize their visions for technically demanding new instruments. At the same time, their ability to work to exacting standards of form and dimension was crucial to disciplinary coherence.

In focusing on the disciplinary level, this chapter highlights the importance of collaborations—in and beyond chemistry—for overcoming the chemical identity crisis. Within the discipline, collaborations between organic chemists and those primarily interested in physical phenomena were decisive. One such collaboration—between Hermann Kopp and Hofmann—features here, revealing Liebig's ongoing role as disciplinary linchpin. It also shows how—despite their common interest in making sense of the physical properties of organic compounds—Kopp and Hofmann approached the relationship between experiment and theory in very different ways.

Around the same time, another former Liebig student, Victor Regnault, began an investigation that would prove crucial to organic chemistry's nascent reliance on physical methods. An eclectic training had left Regnault with a serious interest in organic chemistry and a passion for precision measurement. Commissioned by the French government to study steam engines, he uncovered fundamental problems in thermometry. Regnault's solution, the first reliable air thermometer, was constructed using the considerable resources of Parisian precision science. But as this chapter explains, glassblowers—Geissler prominent among them—responded to Regnault's innovation by constructing standardized mercury thermometers of unprecedented precision and reliability that could be used by chemists working anywhere.

This chapter's shift to consider chemistry as a discipline marks a break, beginning a final section of three chapters that together fulfil this book's claim to offer a disciplinary history. Chapter 8 explains how organic synthesis transformed the laboratory from being merely the place where chemists worked to being a vital material resource. Because it generated industrial wealth, synthesis attracted the support needed to develop laboratory infrastructure. Thus, it introduced a powerful synergy between resources and productivity, reflected in organic chemistry's academic and material dominance ca. 1900.

Chapter 9 concludes this section by using chemists' attempts to synthesize the hemlock alkaloid coniine to illustrate how chemists' synthetic capabilities developed post-1850. Both early failures and Albert Ladenburg's successful 1886 synthesis highlight difficulties in identifying synthetic products. Convincing other chemists that he had made a substance identical to natural coniine was crucial to Ladenburg's claim to have synthesized a natural alkaloid. The present chapter therefore draws from that case study to show how Ladenburg made melting and boiling points into decisive evidence.

Melting and boiling points ultimately enabled chemists to know and, crucially, agree on what their reactions made. These measurements were therefore essential in the synthesis of natural products. But, as we shall see, they were equally vital in grounding the reasoning processes that linked chemists' laboratory work with emerging theories of structure and stereochemistry (atoms in space), creating a novel link between numerical measurements and unseen molecules.

Solving the chemical identity crisis changed how chemists understood pure substances and their molecular identity. At the beginning of this study, they knew a pure organic substance by its origin, smell, taste, and physiological activity. By the end, they knew it by formula, melting and boiling points, and a handful of other physical properties. Organic synthesis therefore redefined the object of its own inquiry. Explaining how chemists and glassblowers accomplished this remarkable feat continues the transformation of practice and material culture that had begun during chemistry's glassware revolution—with equally profound consequences for the discipline.

ORIGINS OF A CRISIS

The chemical identity crisis originated in a shift in how chemists classified substances derived from living sources. Before about 1830, chemists' taxonomic system was primarily based on natural historical classification of the source. Developments in quantitative organic analysis—notably in the work of Liebig—led chemists to seek to redefine organic compounds based on their elemental composition and molecular formula. Thus, analytical innovations went hand in hand with a new understanding of an individual, pure substance, an interconnectedness of experiment and theory that is fundamental to this chapter.[3]

As this book has shown, composition and formula were hard to establish and even harder to agree on. Analytical results depended critically on the sample's identity and purity as well as the techniques used. In addition, many disagreements occurred, because—despite believing they were analyzing the same substance—chemists were in fact working with different substances or with the same substance in different states of purity. By mid-century, chemists knew that distinct compounds might share the same composition and molecular formula, a phenomenon called "isomerism." As analytical methods improved, they identified new kinds of isomerism—including optical isomers that rotated light in opposite directions while otherwise seeming largely identical at the molecular level. Quantitative elemental analysis—no matter how perfected and standardized—could not distinguish isomers. Thus, composition and molecular formula—although they provided the starting point for a new chemical taxonomy—did not uniquely identify substances.

Chemists therefore continued to use established criteria of chemical identity, including origin, taste, and smell—despite a mounting body of evidence indicating their limitations. The case of camphor (chapter 2) illustrates this confusion. Chemists recognized two forms of camphor, both obtained from trees. Natural camphor was extracted from the camphor laurel, while artificial camphor formed when oil of turpentine from pine trees reacted with hydrogen chloride. Because both substances smelled similar and were used for similar purposes, they were believed to be essentially the same. Giessen analysis, however, showed that artificial camphor—unlike the natural compound—contained chlorine.[4] Rising analytical standards therefore challenged the definition of individual substance, revealing the inadequacy of relying on smell and origin.

By the 1840s, origin's limited value was becoming established. Hofmann, for example, showed in 1843 that aniline derived from coal tar was identical to Nikolai Zinin's laboratory product benzidam.[5] Some chemists nevertheless continued to use smell and other physiological properties—even though others, including Hofmann, were increasingly wary. When Adolphe Wurtz produced methyl- and ethylamine in 1849, for example, their pungent smell initially led him to misidentify both substances as ammonia.[6] Hofmann seems to have learned from Wurtz's difficulties. In 1850, he considered the "slight difference" in odor between another pair of similar, foul-smelling substances, aniline and ethyl aniline, "perhaps imperceptible to an inexperienced nose."[7]

Wurtz's error and Hofmann's concern about inexperienced chemists' ability to identify organic bases by smell highlight two aspects of the growing discipline's emerging crisis. First, as we have seen, the turn to synthesis produced increasing numbers of very similar compounds that even experienced chemists struggled to distinguish. Chemists therefore learned that physiological properties were hallmarks of similarity rather than characteristics that uniquely identified single compounds. Although Hofmann, for example, continued to smell and even taste the substances he made, he did so to allocate each substance quickly to its proper family, rather than as a method of identifying it.[8]

The second issue concerned the subjectivity of properties like smell and taste. People's ability to distinguish and identify smells varies widely and cannot be equalized even by intensive training. In addition to being shared by similar compounds, such physiological properties were therefore impossible to distribute from chemist to chemist or between one laboratory and another. As more and more chemists began studying organic compounds, this lack of universality emerged as a critical problem.

Beginning in the 1840s, chemists sought objective criteria of chemical identity that could be shared throughout the community. Solving the chemical identity crisis required new methods and instruments. The remainder of this chapter explains how chemists worked with glassblowers to implement these developments. Just as in the shift to composition and formula, however, we shall see that changes in how chemists identified organic compounds also transformed what they understood an individual, pure substance to be.

ENTER THE SCIENTIFIC GLASSBLOWER

Developments in glass and glassblowing since the 1830s had introduced an important new figure to this story: the professional scientific glassblower. Previous chapters have shown that working in glass gave chemists enormous experimental flexibility, allowing them to produce variant forms of devices such as Liebig's Kaliapparat. Some variants offered improved functionality. But this proliferation ran counter to an increasingly necessary disciplinary standardization. The possibilities and challenges of working in glass led chemists to collaborate with skilled glassblowers. Improving a device often increased its complexity, and hence the skill required to make it—making it preferable or even essential that chemists buy it, rather than trying to make it for themselves. At the same time, professional scientific glassblowers developed chemists' innovations into standard, commercial items.

The widespread availability of standard glassware components transformed chemistry during the second half of the nineteenth century. In novel combinations, its functionality applied in new experimental settings, such glass apparatus continued to support innovation. Mixing and separation, temperature and pressure, change and stability all submitted within glass to the chemist's will as well as their gaze. At the same time, commercially manufactured glassware embodied defined methods and procedures, effectively distributing disciplinary standards.

These issues are illustrated here using Heinrich Geissler's 1851 modified Kaliapparat. Liebig's original, introduced in 1830, measured carbon during organic analysis (see figure 3.5). Dissolving carbon dioxide increased the device's mass, making it possible to calculate the sample's carbon content. As Liebig certainly appreciated, however, dissolving a gas in a liquid was useful in many other experiments. In late 1831, during Friedrich Wöhler's first visit to Giessen, the two friends measured exhaled carbon dioxide using Liebig's Kaliapparat. Thus, Liebig himself began the process by which his device transcended its original context of invention. Twenty years later, Geissler, as we shall see, recognized both the Kaliapparat's wider applicability and its potential for improvement.[9]

Raised among the glass huts of the German state of Thuringia, Geissler is widely celebrated for his contributions to high-vacuum physics during the late nineteenth century.[10] Around 1850, however, Geissler was appointed technician at the University of Bonn, where he founded what rapidly became a highly successful firm for the supply of scientific, and especially chemical, glassware. In fact, Geissler was one of the first professional glassblowers to specialize in producing chemical glassware for retail sale. Well before he became a pioneer of glass vacuum technology, Geissler was breaking new ground in an entirely different field.[11]

Following a youth spent learning all aspects of glass working, Geissler had refined his skill in making instruments such as thermometers and barometers while making his way across Europe during the 1840s. By the time he settled in Bonn, Geissler was ideally placed to support—and to capitalize on—chemists' increasing use of lampworked glassware. This was the community he set out to serve. In early 1851, for example, Geissler developed a new and "unusually sensitive" balance whose beam was constructed from glass tubing. "[O]riginally intended for chemists," this balance was subsequently used by the Bonn mathematician Julius Plücker in his attempts to quantify magnetic phenomena and improve thermometers.[12]

Plücker's ability to produce meaningful results was utterly dependent on Geissler's glassblowing skill. When preparing carbon monoxide, for example, Plücker initially found it impossible to remove contaminating carbon dioxide. Even after passing through Liebig's Kaliapparat, carbon monoxide still contained a tiny amount of carbon dioxide that invalidated Plücker's experiment. Just as Mitscherlich and others claimed, Liebig's Kaliapparat did not always completely absorb carbon dioxide. Although Plücker's goal was to study magnetism, his project relied on practical chemistry—and hence on glass and glassblowing.[13]

Geissler's solution reflected scientific understanding and mobilized superior glassblowing expertise. Incorporating three bubblers in place of the original bulbs increased contact between gas and liquid. Geissler's Kaliapparat absorbed significantly more carbon dioxide than did Liebig's original (figure 7.1). It was also much more complex. No amateur could make this device. Thus, Plücker's collaboration with Geissler transformed Liebig's device into something only a skilled professional could make. Whereas Liebig's analytical project depended on chemists' ability to make his Kaliapparat for themselves, anyone wishing to use Geissler's Kaliapparat would have to buy one. Sold as a standard item, its enhanced performance and increased complexity commanded a higher price.[14] It nevertheless displaced Liebig's original in the analysis of some organic compounds, also marking a significant step in the Kaliapparat's transformation from analytical apparatus to generic gas-absorption device.[15]

The pursuit of improved precision and functionality led chemists, initially in Germany but ultimately throughout Europe and North America, to collaborate with skilled glassblowers. Such collaborations made chemists

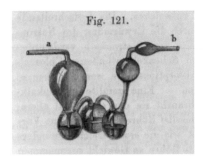

Fig. 121.

FIGURE 7.1

Geissler's 1851 Kaliapparat stood unsupported on three bubblers that formed a tripod base. Thus, it could be weighed on a late-nineteenth-century pan balance. The bubblers' primary function, however, was to improve gas absorption. Because each bubbler fed gas under the liquid surface, bubbles were forced to escape through the liquid, slowing their flow. As a result, Geissler's device absorbed significantly more carbon dioxide than did Liebig's original. Improving functionality, however, also increased the skill required to make the device. No amateur could make Geissler's Kaliapparat, and it was more expensive to buy. It was also harder to clean—with the result that chemists reserved its use for situations requiring superior gas absorption.

Source: Bayerische Staatsbibliothek München, Chem. 102 ic, Carl Remigius Fresenius, *Anleitung zur quantitativen chemischen Analyse,* 5th extended and improved edition (Braunschweig: Vieweg, 1863), 583, Fig. 121, urn:nbn:de:bvb:12-bsb10072595-9. Courtesy of the Bavarian State Library, Munich.

increasingly reliant on expert glassblowing. They also provided new opportunities for glassblowers. What had begun as a movement intended to liberate chemists from dependence on instrument makers thus became the driving force behind the emergence around 1850 of scientific glassblowing as a lucrative professional activity.

A burgeoning trade literature during the second half of the nineteenth century testifies both to the establishment of numerous new firms for the supply of flameworked chemical glassware and to the increasing range, diversity, and complexity of the apparatus being offered for sale. Collaborating with glassblowers gave chemists access to unprecedented functionality. By translating chemists' innovative glassware visions into standard apparatus, professional glassblowers provided a solution to crucial material aspects of disciplinary standardization. As we shall see, this collaboration was how professional scientific glassblowers helped solve the chemical identity crisis.

BOILING POINTS AND CHEMICAL THEORY

An important step in chemists' response to the chemical identity crisis centered on the work of Hermann Kopp. After training with Liebig, Kopp remained in Giessen as Privatdozent in chemistry and physics. His independent research initially had little to do with Liebig's interests. Abandoning organic chemistry, Kopp reverted to a project his former Heidelberg physics professor George Wilhelm Muncke had begun that involved measuring the liquid densities of inorganic compounds.[16] Kopp's goal—following Muncke—was "to prove . . . that all physical constants are to be regarded as functions of the chemical nature of molecules." In 1841, however, Kopp switched focus from inorganic to organic compounds, seeking "a consistent implementation" of atomic theory in the organic sphere. He also began measuring boiling points. Crucially, Kopp claimed that boiling points "might . . . provide important distinguishing characteristics" to help chemists classify organic compounds.[17]

Strong circumstantial evidence indicates that Liebig was behind this change of direction. He actively supported Kopp's project, providing much of Kopp's early boiling point data.[18] When publishing Kopp's papers, Liebig declared them of "great value for the experimenting chemist." Giessen analysis relied on matching analytical data to theoretically plausible formulae. Liebig similarly anticipated that a match between measured boiling points and theoretical values would indicate purity and confirm identity. Thus, Liebig believed that boiling points might serve as numerical characteristics, potentially offering a community-wide solution to the problem of identifying organic compounds.[19]

At this time, chemists sometimes measured boiling points but did not consider them as characteristics. Large differences in reported values for different samples of the same substance did not cause them to doubt that these were, in fact, identical substances.[20] For Kopp, this state of affairs was highly problematic. His "one consistent purpose"—according to his student Thomas Edward Thorpe—was "to establish a connection between the physical and chemical nature of substances."[21] Linking the observable physical properties of substances to their underlying molecules was Kopp's approach to promoting atomism. It also meant that the same substance should boil at the same temperature—provided, of course, it was measured the same way, under the same conditions.

Kopp believed "physical properties [such as boiling point] would gain in significance once it was proved that they followed fixed standards." Anticipating the boiling points of series of similar compounds would show regularities, he also hoped that they might help differentiate organic compounds.[22] The taxonomy of organic compounds was an ideal test bed for this theory.[23] Many chemists, including Liebig and Hofmann, suspected constitution (molecular arrangement) affected boiling point. At this stage, however, Kopp believed it "indubitable" that composition alone determined boiling point.[24] Kopp's conviction ultimately fell victim to a growing body of contradictory evidence. But as we shall see, it had dramatic consequences for how chemists measured boiling points and what they thought those measurements signified.

At first, Kopp's project appeared to make rapid progress. Gathering data from the chemical literature, Kopp calculated an average difference of 18 °C between the boiling points of related methyl and ethyl compounds (today understood as differing by one carbon unit). Kopp anticipated that this figure should apply universally and could therefore be used to predict values for as-yet-unstudied organic compounds. Limited exposure to the complexity of organic compounds was perhaps responsible for his optimism. Moreover, this average concealed wide variations in the underlying data: Kopp had averaged values ranging from 11 to almost 25 °C.[25]

Its insecure empirical basis was not the only problem with Kopp's proposal. There was also a noteworthy circularity at its heart. Kopp intended to calculate boiling points for "analogous compounds." These theoretical values offered what—in language reminiscent of Liebig's—Kopp termed a "control" against which to judge the validity of measured values. At the same time, Kopp intended to use comparisons between theory and experiment to refine his theory—notwithstanding his absolute commitment to composition as the sole factor determining boiling point. Thus, Kopp credited a provisional theory with reliable agency—very different from Liebig and Hofmann's more cautious laboratory reasoning. Indeed, major contemporary criticism of Kopp's work concerned his denial of any role for constitution in fixing boiling point.[26]

Kopp's theoretical preconceptions led him to anticipate a very high degree of regularity in boiling points. But as he gathered more data, and compared theory and experiment, this regularity eluded him. Kopp found different chemists measured "different, in themselves equally reliable, [boiling point] results for the same substance," leaving him unable to confirm his hypothesis, even if it were true. This shortcoming was of great importance, because it alerted him to the unreliability of existing boiling point measurements. Kopp responded by seeking to understand what caused this unreliability, becoming an advocate for improved methods of boiling point measurement.[27] Moreover, this step introduced a crucial convergence between his project and Hofmann's.

Hofmann—no doubt guided by Liebig—quickly accepted that boiling points might help chemists identify organic compounds, and he soon began working to support Kopp's project. When Hofmann and James Sheridan Muspratt published the preparation of toluidine in 1843, they reported toluidine's boiling point. Even more significant, they offered a comparison of the boiling points of benzene, nitrobenzene, and aniline with nitrobenzene, nitrotoluene, and toluidine, respectively, data they claimed showed "undoubtedly that Kopp's rule holds good with regard to the above-cited bodies."[28]

Using boiling points to identify substances became an even more attractive possibility following the turn to synthesis. Synthetical experiments produced large numbers of very similar compounds that were frequently hard to differentiate. By the time he moved to London in 1845, Hofmann was using boiling points to distinguish between his synthetic products and to recognize compounds he already knew. Chemists at the Royal College of Chemistry routinely reported boiling points for substances they made.

In 1855, Hofmann's assistant Arthur Church measured boiling points for a series of aromatic compounds, finding that these showed exactly the kind of regularity Kopp sought. Unsurprisingly, this work was singled out for particular praise by Kopp.[29] Church continued working in this area, and in 1860, he tabulated and analyzed boiling point differences between members of the main families of organic compounds. Church struggled with the "most singular relations" displayed by the amines—one of Hofmann's primary areas of study.[30] But Kopp—aided by Hofmann—successfully rationalized this data by reorganizing it to reflect Hofmann's recent classification of amines as primary, secondary, and tertiary.[31]

By 1860, Kopp considered that "[c]onstant relations . . . between the boiling-points and the formulae of the volatile organic compounds . . . have been most positively established."[32] As the amines illustrated, however, constitution was an essential factor in those relations. Thus, greater experience with organic compounds—largely gained through his collaboration with Hofmann—forced Kopp to abandon his conviction that composition alone determined boiling point. Reformulated to apply only within series of constitutionally similar compounds, what became known as Kopp's Law of Boiling Points was accepted as a useful approximation—rather than a numerically precise law of nature—by chemists during the 1860s.[33]

There was a happy consilience of interests between Hofmann's taxonomic aims and Kopp's goal of discovering how organic compounds' molecular nature determined their boiling points. The "rapid progress of organic chemistry . . . greatly expanded the material available for the discussion of boiling-points."[34] In turn, Hofmann's use of boiling points was an invaluable stimulus to Kopp's work. It was vital to both projects that chemists everywhere should measure the same boiling point for the same substance. As I now show, this was far from being the case during the 1840s—prompting Kopp to seek a consistent definition of boiling point and standard methods for its measurement.

DEFINITION AND STANDARDIZATION

When Kopp began his project, chemists—thanks in large part to Liebig and Giessen analysis—expected the same substance obtained from different sources to have the same formula. But they did not expect boiling points measured for the same substance to be identical. They understood boiling point varied with atmospheric pressure. They also knew—from work done to stabilize the freezing and boiling points of water as thermometric fixed points—that boiling point depended on numerous other factors.[35] Nineteenth-century chemists therefore did not interpret boiling points as bearing on purity and identity. This was the situation Kopp set out to change in the early 1840s.[36]

As we have seen, Kopp relied heavily on boiling point measurements made by other chemists. This reliance alerted him to two fundamental problems. The first was that measurement technique varied widely, which meant that chemists measured different boiling points for the same substance, even at the same pressure. In addition, sample purity would prove increasingly problematic. In the absence of agreed criteria, chemists judged purity in largely arbitrary ways, including appearance and color. Measurement technique and sample purity were in themselves highly technical problems that were difficult to address. They were also—as Kopp recognized—intrinsically linked, making it challenging to decide which was the major source of discrepancy between any two reported values.[37]

Kopp's initial approach to measuring boiling points was far from adequate in these circumstances. This was partly because Kopp—although he
would soon learn otherwise—was initially relatively unconcerned by purity.[38]
Indeed, it was Liebig—rather than Kopp—who first pointed out that boiling
points might usefully indicate purity.[39] Kopp nevertheless developed unique
expertise in boiling point measurement. Over the course of two decades, he
highlighted the wide range of skills—both scientific and technical—needed
to tackle these problems in a concerted manner. Redefining boiling point
and how it should be measured established crucial interconnections between
purity, technique, and instrumentation. It would take immense labor distributed among diverse specialists to standardize boiling points.

One other issue affecting measured boiling points requires emphasis: the
material from which the measuring vessel was made. In 1812, Joseph-Louis
Gay-Lussac fixed the boiling point of water at "100.000 °C" only when measured at a specified atmospheric pressure *in a metallic vessel*.[40] In the wake of
the glassware revolution, however, most chemists measured boiling points
in glass vessels—even though they knew water boiled at a higher temperature in glass than in metal. How much higher was particular to each pair of
vessels. Even worse, the boiling point of water measured in glass previously
exposed to sulfuric acid could be elevated by as much as 6 °C—no matter
how well the flask had been washed. Glass contributed to Kopp's problem—
but it would also play a major part in his solution.[41]

By 1845, supported by Liebig and Hofmann, Kopp was emerging as the
leading expert on boiling point. He used this commanding position to define
"normal boiling point" as the temperature at which "the tension [pressure]
of the vapor of the liquid is equal to the prevailing air pressure." In practice,
however, Kopp remained unclear "what is . . . to be regarded as the boiling
point." A major reason for this gulf between theory and experiment was the
impossibility of rationalizing "readings for the same substance made by two
equally good observers which are not in agreement." It was, he explained,
doubtful whether "as a rule one . . . is entitled to regard each perfectly isolated result for a boiling point as accurate to 1 or 2°." Chemists' failure to
measure reproducible boiling points badly undermined Kopp's project.[42]

Chemists at this time generally measured boiling points in one of two
ways, both of which Kopp believed were flawed. The first and more usual
method involved submerging a thermometer in a liquid contained in an
open glass vessel and then heating the apparatus until a steady temperature

was reached. This method, according to Kopp, typically produced "a too high, an abnormal boiling point." The use of glass was partly to blame—but so was the measurement technique. Stopping the experiment early, for example, prevented the sample from evaporating sufficiently, which Kopp found led to inaccuracy—presumably because the thermometer was subject to a considerable temperature gradient.[43]

Chemists' second method was to measure boiling point during distillation, a technique long used to purify volatile substances, usually liquids.[44] Distillation separated substances by boiling point—removing more volatile substances (usually liquids) from mixtures also containing less-volatile ones (less easily vaporized liquids or solids). When heated, a mixture of two such components produced a vapor richer in the more volatile substance. Carried away from the original mixture through glass tubing, this vapor condensed to give a liquid distillate of the same enriched composition.

Kopp believed that measuring boiling points during distillation compounded the problems of reliable measurement—including those caused by measuring boiling points in glass vessels, since laboratory distillation apparatus was now predominantly constructed using glass components. Moreover, when measuring boiling point, the thermometer was typically submerged in the impure starting mixture, which inevitably compromised the result.[45] Kopp noted that—even when the substance being distilled was relatively pure—the measured boiling point usually differed by "several degrees" from that of the purified distillate.[46] Yet, as I explain below, limited sample availability meant that Kopp would continue to recommend this position for the thermometer in all but exceptional cases.

Kopp's response to these problems mobilized an integrated, glassware-based technology. Constructing his apparatus from standard glass components, he began by seeking to match Gay-Lussac's boiling point for pure water. He chose a thermometer made by the celebrated Parisian instrument maker Charles Félix Collardeau. The next section addresses the importance of thermometry in standardizing melting and boiling points. But it is worth noting here that early-nineteenth-century chemists recognized thermometers made by certain instrument makers as offering superior precision and reliability. Indeed, Liebig's preference for Collardeau's thermometers makes it reasonable to speculate that he may have loaned such an instrument to Kopp.[47]

Kopp's apparatus (reconstructed in figure 7.2) combined a round-bottomed glass flask with a cork fitted into the opening at the neck. A glass tube pierced this cork, connecting the flask's interior to the air outside, thereby maintaining the boiling liquid at atmospheric pressure. Surrounding this tube with a Liebig condenser prevented the escape of vapor. The thermometer also passed through the cork and was positioned so that its thread was always enclosed to the same extent, minimizing the temperature gradient around the thread—a crucial factor in accurate readings.

Except where the substance boiled abnormally, limited sample availability meant that Kopp preferred to immerse the thermometer to a constant depth in the boiling liquid—even though he knew this limited the achievable accuracy.[48] Introducing platinum wire into the liquid minimized "bumping" (*Stossen*), a phenomenon that resulted in "abnormal boiling" and falsely elevated the observed boiling point.[49] Finally, Kopp advised waiting until four-fifths of the liquid had evaporated before recording the temperature.[50] Following this procedure, Kopp measured the boiling point of water in glass at "exactly 100 °C" at 332 mm mercury, the value established by Parisian precision science.[51]

This method enabled Kopp to measure reliable boiling points for numerous organic compounds. By publishing his recommendation, he explained how other chemists could produce equally reliable results—as Hofmann, among others, certainly did. Thus, Kopp established a new connection between boiling point and the purity of substances, thereby inspiring improvements in purification and measurement. Over subsequent decades, fractional and reduced-pressure distillation enhanced separation by boiling point.[52] Chemists also found new ways of measuring reliable boiling points, including those of tiny samples.[53] They used such techniques in combination to prepare substances whose sharply defined boiling points were both hallmarks of purity and recognized as characteristic. Thus, Kopp's project—although it did not fulfil his original goal—marked an important stage in the redefinition of purity and identity that enabled chemists to solve the chemical identity crisis.

STANDARDIZING THERMOMETRY

Kopp was not the only former Liebig student concerned with the physical properties of organic compounds. Victor Regnault—whose analytical

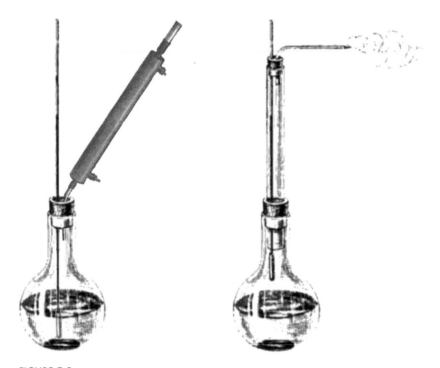

FIGURE 7.2

The apparatus Kopp described in 1845 has been visually reconstructed on the left. It shows the thermometer bulb submerged in the liquid sample, as Kopp recommended at this time. The Liebig condenser (at that time encased in metal) ensured that none of the vaporized liquid escaped while measuring the boiling point at atmospheric pressure. Note the presence of a platinum wire to prevent bumping.

The standard apparatus is shown on the right. Note that by 1857, the thermometer bulb was held above the boiling liquid, as would be usual today, while its mercury thread was encased by a glass tube. This arrangement permitted the escape of vapor while reducing the temperature gradient experienced by the thread—a key source of error in temperature measurement. This composite apparatus was made by joining flask and tube. It was later superseded by a single glass vessel incorporating a long neck that performed the tube's insulating function.

Source: The image on the right is from Thomas Graham and Friedrich Julius Otto, eds., *Ausführliches Lehrbuch der Chemie,* 3rd edition, vol. 1 (Braunschweig: Vieweg, 1857), 190, fig. 254. Courtesy of the University of Wisconsin-Madison Libraries. Composite image on the left courtesy of Andrew C. Warwick.

disputes with Liebig were discussed in chapter 2—shared this interest. Indeed, many of the boiling points that Kopp tabulated during the early 1840s were taken from Regnault's work—which Liebig had perhaps drawn to his attention. Moreover, Kopp was quite critical of the "uncertainty" in Regnault's observations.[54] Under these circumstances, it is unthinkable that Regnault did not closely follow Kopp's project.

As we shall see, Regnault—to an even greater extent than Kopp—saw precision as fundamentally necessary in tackling the problems facing organic chemistry. In fact, it would be Regnault whose passion for precision drove him to do vital work in standardizing thermometry.[55] Only comparable thermometers could enable chemists to standardize boiling and melting points. Producing such devices, however, was a remarkably demanding endeavor, spanning chemistry, physics, and glass.

Regnault's skills in precision measurement and organic chemistry undoubtedly exceeded Kopp's. Regnault was supremely well trained in physics, having studied at two of Paris's most prestigious institutions, the École Polytechnique and the École des Mines. In 1835, he spent several months in Giessen, where his "decided talent" for chemistry impressed Liebig.[56] This encounter would play a central role in shaping Regnault's career. Encouraged by Liebig, Regnault devoted the next 5 years to organic chemistry. As I now explain, this experience fueled Regnault's commitment to precision measurement, ultimately drawing him back toward physics.[57]

At Liebig's request, Regnault investigated so-called "Dutch Liquor" (ethylene dichloride), for which Liebig and Dumas had assigned different molecular formulae. His task was to discover who was right, and it is worth noting that Regnault won Liebig's good opinion despite deciding in Dumas's favor—at least where formula was concerned. A meticulous experimenter, Regnault found Dutch Liquor hard to purify and even harder to analyze—largely because the focus of disagreement between Liebig and Dumas concerned its hydrogen content. Hydrogen's small atomic weight meant that tiny errors in the combustion analysis could have disastrous effects—as could any inaccuracy in the molecular weight determination.

Regnault's response was distinctly Parisian. Despite exposure to Giessen analysis—with its emphasis on replicability over precision—Regnault believed that organic chemistry was just as reliant on precision measurement as was any other science. A "noticeable" divergence between Dumas's and Liebig's boiling points for Dutch Liquor, for example, prompted

Regnault to repeat the measurement—which he did by submerging a thermometer inside a boiling tube filled with sample and adding some metallic wire to prevent bumping. In this case, Regnault agreed with Liebig. His boiling point (82.5 at 756 mm °C) was almost identical to Liebig's (82.4 °C), both being noticeably lower than Dumas's (85 °C).[58]

Regnault also found Dumas's method of molecular weight determination deficient. Based on measuring specific weight (vapor density), Dumas's method proved unsuitable for Dutch Liquor's volatile decomposition products. Thus, Regnault—on returning to Paris, where he was appointed assistant to the geologist Pierre Berthier at the École des Mines—resorted to Gay-Lussac's considerably more complex apparatus to complete his investigation.[59] Specific weight measurements took their place alongside boiling points among the physical data Regnault reported, and which Kopp would later appropriate.[60]

Despite learning Liebig's methods in Giessen, Regnault was convinced that only Parisian precision measurement could solve certain analytical problems. This commitment remained strong as he pursued independent research in organic chemistry, first studying chlorinated compounds related to Dutch Liquor and subsequently shifting focus to the alkaloids. By the late 1830s, as chapter 2 explained, the latter project would bring Regnault into conflict with Liebig, on essentially the same grounds as divided Liebig and Dumas over Dutch Liquor: Regnault and Liebig disagreed about the precision achievable in determining the alkaloids' hydrogen content. Just as Liebig's good opinion surely attracted Regnault to organic chemistry, it seems plausible that his former mentor's dismissal now contributed to Regnault's decision to shift his research toward physics.

By the early 1840s, Regnault—as was common in Paris—occupied two professorial posts: one in chemistry as Gay-Lussac's successor at the École Polytechnique, the other in physics at the Collège de France. It was in this latter role that Regnault was asked by the French government to study the operation and efficiency of steam engines—a commission that led him to investigate the thermal properties of gases. Just as Kopp was launching his quest to standardize boiling points, Regnault began work that would both establish the limitations of existing thermometers and offer a method of overcoming them.

This work was materially as well as mathematically complex.[61] Indeed, measuring temperature was "perhaps the most difficult" problem confronting mid-century physics.[62] Understanding how chemists standardized melting and boiling points, however, is inextricably linked to innovations in thermometer design and standardization. I therefore offer a brief (and in places somewhat simplified) summary of the interaction between chemistry and thermometry in Regnault's work, and of the essential contributions of glassblowers, most notably Geissler, to making standard thermometers. By 1850, Regnault's understanding of the requirements of reliable temperature measurement was transforming thermometer construction.[63]

When Regnault began his investigation, most temperature measurements were made using mercury thermometers. Good-quality instruments were available from skilled makers in metropolitan centers, such as Paris. Yet—as Kopp's work illustrated—it was hard to compare readings from thermometers made by different instrument makers. Regnault recognized this failure of comparability as a fundamental problem, showing that it became much worse at temperatures outside the fixed calibration points, 0 °C (water's freezing point) and 100 °C (its boiling point).[64]

As discussed below, because many organic compounds melt at temperatures well in excess of 100 °C, this would be a significant obstacle in standardizing melting points. The issue, moreover, was intrinsic to the device and was not something that could be overcome merely by improved craftsmanship. As Regnault explained, because mercury's thermal expansion was barely seven times that of glass, variations in the type of glass used, and how the glassblower had worked it, meant that even good instruments were not comparable beyond the fixed points.[65]

Regnault therefore turned his attention to thermometers in which the thermometric substance was a gas. According to Gay-Lussac's law, gases expanded uniformly with heat, which encouraged the assumption that gas thermometers were "perfect" devices, "whose indications would always be proportional to the quantities of heat . . . absorbed, or in which the additions of equal quantities of heat would always produce equal dilatations."[66] But Regnault's experiments showed that neither Gay-Lussac's law ($p \propto T_{abs}$) nor Mariotte's (Boyle's) law ($pV = const$) were "laws of nature, but . . . only approximative, and consequently only true between certain limits." Even gas thermometers were therefore a good deal less accurate than generally believed.[67]

Regnault nevertheless believed that gas thermometers could be made to produce comparable results. Because gases—unlike mercury—expanded so rapidly with increasing temperature, glass's thermal expansion did not noticeably affect the reading.[68] Regnault therefore directed his attention to gas thermometers, selecting air as the thermometric substance. His first decision was whether the air thermometer should operate at constant pressure or constant volume. Finding the more straightforward constant pressure device (figure 7.3, labeled "Fig. 16" in the illustration) inaccurate at high temperatures, Regnault opted for the constant volume device (figure 7.3, labeled "Figs. 13–15").[69]

Working at constant volume, however, introduced new problems: Air's large thermal expansivity caused high pressures inside instruments exposed to elevated temperatures—which was both hazardous and a potential source of error. Regnault responded by filling thermometers with air at reduced pressure, verifying that such devices remained comparable. Because they relied on a widely available thermometric substance and were insensitive to variations in filling pressure, air thermometers were in principle more easily distributable. Regnault's extensive study showed that well-made, properly used constant volume air thermometers were accurate and comparable up to 325 °C. Air thermometers fulfilled Regnault's criteria for a "perfect" or "normal" thermometer, showing equal expansion for equal addition of heat, and he therefore deemed them the only instruments suitable for use in situations demanding great precision.[70]

Air thermometers, however, were not widely available. Regnault's was large, complex, and constructed using the considerable resources of Parisian science. Such an instrument was far beyond the reach of most chemists.[71] It was also impractical for many chemical uses—including measuring melting and boiling points. When an air thermometer could not be used, as Regnault explained, "it becomes necessary to use a mercurial thermometer." In this case, however, "a direct comparison must be made of this instrument with the air-thermometer, in order that its indications may be transformed into those of the normal thermometer."[72]

Thus, Regnault established the air thermometer as a *normal thermometer*— the fiduciary device, against which any mercury thermometer must be calibrated to produce comparable readings. Explaining how mercury thermometers could be calibrated to become normal thermometers, moreover, established glass and glassblowing as paramount requirements of reliable thermometer construction. Unlike the air thermometer, mercury thermometers were made entirely from skillfully lampworked glass tubes into which mercury was introduced. However, glass is not a single substance but a

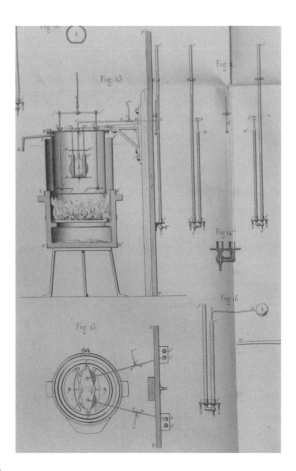

FIGURE 7.3

Regnault's constant pressure air apparatus (Figs. 13–15 in the illustration) incorporated two thermometers side by side, to allow direct comparison of their results. This was an expensive precision instrument, designed to measure the temperature surrounding the thermometer's main bulb A. Constructed of metal as well as glass parts, sealed together using putty, it incorporated complex components such as the three-way tap at the base of the manometer. Heating the air inside bulb A increased the pressure, which was measured by the manometer (two side by side in Fig. 14, again for ease of comparison). The measured pressure was then converted to temperature by calculation. Although the constant pressure air thermometer (Fig. 16) was considerably simpler, Regnault rejected this device, because he found it inaccurate at higher temperatures. As the air inside bulb A expanded, it filled the graduated glass tubes. The higher the temperature, the greater the proportion of the air was at room temperature rather than at the temperature of bulb A.

Source: H. Victor Regnault, *Relation des Expériences Entreprises par Ordre de Monsieur le Ministre des Travaux Publics, et sur la Proposition de la Commission Centrale des Machine à Vapeur, pour Déterminer les Principales Lois Physiques, et les Données Numériques qui Entrent dans le Calcul des Machines à Vapeur,* vol. I (Paris: Didot Freres, 1847), Planche II, Figs. 13–15 (constant volume) and Fig. 16 (constant pressure). Courtesy of the University of Wisconsin-Madison Libraries.

complex mixture, whose composition and properties—including its thermal expansion—vary widely. Regnault's next step was therefore to investigate how the type of glass used affected the performance of mercury thermometers.

Comparing mercury thermometers made "with the flint-glass of Choisy-le-Roi, with common glass, with green glass, and with Swedish glass," Regnault found their behavior varied widely. Mercury thermometers made from "the different varieties of common glass, which are at present used in the manufacture of chemical instruments, do not proceed in strict accordance beyond the fixed points which have been used to regulate their scales." Regnault nevertheless concluded that mercury thermometers could be used in "most experiments," provided they had not been made from glass high in lead—which, he explained, "may be readily detected when . . . worked in the lamp."[73]

By 1850, Regnault had established the possibility that a mercury thermometer could be a normal thermometer, provided it had been expertly calibrated against a well-made normal air thermometer over a wide range of temperatures. Accomplishing this depended crucially on glassblowing expertise and meticulous making. As I now show, however, Regnault's work also introduced possibilities for a material transformation of thermometry that glassblowers such as Geissler would soon exploit.

In fact, there are good reasons why Geissler was particularly well situated to capitalize on this opportunity. The transition from large, complex air thermometers whose construction involved numerous materials and a diverse skill set to mercury thermometers made by lampworking paralleled the displacement of Parisian analytical apparatus by Liebig's Kaliapparat some two decades earlier. Geissler, moreover, was now in Bonn, where—as we have seen—his work with Plücker already included perfecting both a precision balance fashioned from glass and a new and superior form of Kaliapparat. Given Plücker's interests in precision measurement, it is unsurprising that the two now collaborated on some of the mid-century's most important improvements in thermometer design.[74]

Geissler's collaboration with Plücker initially focused on constructing a "compensated thermometer," a device that itself corrected for the thermal expansion of glass—thereby permitting a mercury thermometer to approach "perfect" or "normal" behavior more closely. This remarkable feat required enormous technical expertise as well as extraordinary lampworking skill, and it is far too complex to relate in its entirety. But the first stage of this work amply illustrates my point. Taking only mercury's thermal expansion (which Regnault had determined) as given, Plücker and Geissler set out to determine how a mercury thermometer's glass envelope expanded with heat. They found that this trait was peculiar to each vessel and must therefore be "determined directly for each particular instrument."[75]

Geissler's lampworking skill now came to the fore in managing the many interrelated factors at play. First, he used extremely thin-walled capillary tubes that he made himself, rather than the thicker-walled, commercially available variety (in which temperature gradients built up across the thickness of the glass). Selecting only the most uniform of his home-drawn tubes, Geissler produced a precision instrument that was protected from thermal shock by the thinness of the glass. He then employed his glass balance—sensitive to 0.1 mg—to determine directly the device's thermal expansion. This was but the first of many complex calibration processes required to make a thermometer in which mercury accurately compensated for the glass vessel's expansion.[76]

One of Geissler's later innovations in thermometry was of immediate relevance to chemistry, and especially the measurement of melting points. Introducing nitrogen above the thermometer's mercury thread reduced vaporization of mercury, enabling the thermometer to measure high temperatures reliably. This method enabled Geissler to produce thermometers for use between 360 and 500 °C.[77] His successor Franz Müller found this modification improved accuracy so much that he applied it to the construction of all chemical thermometers intended for use above 200 °C (figure 7.4).

Geissler's thermometers "soon received due recognition," and over the coming decade—years during which he also developed his celebrated vacuum pump and eponymous discharge tubes—Geissler became the acknowledged maker of the world's most reliable thermometers.[78] In 1864, none other than Kopp reported using "thermometers made by Geissler, of Bonn," lent to him by Heinrich Buff, and which Kopp had "repeatedly compared . . . with two normal thermometers of [his] own construction."[79] In 1889, meanwhile, Hofmann measured the boiling points of volatile amines using a Geissler thermometer, which had been tested against a normal thermometer of the Imperial Physical Technical Institute (Physikalischen Technischen Reichsanstalt, PTR).[80] One of the leading advocates for standardization and one of the late-century's most eminent organic chemists both chose Geissler's thermometers over those of any other maker.

(a)

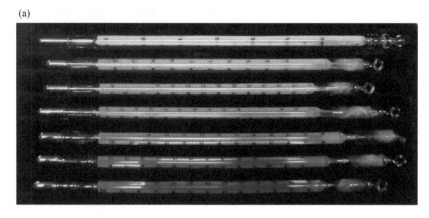

(b)
Normal-Thermometer nach Gräbe-Anschütz. Satz für chemische Arbeiten, bestehend aus 7 Instrumenten von 10 bis 14 cm Länge und ca. 6 mm Durchmesser, aus Jenaer Normalglas, oben mit Ring versehen. Nr. 4, 5, 6, 7 sind mit Stickstoff gefüllt, der Satz ist in ein elegantes Sammetetui eingelegt.

4871 — Der Satz von 7 Instrumenten mit $1/1^0$ Teilung, mit Milchglasskala   . Mark 50,—
4872 — „   „   „   „ 7   „   „ $1/2^0$ „   „   „   . .   „   57,—
4873 — „   „   „   „ 7   „   „ $1/5^0$ „   „   „   . .   „   65,—

Nr. 1, von ca. — 10^0 bis + 50^0 C.
„ 2, „   „ + 40^0 „ + 110^0 „
„ 3, „   „ + 90^0 „ + 160^0 „
„ 4, „   „ + 150^0 „ + 220^0 „
„ 5, „   „ + 200^0 „ + 265^0 „
„ 6, „   „ + 250^0 „ + 310^0 „
„ 7, „   „ + 300^0 „ + 360^0 „

Siehe: Chem. Centralblatt 58 S. 131. Die Destillation unter vermindertem Druck.

FIGURE 7.4
(a) Photograph showing a set of normal thermometers for chemical applications supplied by Heinrich Geissler's successor, Franz Müller; (b) The entry for these thermometers (no. 4871) in Müller's catalog. Made from "normal glass" produced in Jena, Germany's glassmaking center, these instruments were produced in collaboration with the Bonn chemists Carl Graebe and Richard Anschütz. Thermometers like these were essential for accurate melting and boiling point determinations. Those covering the higher temperatures (shown towards the bottom of the photograph and labeled "Nr. 4" through "Nr. 7" in the catalog entry) were filled under nitrogen to minimize vaporization of mercury and consequent inaccuracy.

Source: (a) Courtesy of Collection of Steve Beare; (b) Dr. H. Geisslers Nachfolger (Franz Müller), *Preisverzeichnis: Institut zur Anfertigung und Lager chemischer, bakteriologischer, physikalischer und meteorologischer Apparate*, 9th edition (Bonn: Georgi, 1904), 421. Courtesy of Hathi Trust: https://hdl.handle.net/2027/njp.32101058257674?urlappend =%3Bseq=453%3Bownerid=27021597769408027-459.

Many scientists outside Germany also used Geissler's thermometers.[81] His instruments demonstrated the extraordinary levels of precision achievable in glass. They no doubt inspired makers in other countries to emulate his success—although few achieved similar eminence. In the UK, for example, Louis P. Casella was engaged by Kew Observatory to construct thermometers of improved functionality and sensitivity—a task he considered required "great delicacy."[82] Managed since 1842 by the British Association for the Advancement of Science, one of Kew's main purposes was to serve as "a place for . . . comparison of the various instruments . . . in order that their relative advantages and defects may be ascertained."[83] Just as Geissler's thermometers were standardized against normal thermometers at the PTR, so Casella's thermometers were offered for sale with "verification from Kew."[84]

Regnault's expertise in physics was certainly crucial in making possible the standardization of thermometry. But makers like Geissler and Casella played a central role, operating in synergy with emerging national standards institutions that would later enforce abstractions of best practice that originated in the hands of individual skilled makers. These developments were timely, since it would otherwise have been impossible to standardize the boiling and melting points of organic compounds. They also owed a great deal to organic chemistry—in the impetus Regnault gained from exposure to the field and, as I now show, in the increasing demands organic synthesis placed on melting point measurement.

MELTING POINTS AND NATURAL PRODUCTS

Melting points were essential to solving the chemical identity crisis. Before about 1840, chemists sometimes reported melting points. They did not, however, interpret these as characteristic, accepting large discrepancies between values for the same substance.[85] In the wake of Kopp's work, melting points emerged as potentially even more useful than boiling points for identifying organic compounds. As previous chapters have shown, Hofmann prioritized working with crystalline compounds, because these were easier to handle in the laboratory at a time when recrystallization was chemists' best method of purification.

Where distillation relied on differences in boiling point, recrystallization separated a desired substance from contaminating impurities by using differences in solubility. Skillfully performed, recrystallization was more likely to permit the isolation of a single substance than was distillation. Chemists routinely capitalized on this advantage by transforming volatile organic compounds into crystalline derivatives to purify them prior to analysis.[86]

In principle, melting points were easier to measure and—because they were far less susceptible to variations in pressure—easier to standardize. Just as with boiling points, however, interlinked problems of purification and measurement meant that chemists still struggled to make melting points characteristic. Many crystalline solids once believed to be pure became mixtures when subjected to new definitions of purity—as the work of French chemist Michel Eugène Chevreul illustrates.

Chevreul's study of fatty acids during the 1820s is frequently taken to have established melting points as viable identifying characteristics—largely because Chevreul himself insisted on melting point's value in identifying organic substances.[87] Because many of the fatty acids he isolated (including oleic) were liquid at room temperature, however, Chevreul reported the melting points of just three substances (margaric, stearic, and caproic acids), using melting point to differentiate between two fatty acids in only one instance—when identifying margaric acid as distinct from stearic.[88]

Even margaric acid's melting point, variously reported by Chevreul as 60 °C and 55 °C, did not achieve characteristic status.[89] Thomas Thomson's *System of Chemistry* (1817) and Edward Turner's *Elements of Chemistry* (1833) listed different melting points for margaric acid.[90] In 1838, Berzelius reported its melting point as 58.75–60 °C, attributing the difference between his value and Chevreul's to problems of thermometry. Crucially, the discrepancy did not cause Berzelius to doubt that he and Chevreul had examined the same substance, nor did he speculate on the purity of their samples.[91]

In fact, Chevreul never suggested that melting point could or should displace physiological properties, such as smell, taste, appearance, and origin.[92] Indeed, his own work demonstrated the great difficulty of using melting points this way. Although Chevreul purified fatty acids by recrystallizing them until their melting points were stable, this approach turned out to be unreliable. When margaric acid was subjected to improved methods of purification in 1852, it emerged as a mixture of 10 parts palmitic acid and one part stearic acid.[93]

More generally, chemists' inability to measure replicable values undermined the use of melting points as characteristics—just as with boiling points. Overcoming this problem prompted chemists to redefine melting points in relation to new methods of purification and standards of measurement. Improvements in distillation technique increased the purity of many starting materials and organic reagents so that reactions also produced purer products. Chemists learned to connect sudden melting with purity and slower melting at lower temperatures with the presence of impurities, arriving at reliable melting points that unambiguously identified substances they believed were pure.

These changes went hand in hand with chemists' efforts to prepare compounds in the laboratory that they could show were identical to substances derived from natural sources. By about 1860, so-called "natural product synthesis" invested the chemical identity crisis with new urgency, making it imperative that chemists could agree on whether two substances were different, the same, or merely similar. By this time, they accepted formula and reactivity as indicative—that is, properties that must be the same for two substances to be considered identical. But synthesis, as we shall see, prompted chemists to define natural products—whether extracted from nature or synthesized in the laboratory—as substances whose identity and purity was established by melting points measured according to prescribed standards of procedure and instrumentation.

CONIINE, COMMODITY, AND COMMUNITY

This section draws together the threads of this chapter by anticipating the topic of chapter 9. There we shall see how, in 1886, the Kiel chemist Albert Ladenburg synthesized the first natural alkaloid, coniine. Here, I focus on Ladenburg's efforts to convince his peers he had made coniine. Success required Ladenburg to prove natural and synthetic coniine had the same melting point. As we shall see, this proof entailed managing a new problem in the definition of pure substance. It also relied on the finest thermometers and on using novel glassware to further refine how melting points were correctly measured.

Ladenburg's attempts to synthesize coniine highlighted an important, emerging issue for organic chemists. This issue affected melting point measurements and further challenged chemists' notion of pure substance. Mid-century chemists learned that many natural products showed optical activity, rotating light to either the right (called "dextrorotatory") or the left ("laevorotatory"). Chemists therefore recognized optical activity as an indicative property that any synthetic product must match to be considered an identical substance to the natural target. At the same time, optical rotation measurements had yet to be standardized. Although an individual chemist might compare values for natural and synthetic substances, these had no broader value as characteristics in the 1880s.

Most synthetic products, however, were optically inactive. Unlike living organisms, laboratory chemistry typically made equal amounts of both forms, called "optical isomers," so that their rotations canceled each other out. Optical isomers shared the same composition and molecular formula. But even though their melting and boiling points were identical, they were nevertheless distinct substances. Natural product synthesis typically proceeded by making an optically inactive mixture from which the desired isomer could be separated.[94]

Suspecting coniine was dextrorotatory α-propyl piperidine, Ladenburg separated the dextrorotatory isomer from synthetic, optically inactive α-propyl piperidine.[95] He found that this isomer displayed many of natural coniine's chemical properties, and its effect on light matched natural coniine's. Seeking to prove synthetic dextrorotatory α-propyl piperidine was identical to natural coniine, Ladenburg turned to melting point measurements. Converting both the optically inactive compound and the dextrorotatory isomer to their crystalline hydrochloride salts, he found that these melted at different temperatures.[96] Showing that his synthetic product and natural coniine were identical therefore required further work, ultimately entailing the development of a new, glass apparatus for measuring melting points. This apparatus, devised in 1885 by Ladenburg's assistant C. F. Roth, was given a "lovely, clean design" by Geissler's successor, Müller, who subsequently sold it as a standard device (figure 7.5).[97]

Roth's account of why he produced this new device shows how organic chemistry drove accuracy in melting point measurement. It also confirms that during the 1880s, melting points still suffered from the same problems of standardization that Kopp had identified for boiling points four decades previously. Roth found the two most widely used existing techniques of melting point determination gave different results when applied to coniine. The simplest involved heating the sample and thermometer in a concentrated sulfuric acid bath contained in an open round-bottomed glass flask. The other—developed some 10 years earlier by the Bonn chemists Richard Anschütz and Robert Schulze in collaboration with Geissler—was

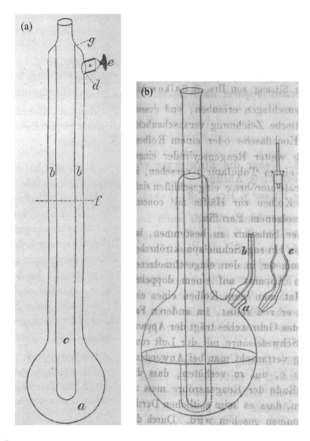

FIGURE 7.5

(a) Roth's apparatus (1886, left); shown alongside (b) the earlier apparatus of Anschütz and Schulze (1877, right). Both items were made in Bonn: the earlier one by Heinrich Geissler, Roth's by Geissler's successor Franz Müller. In each case, sample and thermometer sat within an inner glass tube, surrounded by concentrated sulfuric acid contained in the outer round-bottomed flask. A closed tap or drying tube prevented the acid from absorbing atmospheric moisture when the apparatus was not in use; a small opening maintained atmospheric pressure during use while simultaneously minimizing exposure to corrosive acid fumes.

Source: (a) C. F. Roth, "Ein neuer Apparat zur Bestimmung von Schmelzpunkten," *Berichte der deutschen chemischen Gesellschaft* 19 (1886): 1971; (b) Richard Anschütz and Robert Schulze, "Ueber einen Einfachen Apparat zur bequemen Bestimmung Hochliegender Schmelzpunkte," *Berichte der deutschen chemischen Gesellschaft* 10 (1877): 1801. Both images courtesy of the University of Wisconsin-Madison Libraries.

similar except that the flask containing the sulfuric acid was closed. This reduced chemists' exposure to corrosive fumes but also tended to give higher readings than "free sulfuric acid."[98]

This failure of comparability was a serious problem for Roth. Where the chemist "wishes to use the melting point later as an aid to the identification of compounds," he explained, "a perfect agreement of his found melting point with that previously stated is most highly desired, indeed necessary for him."[99] The difficulty, he believed, originated in temperature variations around the thermometer's mercury thread—which required every measurement to be corrected by a calculation involving mercury's "apparent expansion coefficient," a value that varied for each type of glass and every individual thermometer.[100]

Roth addressed this problem by elongating the neck of the device. On heating, the rising sulfuric acid level surrounded the thermometer, minimizing variation of temperature around its thread. Roth tested his device on six substances, including two well-characterized, simple organic substances, benzoic acid and urea, as well as the mercury double salt of α-picoline (starting material for Ladenburg's coniine synthesis) and nicotinic acid—whose melting point was around 235 °C. He compared melting points for each substance measured using "one and the same thermometer"—almost certainly a Geissler normal thermometer—in his new apparatus and with the more usual open round-bottomed flask.[101] In each case, he corrected the latter value using a standard correction formula.[102] The largest discrepancy between the corrected result and the melting point measured by his new apparatus was 0.16 °C, well within the accuracy of correction. Roth therefore concluded that his apparatus eliminated the need for melting point correction.

Ladenburg now used Roth's apparatus to identify coniine as dextrorotatory α-propyl piperidine. Having separated the desired optical isomer from his optically inactive laboratory product, Ladenburg converted both his synthetic product and natural coniine to their hydrochloride salts. Assisted by Roth and using his apparatus, Ladenburg measured the melting point of dextrorotatory α-propyl piperidine hydrochloride at exactly 217.5 °C, while natural coniine's hydrochloride melted at 217.5–218.5 °C.[103] Melting points within a degree of each other convinced Ladenburg—and the chemical community—he had indeed synthesized coniine.

Together, Ladenburg, Roth, and Müller redefined coniine's identity in terms of a sharp, accurate melting point measured according to a set procedure in standard apparatus. This accomplishment embodied every aspect of chemists' response to the chemical identity crisis. Roth's apparatus capitalized on both the innovative possibilities of working in glass and the potential for glass apparatus to embody crucial aspects of laboratory standardization. Developed to enable Ladenburg to demonstrate his synthetic product was identical to natural coniine, it was made commercially available by Müller's Bonn firm. Ladenburg's synthesis, moreover, entailed operating at a new standard of purity—with respect to natural coniine's optical rotatory power and, as chapter 9 explains, by means of improved methods of distillation and crystallization. Its laboratory synthesis transformed coniine into a substance knowable by its molecular formula and melting point alone.

CONSTANTS OF NATURE

Chemists' redefinition of pure substances according to their melting and boiling points constituted a major advance in the science—as Kopp, Hofmann, and Liebig clearly appreciated. More than merely an alternative approach to identification, sharp, precise melting and boiling points offered a new, shared definition of substance and purity that was necessary for the development of synthetic organic chemistry. As a growing community attempted to replicate and develop the work of Hofmann and others, reliable criteria enabling chemists everywhere to agree on what they had made became essential. Inseparable from the instruments and techniques used to produce them, standard melting and boiling points remained chemists' only reliable way of identifying synthetic products until well into the twentieth century.

Highlighting the intrinsic link between purity and identity reinforces the limitations of using modern names and formulae in historical contexts. Because distillation and crystallization were known techniques, it is tempting to assume that nineteenth-century chemists worked with compounds of similar purity to those found in laboratories today. In fact, as many examples in this chapter and elsewhere in this book attest, nineteenth-century chemists understood the relationship between purity and identity in very different terms from their twentieth and twenty-first century successors. Effacing that distinction makes past chemistry easier to grasp—but it does so at the expense of obscuring the very struggles that produced the modern, molecular understanding of organic compounds.

In highlighting how chemists sought to stabilize and rationalize properties and reactivity, this study has begun to expose the origins of the possibility of a molecular understanding of organic substances.[104] Such an understanding required a convergence between experiment and theory: on one side, the ability to prepare substances with clearly defined physical properties; on the other, a developing conceptualization of molecules in terms of their constituent elements. Here, as elsewhere, successful laboratory reasoning prioritized experimental knowledge over prior theoretical commitment. Thus, Hofmann's rejection of premature theoretical allegiances in his widely respected studies of the amines (see chapters 5 and 6) provides a useful comparison point for Kopp's conviction that molecular composition alone determined boiling point—a belief that led him to handle numerical evidence in ways that met with skepticism from other chemists.

Established theories—which by 1850 included theories that constrained molecular composition—had their place in laboratory reasoning. But nascent theories of constitution, structure (and ultimately, stereochemistry) were the outcome of laboratory reasoning, not its basis. For example, Ladenburg's identification of natural coniine as dextrorotatory α-propyl piperidine confirmed and completed a partial, provisional constitution. It was also, as chapter 9 relates, an important step in Ladenburg's acceptance of aromatic structural theory, including the benzene ring he had hitherto vehemently opposed.[105]

Solving the chemical identity crisis and producing reliable thermometers increased chemists' reliance on glassblowing expertise. Geissler, Müller, and others were well placed to serve this new niche, providing technical support in universities and founding independent businesses for the commercial supply of scientific glassware. This was the birth of the scientific glassblower—a new profession marrying absolute command of glass as material, consummate lampworking skill, and the capacity to develop scientific understanding. In working with Plücker to create a compensating thermometer, for example, Geissler acquired knowledge of glass's thermal properties that would prove crucial in his development of the electrical discharge tubes for which he is most famous. Making vacuum-tight glass-to-metal seals relied on understanding, exploiting, and minutely managing the differing thermal expansivities of different kinds of glass.[106] Thus, Geissler's emergence as one of the foremost instrument makers pioneering high-vacuum physics relied on prior experience of making thermometers and other chemical glassware.

Little wonder that, following Geissler's death at age 65 in 1879, Hofmann celebrated his combination of utterly unprecedented, "astonishing"
lampworking skill with "profound physical understanding." Scientists frequently struggled to recognize their ideas in the working instrument, such
was Geissler's ingenuity in fulfilling their requirements. Hofmann especially
respected the unceasing quest for perfection and will to self-improvement
that had enabled Geissler to transcend his humble beginnings and lack of
formal education. The "self-made" Geissler, according to Hofmann, was no
mere instrument maker but an active participant in scientific and technical
progress.[107]

Locating the origins of scientific glassblowing in chemistry indicates the
need to reevaluate accepted historiography of nineteenth-century science.
It points to chemistry as the original precision science in glass, and therefore
the primary site of nineteenth-century science's nascent ability to control
and manipulate nature. Organic chemistry—and especially the chemical
identity crisis examined in this chapter—was fundamentally implicated in
that project. Toward the end of the century, the melting points of organic
compounds including aniline and quinoline supplemented the freezing
and boiling points of water as fixed points for calibrating thermometers.[108]

Indeed, chemists' use of glass apparatus and precise procedure to stabilize
melting and boiling points was ultimately so successful that sharp melting
and boiling points ceased to be the product of chemists' work and became
what Joseph Henry, Secretary of the Smithsonian Institution, in 1873 called
"Constants of Nature."[109] In fact, Henry's term remained heavily aspirational, because details of procedure, including who had carried out the
measurement, continued to be necessary to make sense of physical data.
Melting and boiling points only became constants of nature when measured
according to precise methods in expertly manufactured instruments. They
did, however, signal a widespread shift from physiological to physical properties as identifying characteristics.[110]

The 1880s saw disciplinary acceptance that melting and boiling points
(and other physical properties) were real constants in the sense that every
laboratory should be able to reproduce them, any discrepancy indicating erroneous results. Tabulated in volumes of physical and chemical
constants—notably the (1883) *Physical Chemical Tables* published by Swiss
chemist Hans Landolt and German physicist Richard Börnstein—standard
melting and boiling points appeared alongside the corrections needed to

convert the readings of mercury thermometers into those of Regnault's standard air thermometer.[111] Late nineteenth-century science relied on a growing apparatus of precision and standardization to produce and maintain these constants. By summarizing this framework in numerical tables, scientists effectively concealed the vast labor expended in its creation.

Chemists' reliance on glass and glassblowing to develop and distribute standard procedures and outcomes extended beyond the methods of purification and characterization described in this chapter. Virtually all operations of organic synthesis were carried out in glass. Chemists made nature submit to the demands of synthetic organic chemistry by altering the material basis of the tools they used. As chapter 8 explains, chemists also transformed their working environment to make this new world in glass conformable to its academic setting—something they accomplished by developing the laboratory as an essential component of chemistry's material culture.

8 LABORATORY LANDSCAPE

This chapter explains how my study's leading actors shaped the laboratory, first making it capable of supporting collective training and research in organic analysis, and later developing it to manage the new methods and risks of synthetic organic chemistry.[1] As we shall see, the rise of analysis and the turn to synthesis transformed the laboratory from merely a place where chemists worked into an essential tool in their struggle to understand and manipulate organic compounds. From about 1860 onward, purpose-built laboratories were an increasingly necessary component of disciplinary material culture.

Although it focuses on a small number of laboratories, this account may safely be generalized. Having established the far-reaching disciplinary consequences of Justus Liebig and August Wilhelm Hofmann's chemistry and signaled the exemplary value of Albert Ladenburg's 1886 coniine synthesis, their laboratories may also be taken as indicative. Abundant evidence confirms that the laboratories built for Liebig in Giessen in the late 1830s and for Hofmann in Bonn and Berlin in the mid-1860s both reflected contemporary state of the art and served as inspiration for future developments—in chemistry and beyond. Moreover, Hofmann's experience in London explains why the turn to synthesis pushed the boundaries of existing facilities, prompting further changes in laboratory design. By 1878, when Ladenburg's Kiel laboratory was completed, the transformed institutional chemical laboratory existed as a powerful model of remarkable reach and durability.

The present chapter explains how—and, crucially, why—organic chemists spent years on the design and construction of new laboratories.[2] In doing so, it extends our understanding of chemistry's nineteenth-century institutionalization. Situated in the context of dominant exemplars drawn from physics, and based on a change during the 1870s that historians have called the "institutional" or "laboratory revolution," existing accounts have urged state and nation building as primary causes of chemical laboratory development.[3] Yet, as I show below, changes in what chemists could do and sought to accomplish, and in how they worked and were trained, were shaping the chemical laboratory from the inside out even before 1850.

This chapter identifies equipment, infrastructure, training, and organization as the fundamental units of laboratory transformation. By 1880, Hofmann would refer to Germany's chemical laboratories as "palaces and temples," envied the world over.[4] This splendor, however, did not arise simply because governments willed it so. Nor can chemistry's institutionalization be understood as merely a change of scale.[5] Instead, organic chemistry's accomplishments and possibilities simultaneously required a new laboratory landscape and equipped chemists to attract crucial support for its creation. As chemists themselves will tell us, chemistry led the institutionalization of the sciences.

GIESSEN MODEL

Liebig is famous for instituting a new and much-emulated system of laboratory-based chemical training in Giessen. In developing the so-called "Giessen Model," however, Liebig changed much more than how chemists were trained.[6] His new analytical apparatus—the Kaliapparat—prompted Liebig to refashion organic analysis as a large-scale, collective endeavor in which students performed cutting-edge research using apparatus built from small glass components. Students learned—by doing and by watching more experienced chemists at work—how such experimental work could be turned to scientific account, supplying essential manpower for Liebig's analytical project in the process. The rise of Giessen analysis drove a novel organization of labor as well as innovations in equipment. It also, as we shall see, produced a new kind of laboratory.

Earlier chapters have shown that Liebig's project involved standardizing analytical practice—beginning inside his own laboratory. Just as Liebig sought to control organic analysis elsewhere through his roles as author and editor, so he imposed rigid disciplinary norms in Giessen. Liebig is celebrated for Giessen's structured system of analytical training and his individual interactions with students, especially those undertaking research close to his own interests. But growing student numbers during the mid-1830s meant that work in his laboratory increasingly proceeded without Liebig's immediate involvement. Laboratory organization—material as well as social—thus became essential to his quest.[7]

The importance that Liebig placed on laboratory facilities is reflected in the energy he devoted to their provision—at a time when most universities offered little to no support in this regard. When Liebig arrived in Giessen, he found that limited resources were available to support his research. After establishing a private laboratory in the guardroom of a former barracks, Liebig used student fees to buy equipment and supplies, acquiring crucial apparatus from Paris.[8]

His laboratory remained privately funded until 1833, when Liebig first sought university assistance in addressing its limitations. Working environment was a major concern. Rapidly increasing student numbers resulted in severe overcrowding, while the laboratory's combination of open coal-burning furnaces and poor ventilation meant that air quality was dreadful.[9] Even in the dead of winter, students frequently worked with doors and windows wide open.[10] These problems affected the laboratory's efficiency. They also represented a serious threat to the safety of those working in them.

Despite this situation, the university was unwilling to help. Characteristically tenacious, Liebig persisted. A small 1835 addition offered temporary relief—but Liebig's rapidly expanding school soon outgrew these extended quarters. In 1839, funding was at last granted for further building work and remodeling. This extension, designed by the architect J. Paul Hofmann (referred to hereafter as Paul Hofmann to distinguish him from his son August Wilhelm Hofmann), substantially increased the institute's floor area, adding two large rooms to its existing footprint: a lecture theater and a new laboratory for organic analysis. This latter was the centerpiece of Liebig's new laboratory (figure 8.1).[11]

A brute expansion of the laboratory's capacity was nevertheless not Paul Hofmann's primary goal. Increased working efficiency was important, and it was crucial that the laboratory space be organized to allow Liebig to retain "an easy overview of the whole." Accomplishing this depended on an "intimate collaboration" between chemist and architect. While the chemist understood the technical demands, these were best realized by

FIGURE 8.1

Wilhelm Trautschold's 1842 woodcut showed advanced students and assistants, including August Wilhelm Hofmann (far right), working in Liebig's new analytical laboratory. This was where they performed analyses using the Kaliapparat, held in this image by Mexican student Manuel Ortigosa (far left). The figure surveying the laboratory from the lecture theater next door (center, back) might plausibly be supposed to be Liebig. Ventilated glazed enclosures on the back wall helped reduce chemists' exposure to noxious fumes.

Source: J. Paul Hofmann, *Acht Tafeln zur Beschreibung des chemischen Laboratoriums zu Giessen* (Heidelberg: Winter, 1842). Courtesy of the Duveen Collection, University of Wisconsin-Madison Libraries.

the architect—especially when working within the constraints of an existing building rather than starting from scratch.[12]

The new laboratory incorporated some noteworthy features, and it significantly reorganized how interior space was used. Each activity was allocated to a separate room or distinct physical space, something that had not been possible in the laboratory's previously cramped quarters (figure 8.2). Analytical balances were housed in a dedicated weighing room separated from the main laboratory, which protected them from air currents, improving precision and reliability. Another room housed a small library, while the glassblower's torch stood in the storeroom, lit by a skylight.[13]

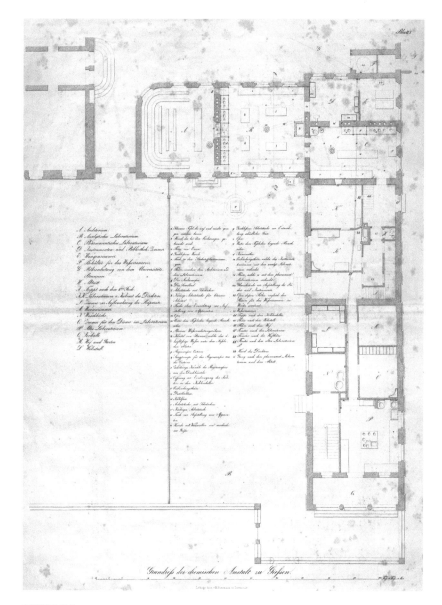

FIGURE 8.2

Ground plan of the chemical institute in Giessen designed for Liebig by the architect Paul Hofmann in 1839. Hofmann's addition to the building consisted of the larger lecture theater (A) and new analytical laboratory (B). Elsewhere in the existing building, rooms were newly designated for specific uses. The building work also improved ventilation.

Source: J. Paul Hofmann, *Acht Tafeln zur Beschreibung des chemischen Laboratoriums zu Giessen* (Heidelberg: Winter, 1842). Courtesy of the Duveen Collection, University of Wisconsin-Madison Libraries.

The new organization not only reflected changes in what was being done, but also by whom. Beginners studied the basics of analysis in the pharmaceutical laboratory, for example, while advanced students learned to use the Kaliapparat in the large analytical laboratory. More experienced chemists guided their junior colleagues—as when August Wilhelm Hofmann, then Liebig's assistant, took the newly arrived Adolphe Wurtz under his wing in 1842.[14] Thus, physical organization embodied a social order intended to foster productivity. The analytical laboratory, moreover, was subject to Liebig's panopticon-like gaze: even when teaching in the next door lecture theater, Liebig might observe through a connecting hatch (figure 8.1).

Safety was an overriding concern. The new science of organic chemistry was a smelly, hazardous business. This was especially the case where many junior chemists worked in close proximity. Paul Hofmann exercised considerable ingenuity in minimizing the dangers inherent in Liebig's collective approach to chemical training and research, at both the level of building design and through the incorporation of new fittings. Glazed, ventilated enclosures were a significant innovation, the first time such installations were provided in a teaching laboratory. "[R]equired" to minimize exposure to damaging fumes, while allowing easy access to the apparatus being used, these enclosures were fitted in the analytical laboratory but not in the pharmaceutical laboratory, which was therefore suited only to less hazardous operations.[15]

Other amenities, including building ventilation and pipes distributing Giessen spring water, improved conditions throughout. Easy access to clean water enhanced working efficiency. It also addressed concerns about safety, equipping chemists to respond more swiftly to spills, accidents and—crucially—fire.[16] Controlled fires were everywhere in Liebig's laboratory, used to heat ovens, sand baths, and distillation apparatus. Paul Hofmann's design supplied these fires with essential air, while minimizing chemists' exposure to harmful fumes.[17] Fire combined with substances that were mainly flammable and frequently toxic constituted a hazardous working environment, which the new laboratory was designed to mitigate.[18]

Liebig's Giessen laboratory was an important milestone in chemistry's institutionalization. No longer merely "a small kitchen with almost ordinary ovens," it was one of the first institutional laboratories including purpose-built spaces and custom fittings.[19] It fundamentally altered what chemists expected a laboratory to be and do, setting a new agenda for how they approached laboratory building and equipment. From this point on, chemists actively collaborated with architects in laboratory design. Paul Hofmann's book was likewise widely imitated—as both a commemorative volume and a means of documenting best practice.[20]

As the remainder of this chapter will show, however, future laboratories did not merely emulate Giessen. Paul Hofmann and Liebig had identified and responded to what would become intensifying concerns for chemists and laboratory architects everywhere: the provision of clean air; protection from fire, explosion, and noxious chemicals; effective pedagogical and research organization; and how the chemical laboratory could conform to its academic setting. These are central issues for the remainder of this chapter.

LONDON ADAPTATION

The Giessen laboratory was home in the mid-1840s to a major advance in chemistry: the introduction of "synthetical experiments" by Liebig's star pupil August Wilhelm Hofmann.[21] Like analysis, synthesis had far-reaching consequences—for training and practice, safety, and the laboratory itself. As the British architect Edward Cookworthy Robins would later remark, the "fittings are not independent of the structure in science laboratories, but in most cases are an integral part of it."[22] Chemical and architectural concerns became entwined in the provision of facilities for training and research that were suitably appointed, effectively ventilated, and well lit, promoting efficiency and productivity while reducing risk.

Thus Hofmann, while pioneering organic synthesis, also became a leading contributor to laboratory architecture, beginning with the new premises of London's Royal College of Chemistry (RCC). Twenty years in London were to teach Hofmann a great deal about the demands and dangers of doing and teaching organic synthesis, and about the laboratory's essential role in managing the new science. Indeed, the difficulties he experienced in funding adequate laboratory facilities were no doubt especially instructive—even as several German states provided increased support for institutional chemical laboratories.[23] By the 1860s, his world-leading reputation secured, Hofmann was painfully aware of the limitations of his situation when compared to peers in Germany. He was also, as we shall see, uniquely equipped by his London experience to build state-of-the-art laboratories in Bonn and Berlin.[24]

Hofmann's commitment to the laboratory was shaped by his background and Giessen training. Son of Liebig's architect, J. Paul Hofmann, he was originally intended to follow in his father's footsteps before being drawn to chemistry by an encounter with Liebig. Thus, Hofmann—perhaps more than any other chemist of his day—understood architectural as well as chemical concerns in laboratory building. But he did so from a distinctly German perspective. When he moved to London, Hofmann initially drew heavily on Liebig's expertise and German resources. One of the RCC's main purposes was to provide instruction in using Liebig's Kaliapparat. Thus, Hofmann relied on glassblowing skill, brought to London in the person of his first assistant, Hermann Bleibtreu, along with a quantity of German glassware.[25] Yet the RCC's London location and distinct institutional identity, combined with Hofmann's research in synthesis, meant that the new laboratory would be no mere copy of Giessen.

Hofmann's first-hand experience of laboratory building began in September 1845, when he arrived in London as director of the newly founded RCC. He quickly became an active member of the Building and Laboratory Committees, created to oversee plans for the new College and operation of its laboratory. Teaching began in October 1845 in temporary Mayfair premises secured for the College by one of its founders, John Lloyd Bullock. By the end of the year, the governing Council had taken a lease on a nearby plot at 16, Hanover Square.[26]

In selecting this location, the RCC's Council followed the precedent set by other scientific institutions. Both the Zoological Society and the Royal Agricultural Society occupied premises in Hanover Square, while the Royal Institution (RI)—where Bullock and his associate John Gardner had originally hoped to situate their "Davy College of Practical Chemistry"— was less than a half-mile away in Albemarle Street.[27] Indeed, some members of Council sought to model the RCC explicitly on the RI, envisaging "a kind of panopticon" that would attract additional subscribers "by means of magnificent exhibitions and entertaining and educational evening lectures."[28]

As we shall see, such ambitions were to be thwarted—by the kind of chemistry the RCC was created to promote, the limited resources it commanded, and the divergent goals of its Council, donors, and director. Largely supported by private subscriptions, the RCC's financial situation remained precarious until its 1853 merger with the state-funded School of Mines.[29] Donors among the landowning aristocracy anticipated practical agricultural benefits, while medical men sought novel therapeutic agents.[30] Hofmann's aims, as chapter 4 explained, were also about to change. Initially attracted to London by the prospect of profiting from Liebig's "amorphous quinine," that scheme's near-scandalous outcome prompted Hofmann to redirect his energies toward academic research. By the end of 1845, Hofmann once again took up the investigation of natural alkaloids by means of synthetical experiments on aniline.[31]

Practical organic chemistry was out of place in Mayfair, synthesis even more so than analysis. Residents of this refined, wealthy district had welcomed the RI's celebrated public lectures—probably unaware of what went on in the small, subterranean laboratory. Although it could not accommodate Bullock and Gardner's proposed "Davy College," the RI's success perhaps lulled the RCC's founders and supporters into a false sense of security. But as I show below, there was good sense in University College professor of chemistry Thomas Graham's view that "[t]he best locale for a Giessen school is not London, but some country town."[32]

Hofmann nevertheless displayed considerable ingenuity in responding to the challenges of funding, purpose, and location—while shaping the new institution to serve his academic ambitions. The early RCC hosted some public events, including a lecture delivered on July 3, 1846 by Gardner. As chapter 4 related, Gardner was criticized by pharmacists for promoting Liebig's amorphous quinine shortly after his colleague Bullock was granted a patent on its preparation. Amorphous quinine was one of several examples that Gardner used to celebrate pure science's ability to produce practical benefits. Hofmann subsequently preached the same message in a similar forum, a strategy intended to attract much-needed additional funding, while placating existing donors concerned that the RCC had yet to produce much of agricultural or other value.[33]

Hofmann's attention during the RCC's early years focused on practical training and research to the exclusion of almost everything else. For at least the first year, there were no formal lectures, Hofmann instead delivering impromptu instruction in the laboratory.[34] Discovering the lamentably poor chemical knowledge of many of his students, he adopted as their guide the elementary textbook recently published by George Fownes, who had recently returned to London after studying with Liebig in Giessen.[35] Textbooks were nevertheless of limited value in practical chemistry. As Hofmann explained, the "most perspicuous and minute description of an experiment offered . . . in a manual does not secure [the student] from a host of difficulties which beset him as soon as he attempts to imitate it." Such problems, however, were "readily overcome under the eye of a teacher."[36]

Hofmann dedicated himself to the laboratory, in terms of both social organization and material environment.[37] As we saw in chapter 5, he was delighted to find two already-skilled chemists, Edward Chambers Nicholson and Frederick Abel, among the college's early enrollees, training them to manage the teaching laboratory and deliver elementary instruction—much as Hofmann himself had done in Giessen.[38] Even in the temporary laboratory, he sought Liebig's advice concerning lighting and the appropriate arrangement of furniture, seeking to make best use of what he had already recognized as cramped premises.[39]

Despite Hofmann's efforts, ventilation soon emerged as a serious issue affecting both the laboratory's interior and the surrounding area. In fact, the matter became so troubling that it seems Hofmann again consulted Liebig, this time concerning plans for the RCC's permanent laboratory. As the RCC's General Meeting heard in August 1846, "the most eminent Chemists were also consulted as to the arrangements and plan of ventilation alike for securing the health of the students, and the prevention of annoyance to the vicinity."[40]

Most problems with ventilation stemmed from two causes. The first—called "sulphuretted hydrogen" (hydrogen sulfide)—mainly related to teaching, while the second—aniline—was a direct consequence of Hofmann's research. Sulphuretted hydrogen—a highly toxic gas with the characteristic smell of rotten eggs—was widely used in analysis. Aniline—the starting material for Hofmann's synthetic investigations of organic bases—was obtained from indigo by large-scale destructive distillation, a process that frequently released foul smelling substances into the atmosphere. These issues were major considerations in designing the new building.[41]

As planning proceeded during Spring 1846, regular meetings of the Building Committee reviewed alternative approaches to ventilation. The Committee was initially disposed to adopt a culvert and shaft system put forward by Dr. Reid, a Council member involved in the building project. Reid's method involved considerable initial outlay but—unlike the alternatives—would avoid "the great disadvantage of exposing the institution to the complaints of the neighbours." At its next meeting, the Building Committee nevertheless referred Warren de la Rue's alternative "plan for ventilating the laboratories by means of a steam engine" to the architect James Lockyer for evaluation.[42] De la Rue—who, besides his industrial activities and interest in photography, was a keen chemist—took an active role in the new College, learning how to analyze organic compounds in its laboratory.[43] In the absence of final plans or other evidence concerning the finished building, it is impossible to know which system ultimately prevailed.[44] But, as we shall see, ventilation continued to be problematic throughout the RCC's life.

In addition to general ventilation, the RCC's new laboratory was equipped with specific measures to remove sulphuretted hydrogen, which Hofmann considered "the most troublesome agent of laboratories." Again assisted by de la Rue, Hofmann developed a method by which "the whole of this gas is converted into innoxious sulphurous acid before it passes into the external atmosphere," and in 1848 he hoped "it would gratify the Council to learn that no complaints have been made by our neighbours on the ground of chemical nuisances."[45]

Inside the laboratory, the situation remained problematic. Despite "an additional enclosure for the evaporation of acids . . . whereby the atmosphere of the Laboratory has been improved," a growing student body continued to find the facilities wanting.[46] Just before Christmas 1849, Hofmann experimented with "an Arnott ventilator in the higher and the lower laboratory," but without success.[47] The following May, students subscribed £12 toward a ventilation system of their own design that used burning gas to create a current of air.[48] Despite an additional dedicated £20 donation received in July, the cost—as estimated by de la Rue—proved prohibitive, and the plan lapsed.[49]

Thus, the RCC's students continued to work in smelly, hazardous conditions. William Henry Perkin—who entered Hofmann's laboratory in 1853— later recalled making sulphuretted hydrogen "in a small square chamber connected with the chimney flue," and lamented the laboratory's almost complete lack of "stink closets."[50] Gersham Henry Winkles' large-scale isolation of trimethylamine from the brine of salted herrings in 1855 (which produced a boiling point conundrum that Hofmann helped Hermann Kopp resolve; see chapter 7) no doubt tested his fellow students' resilience and collegiality.[51]

Some improvement in air quality was nevertheless achieved in 1854, when Hofmann developed a gas-fired furnace to replace existing charcoal-burning furnaces for organic analysis.[52] The new furnace further accelerated the rate at which analyses could be done, each requiring "scarcely more than an hour" if performed by experienced hands. But it offered additional advantages. Gas burned more cleanly than charcoal did, and the fire was easier to control. Hofmann's innovation therefore made analysis safer, significantly reducing the number of accidents caused by cracked tubes and the like.[53]

Meanwhile, Hofmann's return to research had intensified the ventilation problem. His synthetic investigation of organic bases started from aniline, chosen because of its similarity to natural alkaloids (especially coniine and nicotine) and its tendency to form crystalline compounds. But aniline—like almost all nitrogen-containing organic compounds—was also very smelly. One of the RCC's early students James Campbell Brown compared the "disagreeable" smell of dilute aniline with the pure substance's "sweetish and by no means offensive smell." In August 1846, large-scale preparations of aniline in the basement kitchen provoked repeated complaints from a clothier on the north side of Oxford Street.[54] The Council responded by instructing Hofmann to move chemicals from the basement kitchen to the upstairs laboratory, presumably believing that the latter was better ventilated.[55]

Complaints like this perhaps contributed to Charles Blatchford Mansfield's decision to tackle a large-scale distillation of coal tar at his St John's Wood lodgings while preparing benzene for the 1855 Paris Exhibition. It was an ill-fated choice. Despite Mansfield's world-leading expertise in this operation, the still caught fire.[56] To save the building and its residents, Mansfield carried the burning still downstairs and into the street, sustaining terminal burns in the process.[57]

The RCC's confined plot also impacted its organization. As in Giessen, the teaching laboratory separated beginners studying the basics of inorganic analysis from more advanced students learning how to use the Kaliapparat. In London, however, both groups—numbering several dozen in total—worked on opposite sides of the same room. This cramped arrangement, which included very inexperienced chemists, enormously magnified the potential for harm. Accidents were more probable, and their consequences likely to be more serious, where many were working in close proximity—as the following incident confirms.

Perkin recollected a student standing near Hofmann "pour[ing] concentrated sulphuric acid into a thick glass bottle . . . which contained a small quantity of water." The bottle cracked, splashing acid into Hofmann's eye.

Fortunately, Hofmann's vision was not permanently damaged—although he "had to be kept in bed in a dark room during several weeks." Even though Hofmann continued throughout to receive progress reports from his students and direct their work, this experience was surely a vivid and very personal reminder of chemistry's inherent dangers.[58]

Hofmann's research changed what the RCC's more advanced chemists were doing. Analysis was increasingly routine—even for substances containing nitrogen. Relatively inexperienced chemists—de la Rue, for example—successfully learned to use Will and Varrentrapp's modified Kaliapparat.[59] By about 1860, most combustion analyses connected to Hofmann's research were delegated to "a sort of under-assistant" named Hadrell, who worked in the basement.[60] As Hofmann reported in 1848, "the greater variety of operations performed by the more advanced students" was a serious contributor to problems with ventilation.[61]

Those conducting original research worked either at places in the main teaching laboratory "along the wall with windows overlooking Oxford Street, very light and not too cramped," or in Hofmann's private laboratory—a spatial separation that reflected institutional hierarchy while minimizing risk. Hofmann's investigations of organic bases routinely required working with highly toxic, noxious substances, including cyanides. The products of synthesis were frequently even worse. As described in chapter 5, anilocyanic acid—a crucial substance in Hofmann's route to the ammonia type—filled his laboratory with a "most powerful odour, recalling at once the odour of aniline, of cyanogen, and hydrocyanic acid [hydrogen cyanide], provoking lachrymation in a most fearful manner, and exciting too in the throat, the suffocating sensation produced by the latter."[62] Hofmann evidently knew what it was like to breathe hydrogen cyanide.

Hofmann's assistants performed much of this difficult, dangerous work. In 1850, he acknowledged Nicholson and Abel's "unremitting zeal, and . . . remarkable experimental skill" in studying anilocyanic acid.[63] The following year, Hofmann highlighted senior assistant James Smith Brazier's help during his investigation of organic bases. Smith Brazier helped Hofmann promote the use of a glass tap-funnel in purifying synthetic products such as ethylaniline (chapter 6), performing numerous sealed tube reactions during synthetic studies of the aniline bases.[64]

New methods—notably the sealed tube reaction—further increased the hazards associated with novel, unknown compounds. Heating starting materials and reagents together inside glass was effective—as Hofmann's synthetic studies showed. But it was also dangerous, sometimes extremely so, posing a particular challenge to laboratory safety in this period. Developed by the English chemist Edward Frankland while in Robert Bunsen's Marburg laboratory during the 1840s, this important new technique was fundamental to Hofmann's synthetic method.[65] Hofmann's investigations of organic amines and phosphorus bases relied on sealed tube reactions, and—as we shall see below—they remained a staple of his chemistry in Berlin.[66]

The process involved heating reactants and solvent together inside a sealed glass tube. Elevating pressure as well as temperature often successfully promoted reaction, but it was also dangerous—especially because heating generally involved naked flames. Sealed tube reactions frequently resulted in explosions, sending glass fragments and hot chemicals flying. Even when things went to plan, it was impossible to be sure what was inside before opening the tube. Most organic compounds are flammable, and many are toxic or carcinogenic. Moreover, such risks are impossible to evaluate without considerable knowledge of the substance concerned. These features made sealed tube reactions inherently hazardous, even for experienced chemists. Frankland—an expert glassblower as well a skillful experimenter—suffered zinc poisoning while opening a sealed tube. Fortunately, despite Bunsen's fears that his student "might be already irrecoverably poisoned," Frankland recovered.[67]

As with ventilation, the available resources limited Hofmann's response to the dangers of sealed tubes. Chemists at the RCC—including Nicholson, Abel, and Smith Brazier—carried out sealed tube reactions in "air- and oil-baths" located in the basement.[68] Hofmann was certainly wise to confine this hazardous process below ground, far away from both his private laboratory and the upstairs teaching laboratory—a pattern later laboratories would follow. But, as those laboratories also show, sealed tube reactions warranted even greater protection in the form of specially designed enclosures and dedicated rooms. The RCC's cramped urban site, however, effectively prevented any alteration to the building, while Hofmann's persistent inability to secure funding for even basic ventilation measures surely made it impossible to justify more specialized expenditure.[69]

Unable to improve safety by material means, Hofmann seems to have adopted the only possible alternative: more highly skilled personnel, in the form of German-trained research assistants. Although Germans were barred from employment in the college, many came to London during the 1850s, contributing to Hofmann's private research while supporting themselves with paid work elsewhere—for example, as analysts for the Royal Mint.[70] Such experienced and committed chemists could be allowed to perform even dangerous manipulations in less than ideal circumstances—partly because they were less likely to have accidents, but also because they better understood the risks they were taking.

By the 1860s, this German "inner ring" became the engine of Hofmann's research productivity, its separation from the surrounding English institution cemented in social as well as organizational terms.[71] Hofmann's private laboratory operated much as though it were in Germany: Its researchers, primary language, social norms, methods, and equipment were almost entirely German. This situation no doubt strengthened Hofmann's professional allegiance to his homeland, pitting the technical superiority of German chemistry against the limited resources available in Britain—even as Hofmann's reputation was reaching its zenith, and chemically produced synthetic dyes such as mauve were beginning to make serious money. The main opposing force—Hofmann's personal loyalty to his long-term supporter in London, Prince Albert—was removed by the Prince's death in 1861. From this point on, it is reasonable to conjecture, Hofmann was ready to return to Germany.

CHEMICAL ATMOSPHERE

Such an opportunity was not long in coming. In the 1860s, two leading universities sought to lure Hofmann back to the wealthy German state of Prussia. The first approach came in 1861 from Bonn, the institution that Hofmann had left to come to London some 15 years earlier; the second—following the demise in 1863 of Eilhard Mitscherlich, Liebig's erstwhile analytical antagonist—from the capital, Berlin. Both came with generous guaranteed funding for new purpose-built chemical institutes, testament to Prussia's support for the rising science. Although he ultimately decided on Berlin, Hofmann would play a leading role in both projects, overseeing the construction of buildings that reflected the state of the art in laboratory design ca. 1865.

The Berlin laboratory has more immediate significance in this book, because this was where Hofmann spent the final decades of his career. The

Bonn laboratory—the first to be built—nevertheless warrants attention. "[U]nfettered by narrowness of space," its greenfield site on the edge of the city meant that Hofmann could fully implement an approach to teaching and research developed over 20 years in London.[72]

Although Hofmann's Berlin laboratory would set a new standard for grandeur, elegance, and cost—key markers of existing historiography—the Bonn institute, as we shall see, more fully reflected Hofmann's views concerning the laboratory's role in enabling teaching and research in organic chemistry (see figure 8.3 [Bonn]; figures 8.4 and 8.5 [Berlin]). Combined with experience gained at London's RCC, planning the Bonn laboratory uniquely qualified Hofmann for Berlin. By the time he was approached about Berlin, no one knew more about laboratory design than Hofmann—including how to make the most of a confined city-center plot in the Prussian capital.[73]

Hofmann took these commissions seriously from the outset. Reporting to an interested British government in 1866, while the two new institutes were under construction, he also had powerful arguments for why others should do the same. The new institutes would provide an "immediate impetus . . . to the prosecution of chemical studies" in Bonn and Berlin, their "unusually large" funding "a tribute . . . to the influence of Chemistry on the modern aspect of the world." Nor "could they remain without effect upon other departments of physical science." Hofmann proposed the chemical laboratories in Bonn and Berlin as models, not only for chemistry, but also for "the two other great branches of natural science, Physics and Physiology, to which, as well as to Chemistry, the future belongs."[74]

However, such expectations would not be realized merely by spending money. While planning the Bonn institute, Hofmann reviewed "drawings and plans of nearly every existing laboratory." He also spent several months in the fall of 1863, accompanied by the University's architect Augustus Dieckhoff, visiting laboratories in Karlsruhe, Munich, Zurich, Heidelberg, and Göttingen, as well as "the splendid institution just completed in the University of Greifswald"—also in Prussia. The result was a design occupying roughly two-thirds of the 45,000 sq. ft plot and projected to cost 123,000 Thalers (equivalent to approximately $50 million today).[75]

Dieckhoff's design reflected chemistry's rapidly increasing scale and rising status, embodying "the dignity of a great public building dedicated to science."[76] As was then usual, the director was expected to live on the premises in what Hofmann described as a "spacious" and "tastefully ornamented" residence. Other aspects of the director's residence were unprecedented—most notably, the "splendid ball-room," a space approaching 600 sq. ft with a 19 ft ceiling. This Hofmann described as "amply satisfying the social requirements of a chemical professor of the second half of the nineteenth century."[77] By far the grandest building on the University's Poppelsdorf site, the new institute's social function was evidently important.

These extraordinary elements should not distract us—as they assuredly did not distract Hofmann—from the new institute's more essential features. Hofmann's report prioritized safety and productivity, explaining how every aspect of the new building promoted teaching success and research output. Light, clean air, and access were vital throughout and were facilitated by building around four open courtyards. "Especial attention" was devoted to the mains supply of "water, steam, gas, and sulphuretted hydrogen"—and to measures for the "speedy and safe removal of all by-products," including "flues . . . for ventilation and carrying off injurious fumes."[78] Hofmann was also proud of the "handsome garden," intended to provide hard-working chemists easy access to much-needed fresh air and exercise.[79] Within, Hofmann detailed everything from the work tables, cupboards, and equipment that constituted each student's "chemical estate" to the 250-seat lecture theater at the building's center (figure 8.3).[80]

Hofmann's design reflected his pedagogical philosophy and served his research goals. Institutional effectiveness was founded on "teaching power" and "maintaining discipline"—which translated into three teaching laboratories, each housing twenty students and supervised by a dedicated assistant. The three laboratories, each "profusely lighted by ten windows" and with its own assistant overseer, were segregated according to seniority. The first was for "beginners" learning inorganic analysis, while more "advanced students" learned quantitative analysis (including how to use the Kaliapparat) in the second. "Young chemists" embarking on research—usually closely connected to the director's interests—occupied the third laboratory, nearest to the director's study.[81]

Hofmann organized the institute this way to promote what he called its "chemical atmosphere"—an eloquent shorthand for a learning environment that facilitated students' rapid acquisition of manipulative skills and research expertise. Twenty, he believed, was the largest number of students an assistant could "superintend, for any length of time and with satisfactory

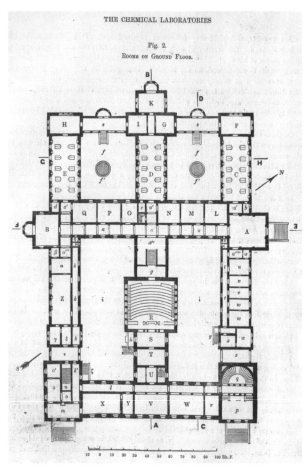

FIGURE 8.3

Elevation to the street (above) and ground floor plan (below) of the laboratory built for Hofmann in Bonn. Built on a greenfield site, this laboratory was designed with light, air, and safety in mind. The large courtyards provided good natural light to the main teaching laboratories (C, D, E) and lecture theater (I) as well as allowing for easy entry and exit to the building. The operation rooms (F, G, H) were connected by a covered colonnade (*e*) for outdoor work.

Source: August Wilhelm Hofmann, *Report on the Chemical Laboratories in Process of Building in the Universities of Bonn and Berlin* (London: Clowes and Son, 1866), figures 9 (above) and 2 (below). Courtesy of the University of Wisconsin-Madison Libraries.

results." Small enough to maintain that "personal supervision," without which "the chemical institution, however excellently it may be organized in other respects, will yield very indifferent results indeed," it nevertheless retained the advantages of communal laboratory practice. Working alongside his peers enabled the student to gain "in a comparatively short time, an amount of experience which, working alone or in company with only a few, he could scarcely gather during years of diligent labour."[82]

The main laboratories were fully equipped for "all ordinary chemical work"—but this involved the provision of new facilities, including the small evaporation niches whose primary purpose was to ensure "speedy and certain withdrawal of obnoxious gases." A "great number of chemical operations" produced "vapours detrimental to health . . . which should be removed as quickly as possible from the working rooms." This problem, moreover, "recurs very often," especially when "a great number of operators are at work." Existing chemical laboratories, as Hofmann explained, were typically provided with "large places ventilated by chimneys." Students, however, often found it "scarcely worth while" to leave their workplace to use them. But, by working at the open bench, they produced fumes in sufficient volume to overload the general ventilation system.

Hofmann's response was both pragmatic and material. Rather than seeking to enforce the use of distant facilities, his design instead encouraged use of evaporation niches by positioning them closer to students' normal workplaces. Every laboratory in Bonn included "a considerable number of smaller [niches] . . . fitted in the spaces between the windows." Ventilated by means of gas burners and closed at the front by a glass pane, these improved enclosures served as models for many later institutions—including Berlin and the laboratory built for Hermann Kolbe in Leipzig at around the same time.[83]

Despite these measures, there were "certain operations which cannot be well conducted in [the main laboratories]." Hofmann therefore incorporated additional facilities to promote safe working. Three operation rooms, adjacent to each laboratory, contained benches for "large and complicated apparatus" and additional "evaporation niches" to carry away acidic and ammoniacal fumes. The three rooms differed in size and fittings, according to each group's requirements. Advanced students learning quantitative analysis in the second laboratory needed standard equipment associated with using the Kaliapparat, and their operation room was therefore smaller.

But the young chemist "engaged in actual research m[ight] at any moment have to fit up new or reconstruct old apparatus." This required more space and a greater variety of tools, including a "blowpipe-table, which is scarcely ever at rest." Essential to research in organic chemistry, glassblowing had its own place in Hofmann's laboratory.[84]

Even these provisions were sometimes insufficient. Each operation room therefore opened onto a covered outdoor "Colonnade," equipped with gas and water and intended for procedures likely to produce "noxious vapours" and those involving the preparation of organic sulfur compounds "like mercaptan," whose "disgusting smell" (like rotting cabbages or smelly socks) might "infect the air of the operation rooms and of all the adjoining quarters."[85] More dangerous still were sealed tube reactions, which were relegated to "protected niches" in basement furnace rooms.[86] Hofmann did not specify the nature of this protection, but when Robins visited during the 1880s, he noted the Bonn laboratory's "specially arched niches, let into the walls and provided with enclosing iron doors, for the protection of the manipulator when experimenting with substances at high temperature in sealed tubes."[87]

A significant portion of the building was dedicated to the director's research, its organization broadly mirroring the range of facilities provided in the institute as a whole. For example, in addition to a private laboratory "lighted by four windows" and associated rooms for weighing and combustion analysis, there was a *"Portico,* for experiments requiring to be carried out in the open air." Connections to other parts of the institute were important—enabling both the director and his assistants to move swiftly between their various responsibilities. And a safe, speedy exit to the courtyard was essential in case of fire or explosion.[88]

The changing nature of research was evident in the library facilities as well as the building. Where students had been mainly responsible for the ad hoc creation of a library at the RCC, such a facility was now essential.[89] Even beginning researchers had to be able to consult original publications, rather than the "condensed" and "often more or less garbled" descriptions provided in manuals and treatises. The "literature of chemistry, though the youngest of the sciences," Hofmann explained, "has already attained to very considerable dimensions." As a result, "collect[ing] the works, which have to be consulted in the prosecution of even limited investigations, in most cases far exceeds the power of any single individual."[90] The turn to synthesis meant that it was increasingly difficult for researchers to work in isolation. In the era before the introduction of research aids such as Friedrich Konrad Beilstein's *Handbook* to facilitate literature searches, working in association with others was essential to mastering existing research and planning effective next steps.[91]

In the end, although he invested so much in its creation, Hofmann would never direct this institute—which instead passed into the hands of another eminent organic chemist, August Kekulé. What was built in Bonn nevertheless makes clear how organic synthesis had changed Hofmann's view of what a good laboratory should provide. Just as in London, pedagogical approach and research efficiency were key. But by the 1860s, attaining these goals required both far greater resources and vastly increased measures to protect inexperienced chemists and expert researchers alike from the hazards of learning and doing organic synthesis.

MATERIAL DEVIATION

Many similarities between the two buildings show that Bonn served as a model for Berlin. But there were also important differences. In Bonn, a generous budget and greenfield site meant that Hofmann was able to give free reign to his institutional vision. In Berlin, by contrast, he was confined by the "limited dimensions of the site, and close proximity of lofty buildings on either side." Located "in one of the best and busiest quarters of the city," Berlin's new chemical laboratory was surrounded by "many of the great institutions of Berlin, devoted to art and science." It certainly did not require a ballroom. But "the limited area available"—under 22,000 sq. ft, "somewhat smaller" than the Bonn building—meant that its practical requirements "had to be satisfied in a very different manner" from Bonn.[92]

As Berlin boomed, acquiring even this limited site was tricky. Part of the plot was provided by the Berlin Academy of Sciences, while the remainder was purchased—at a cost Hofmann found "astonishing." The Academy, which relinquished its own chemical laboratory in this process, received 24,000 Thalers as compensation, but also had to be persuaded of the merits of an institution with "purposes somewhat foreign to its immediate object." Despite funding amounting to 318,000 Thalers (equivalent to approximately $125 million)—of which site acquisition consumed roughly one third—designing and building Berlin's new chemical institute proved a demanding task.[93]

Berlin's chemical laboratory certainly took longer to complete than that in Bonn. Planning started in 1864, and construction began the following spring. Soon afterward, following a "magnificent" farewell dinner at London's Albion Tavern chaired by de la Rue, Hofmann left for Berlin. Those present, including Nicholson, Abel, and Smith Brazier, heard that the British government "declined to release" Hofmann from his appointment, in hopes that he might one day return to England.[94] It seems that the British underestimated the new facilities Hofmann was being offered in Berlin—and their significance for chemistry's future development. When Hofmann took up his new post that autumn, however, the new laboratory remained incomplete. It took 2 years to finish the building—at which point many of its interior fittings were still in planning. Hofmann's Berlin laboratory was not completed until May 1869, 4 years after its groundbreaking.[95]

The main difficulty Hofmann identified concerned the all-important provision of light and air. The plot's "confinement between two high walls" required a "material deviation" from what had been built in Bonn. Hofmann's design—realized with the University's architect Albert Cremer, who had recently built a new Anatomical School nearby—located "those rooms in which light and air are of primary importance in the story overlying the ground floor." These included the institute's "two magnificent laboratories"—each intended for the use of 24 students but capable of accommodating a greater number. As in Bonn, arranging the building around inner courtyards meant that the laboratories—on the institute's north front on Georgenstrasse—could be flanked on both sides by "a row of colossal windows" (figure 8.4).

Other provisions were likewise familiar from Bonn. Small "evaporation niches" (*Abdampfnische*) were fitted between all the window pillars in both the main laboratories and the shared operation room between them. Enclosed between the external windows and internal sashes, these prototype fume hoods were provided with excellent natural light. Elsewhere, wider "combustion niches" (*Verbrennungsnische*) of similar design were provided for organic analysis, remarkable for enabling students to "keep all their apparatus entirely in the closet when at work."[96] Stools were provided, but—since chemists routinely worked standing up—these were mainly "for standing on, in order to bring down bottles, etc. from on high."[97]

Elsewhere, Hofmann's design incorporated significant compromises. Those conducting original research worked in a smaller laboratory toward

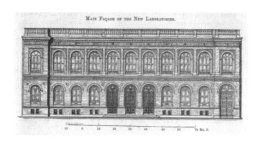

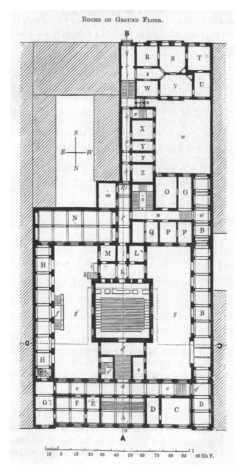

FIGURE 8.4

Elevation to street (above) and ground floor plan (below) for Hofmann's Berlin laboratory. The building set a new standard in grandeur and luxury, but its cramped city-center location imposed significant constraints on its design and interior layout. Originally intended to house 70 working chemists, Hofmann's laboratory soon proved inadequate to disciplinary demands—in terms of both its size and the kind of facilities it provided.

Source: August Wilhelm Hofmann, *Report on the Chemical Laboratories in Process of Building in the Universities of Bonn and Berlin* (London: Clowes and Son, 1866), figures 22 (above) and 17 (below). Courtesy of the University of Wisconsin-Madison Libraries.

the building's rear, distant from the main teaching laboratories. Both this and the director's private laboratory—although they were on the first floor—had windows on only one side, facing internal courtyards (figure 8.5). The far wall of the research laboratory flanked existing buildings, while the director's private laboratory adjoined an internal corridor. In addition, although Berlin University enrolled many more students than Bonn, the institute's central lecture theater was restricted to the same 250 seats—a fact that Hofmann, who was by this time skilled in delivering well-received lecture-demonstrations, evidently regretted.[98]

The design maximized limited space by making many of the smaller rooms serve dual functions as corridors.[99] While touring Hofmann's laboratory, Robins observed that, "even the balance-room is not sacred." People passing through the balance-room made accurate weighing difficult, but there were other, potentially much more serious problems, with this use of space. During Robins's visit, a fire "ran along upon the ground of the balance-room, and though speedily extinguished, proved the danger of making the subsidiary rooms passages between the main laboratories."[100]

Research in organic synthesis exposed chemists to increasingly serious risks. Greater reliance on sealed tube reactions was an important case in point. A series of ground floor furnace rooms were equipped for "experiments made under great pressure, such as digestion of substances at high temperatures in sealed tubes." Appreciation of the dangers of sealed tube reactions, meanwhile, had also grown. In the first instance, sealed tube reactions were to be carried out in "special arches . . . let into the walls, provided with strong iron doors for the protection of the manipulator in case of explosion."[101] Some sealed tube reactions were nevertheless so hazardous that they had to be performed outside. The Berlin laboratory's "magnificent Colonnade"—a space 18 feet high and 100 feet long for open-air working—therefore offered benches to which strong iron boxes for sealed tube reactions were fitted.[102]

The changes driven by synthesis were nowhere more visible than in Hofmann's private laboratory. During the summer of 1867, even as he and Cremer finalized the Berlin laboratory's interior, Hofmann—working in his private laboratory at the RCC—received a vivid reminder of the hazards of synthetical experiments. Reacting chloroform with organic bases including ethylamine produced a substance whose "penetrating odour . . . surpassed

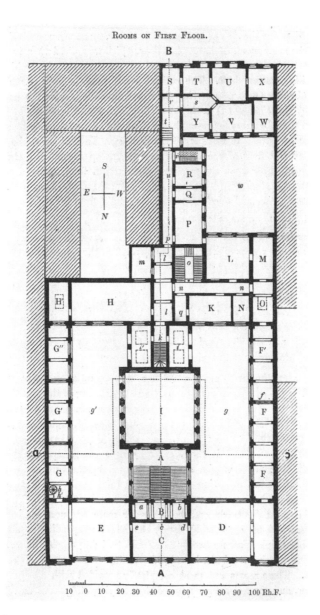

ROOMS ON FIRST FLOOR.

10 0 10 20 30 40 50 60 70 80 90 100 Rh.F.

FIGURE 8.5

Although Hofmann's Berlin laboratory reflected similar design principles to the Bonn laboratory (cf. figure 8.3), the presence of existing buildings on either side of the narrow plot available in Berlin placed severe constraints on its layout. The main teaching laboratories were D (beginners) and E (advanced students), while those conducting original research worked in H. Hofmann's private laboratory was L. Many rooms served double duty as corridors, including G', the "portion of left gallery for balances." This design made it possible to accommodate more chemists but at significant cost to their safety.

Source: August Wilhelm Hofmann, *Report on the Chemical Laboratories in Process of Building in the Universities of Bonn and Berlin* (London: Clowes and Son, 1866), figure 18. Courtesy of the University of Wisconsin-Madison Libraries.

anything . . . it [wa]s possible to conceive."[103] The product of Hofmann's reaction was not only noxious; it was also acutely toxic and highly flammable. It is no wonder that he was so concerned about ventilation and protection from hazardous fumes.

Hofmann responded to this and similar episodes by requiring the best possible protection in his new private laboratory. Most notable was a working niche (*Arbeitsniche*), similar in function but considerably larger than the niches supplied elsewhere in the building. Some 8 feet across, its worktop formed of slate and lit from three sides, the front opening was fitted with glass panes that could be raised and lowered. Within, an air current driven by a gas burner carried away fumes. This design was the state of the art in managing practical hazards (figure 8.6).

This larger working niche addressed another emerging issue for synthetic organic chemists. As their repertoire of known reactions increased, chemists sought to synthesize specific molecular targets. Initially, these attempts were based on following extended, linear reaction sequences. Many reactions, however, produced very little product, especially following purification. Rapid diminution of material from one step to the next frequently

FIGURE 8.6

This working niche offered chemists working in Hofmann's private laboratory maximum protection, even when doing hazardous work on a large scale. About 8 ft wide, it could contain experimental apparatus entirely within its well-lit and ventilated domain.

Source: Albert Cremer, *Das neue chemische Laboratorium zu Berlin* (Berlin: Ernst and Korn, 1868), Blatt 9, Figs. 1–4. Courtesy of the National Library of Medicine. NLM ID 101434380: http://resource.nlm.nih.gov/101434380.

made it imperative to begin on the kilogram scale. Chemists therefore per-formed large-scale preparations of their chosen starting materials.[104]

Starting materials were frequently obtained from industrial facilities—as when Hofmann received 20 kg of ethylamine from his former assistant Nich-olson, proprietor of the London dye firm, Simpson, Maule and Nicholson. But—as we shall see in chapter 9—such substances generally required fur-ther purification before being viable synthetic starting materials. The work-ing niche therefore joined existing provisions for large-scale working, enabling chemists to prepare and purify starting materials in safety while retaining access to specialized facilities available only in a well-equipped laboratory.[105]

Hofmann worked in this laboratory from its completion until his death in 1892, directing an exemplary institution and leading a remarkable research program in synthetic organic chemistry. Hofmann trained hun-dreds of chemists during this period. His students learned how to analyze; how to blow glass; how to react, distill, and crystallize; how to measure melting and boiling points; and how to write research papers—thousands of which appeared in the *Reports* published by the German Chemical Soci-ety, founded by Hofmann in 1867 on the model of its London prede-cessor. But this was not all they learned. As one such student, Ferdinand Tiemann, recalled in 1892, the Prussian government had asked Hofmann "to re-organize its much-neglected [system of] chemical instruction" by build-ing "new laboratories that fulfilled all the requirements of modernity. We all know," Tiemann concluded, "how he fulfilled these tasks." A much-loved and inspiring teacher, Hofmann's students also came to understand why a good laboratory was an essential tool of their science.[106]

LABORATORY LEARNING

Around the time Hofmann was leaving London for Berlin, Albert Laden-burg embarked on a journey of similar significance—one that would shape both his chemistry and his attitude toward the laboratory. Early in 1865, he left Heidelberg—where he had trained with Bunsen and Ludwig Carius—for Kekulé's laboratory, then in Ghent. This was the start of a 2-year European tour culminating in Ladenburg's return to Heidelberg in 1868, where he set up a private laboratory for research and teaching. Working in less-than-ideal circumstances taught Ladenburg a great deal about the importance of a proper laboratory.[107]

Initially trained in inorganic chemistry, Ladenburg's first foray into the organic realm took place when he joined Carius's small private laboratory and began developing the analytical method that Carius had published in 1860. Carius's procedure improved the determination of several elements, including chlorine, bromine, iodine, sulfur, selenium, phosphorus, and arsenic—elements whose importance was increasing as organic chemists' field of study expanded. It worked by complete oxidation of the sample, achieved by repeated treatment with hot nitric acid at high pressure in two sealed tubes, one inside the other (figure 8.7).[108] Ladenburg now sought to use a similar method to measure oxygen content, then routinely calculated rather than experimentally determined.

Briefly considering Carius's sealed tube method offers valuable insight into conditions in his laboratory and the kind of skills Ladenburg learned there. Glass and glassblowing were key. Both inner and outer tubes were made from Bohemian potash glass, which withstood high pressure and oxidative conditions. Sealing the small, inner thin-walled tube relied on "practice and the use of a sharp flame" to avoid overheating the sample inside. The sample tube was then introduced into the outer tube, which was half-filled with nitric acid. The analyst expelled the remaining air by boiling the nitric acid before rapidly sealing the tube. This crucial step exposed the analyst's skill in preparing the apparatus. If the proportion of air was greater than half, it was difficult to expel completely; if it was much less, then the tube was likely to explode.

The analysis was initiated by shaking the outer tube until the ends of the small inner tube broke off, exposing the sample to nitric acid. Finally, the sealed tube was heated in an air-bath. Carius recommended pointing the air-bath's open end toward a corner of the room. "Should a well-made tube explode, which seldom happens," he assured readers, "it would be hurled into the corner as powdered glass with a big bang, and even the most violent explosion is [sic] completely safe."[109]

The nature of the substances being treated this way indicates that the conditions in Carius's laboratory were hazardous in the extreme. Organic compounds containing halogens, sulfur, phosphorus, and arsenic are almost invariably toxic and flammable, and they are certainly unpleasant. Carius's laboratory occupied cramped, private premises, and therefore lacked all but the most basic facilities. Ladenburg's decision to try organic chemistry immediately plunged him into a highly dangerous environment.[110]

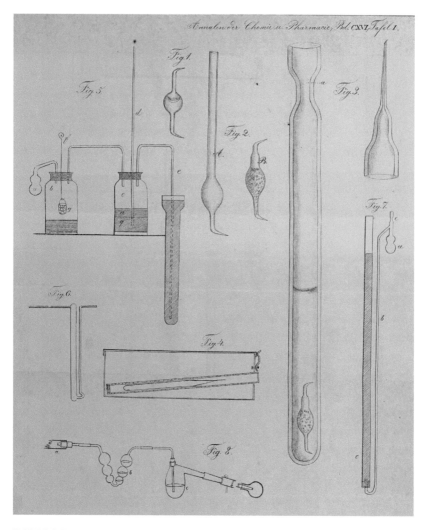

FIGURE 8.7

Carius's sealed tube method of analysis involved sealing the sample inside a small, thin-walled glass tube (labeled "Fig. 1" and "Fig. 2" in the illustration), which was then introduced into a larger, thick-walled tube, of 10–12 mm inner diameter, half-filled with nitric acid (labeled "Fig. 3"). Once the outer tube was sealed, the ends of the inner tube were broken by shaking, so that nitric acid reacted with the sample inside. The tube was then heated in an air bath (labeled "Fig. 4").

Source: Ludwig Carius, "Ueber die Elementaranalyse organischer Verbindungen," *Annalen der Chemie und Pharmacie* 116 (1860): Taf. 1. Courtesy of Hathi Trust: https://hdl.handle .net/2027/uc1.c036497440?urlappend=%3Bseq=403%3Bownerid=115420891-0.

Despite the dangers, Ladenburg was won over to organic chemistry. It also seems likely that his decision met with family support. In April 1865, Ladenburg's father Leopold and uncle Seligmann Ladenburg—both senior figures in the family's Mannheim private bank—became significant shareholders in an exciting new venture. Founded by gasworks-owner-turned-dye-manufacturer Friedrich Engelhorn, the Badische Anilin- und Soda-Fabrik (BASF) would capitalize on the possibilities of profit from coal tar—furthering the synergistic transformation of chemistry and industry launched 20 years earlier in Giessen. The Ladenburgs were so confident of success that they acquired land across the Rhine in Ludwigshafen even before the contracts were signed. Engelhorn had already employed two of Liebig's former students, brothers Carl and August Clemm—whose chemical training was about to catapult them to successful careers in business.[111] That same spring, Ladenburg—presumably funded by his parents—left Heidelberg for Ghent, to work in Kekulé's laboratory.

Although he stayed only 6 months, Ladenburg's encounter with Kekulé would be highly significant for his career.[112] This was where Ladenburg began studying benzene derivatives, which he initially conceptualized in terms of Kekulé's recently published theory of aromatic structure.[113] Further investigation, however, would lead him first to criticize and then to reject Kekulé's hexagonal ring structure for benzene, to which he became vehemently opposed.[114] Commitment to the primacy of experiment over theory left Ladenburg unable to accept Kekulé's proposal, which he regarded as incompatible with the available evidence. As chapter 9 explains, this commitment would also shape Ladenburg's approach to organic synthesis.

In terms of the laboratory, however, Ladenburg's most formative experiences still lay in the future. From Ghent, he traveled via London—where he met Edward Frankland—to Paris. Initially intending to work with Berthelot, Ladenburg rapidly changed his mind when offered an empty room as his laboratory. Instead, he sought a place in Wurtz's laboratory, where—despite encountering what he later described as a surprising ignorance of Kekulé's recent work—he continued experimenting on benzene derivatives.[115]

Wurtz's laboratory in the Faculty of Medicine was certainly an improvement on Carius's. Following Liebig's example, Wurtz—who was determined to "create a modern laboratory sufficient to the demands of the science"—used income from teaching to supplement the limited resources provided by the faculty. Occupying three rooms carved out of the anatomical lecture theater, his laboratory received good natural light. By the 1860s, it was fitted with work benches and supplied with gas. A couple of stonework tables for more dangerous work stood in the outside courtyard. One of these was provided with a "good hood and chimney." This was where Ladenburg joined some 15 others, including the young Charles Friedel.[116]

Ladenburg's first experience of a reasonably well-equipped research laboratory was soon complemented by a near-disastrous exposure to the consequences of inadequate laboratory facilities. After a brief interlude back home in Germany, Ladenburg returned to Paris in the winter of 1866, this time joining Friedel's research project on organic silicon compounds in the latter's laboratory at the School of Mines—then housed in "the fine palace in the Luxembourg Gardens." According to James Mason Crafts—already Friedel's collaborator, a decade before publication of their eponymous reaction in 1877—Friedel "never had a well-equipped laboratory provided with labor-saving contrivances." Organic chemistry was done in a "small room," while hazardous attempts to produce new minerals using high temperature and pressure in a platinum-lined heavy steel tube nick-named "Jacob" took place in "a dark, vaulted cellar."[117]

Soon after entering Friedel's laboratory, Ladenburg was seriously hurt in an explosion—probably caused by the ignition of silicochloroform (trichlorosilane) with air. This "temporary disablement" did not dampen Ladenburg's enthusiasm for organic chemistry. By the following summer, he was back in the laboratory, working hard to make up for lost time. But this experience no doubt heightened his awareness of organic chemistry's dangers.[118] Returning to Heidelberg, where he became a Privatdozent in 1868, Ladenburg knew from painful personal experience just how important adequate laboratory facilities were.[119]

FIT FOR PURPOSE

Understanding the importance of a good laboratory did not mean that it was straightforward to acquire one. As we have seen, developments in organic synthesis increasingly tested the limits of private laboratories, driving a growing requirement for purpose-built laboratories. Adequate facilities were fundamental obstacles facing junior chemists, and they were paramount considerations in employment negotiations—illustrated here by

Ladenburg's experiences as a Privatdozent in Heidelberg and, later, as a professor in Kiel.

Ladenburg's story reveals much about what it took to make a late-nineteenth century chemical laboratory "fit for purpose."[120] When he moved to Kiel in 1872, the university built him a new institute. But Ladenburg did not command resources of the kind made available to Hofmann. His new laboratory—like many others elsewhere—was therefore smaller and far less richly equipped than those built for Hofmann in Bonn and Berlin. Completed in 1878, it nevertheless proved sufficient for synthesis. This laboratory was where Ladenburg synthesized coniine in 1886.

Ladenburg's laboratory struggles began almost as soon as he became a Privatdozent. In Heidelberg at that time, where Bunsen was professor, junior chemists worked and taught in private laboratories they set up, and paid for, themselves. Thus, Ladenburg spent the summer of 1868 finding and equipping rented premises, and hiring a servant—all while completing the manuscript of his *Lectures on the History of Chemistry,* delivered for the first time that autumn.[121] Clearly unhappy at this state of affairs, Ladenburg—together with his young colleagues—petitioned the government for better resources, including a fit-for-purpose laboratory. The answer, when it eventually came, was roundly negative. Indeed, Ladenburg later held this attempt at activism responsible for the fact that it took him 3 years to gain the right to use the title of "professor."

Things went from bad to worse when Ladenburg's landlord terminated the lease on his laboratory. Ladenburg—with financial assistance from his father—responded by buying a new property. But when he attempted to remodel and refit these new premises, Ladenburg confronted local residents' resistance to having a chemical laboratory in their neighborhood. His neighbors took him to court—where only a good lawyer, combined with his father's backing and excellent connections, saved him. Thus, it was not until around the end of 1870 that Ladenburg finally acquired an adequate laboratory.[122]

In the spring of 1872, Ladenburg was invited to become professor of chemistry in Kiel, on Prussia's Baltic coast. Initial astonishment soon gave way to reluctance prompted by concerns about laboratory facilities. After considerable struggle, Ladenburg had finally acquired a good laboratory, and he was understandably eager to remain in Heidelberg. But when he approached the university, hoping to negotiate his position, no improvement was forthcoming. Forced to contemplate the move, Ladenburg visited Kiel.

His visit was far from encouraging. The city made a "terribly poor" impression; the university and its institutes were no better. Chemistry in Kiel at that time was merely a "pharmaceutical branch" in the medical faculty.[123] The university's chemical laboratory, which dated from 1802, was extremely limited. Indeed, this laboratory—situated in a rented house near the castle—was destined to remain in the possession of the present incumbent, Carl Himly. Moving to Kiel would entail starting from scratch where the laboratory was concerned.[124]

For Ladenburg, this was a considerable obstacle: As we have seen, he was well acquainted with the dangers of provisional laboratories. Even the promise of a new institute—part of Prussia's plan to revive Kiel as a university town—was initially insufficient inducement. The plan was that the new laboratory would be ready within 3 years—which turned out to be a woeful underestimate. In the interim, Ladenburg would have to make do as best he could. It was not an attractive prospect.

Ladenburg's family again intervened, ostensibly on the advice of senior Prussian officials among their acquaintance. According to Ladenburg, they were adamant that such a prestigious appointment was not to be refused.[125] It nevertheless seems probable that family investments in the chemical industry were involved. By this time, BASF's Ludwigshafen works were generating huge profits from aniline dyes—mainly manufactured under license from British and French companies. BASF had also recently hired Heinrich Caro, an experienced dye chemist whose association with Adolf Baeyer was to become a model of academic-industrial research collaboration.[126] Aware of chemistry's commercial possibilities, Ladenburg's family surely guided his career at this crucial juncture.

At last, Ladenburg accepted—but he delayed moving to Kiel by 6 months to allow for the arrangement of a temporary laboratory in rented rooms.[127] His caution was wise. In fact, it would be almost 6 years before the promised purpose-built institute was ready for occupation. Meanwhile, Ladenburg set up an interim laboratory and began teaching. Initially disappointed by the small number of students, he soon realized that this had advantages. As in Heidelberg, fewer students meant more time for research. Ladenburg

continued investigating aromatic chemistry, becoming increasingly interested in aromatic compounds containing nitrogen—a preliminary step
toward the natural alkaloids, whose synthesis would become the focus of
his later research.[128] He also discussed plans for the new institute with the
architects, Martin Philipp Gropius and Heino Schmieden.[129]

Gropius and Schmieden designed five new institutes in Kiel, for physiology, anatomy, zoology, and the university library as well as for chemistry. At 228,000 Thalers (equivalent to ca. $75 million) Ladenburg's was
the most expensive and spacious. Occupying a prime site on Brunswick
Street, with an adjoining garden, it was also the first to be completed (see
figure 8.8). The scale and form of the building reflected local constraints—
geographical as well as financial. Sloping terrain forced a compressed, linear
layout, very different from those in Bonn and Berlin, while the institute
was much smaller overall, intended for just 16 students (figure 8.9).[130]

Despite these differences, Ladenburg's institute reflected organizational
ideals similar to those that Hofmann had realized in Bonn and Berlin.[131]
Managing a chemical institute remained more a lifestyle than a job, with
most directors continuing to live on the premises—even as laboratory
hazards increased. Thus, the new building housed not only Ladenburg's
private laboratory but also his living quarters. As in Bonn and Berlin, three
distinct laboratory spaces separated students at different stages of training.
Several of the institute's rooms were assigned to specific activities, including
a preparation room (*Vorbereitungs-Raum*), dedicated balance room (*Waage-
Zim[mer]*), and storerooms for supplies and equipment (*Vorräte, Präparate,
Apparate*). The room for organic analysis (*Organ[ische] Analyse*) was sandwiched between the director's laboratory (*Laboratorium des Directors*) and
one of the smaller research laboratories (*[Saal für] Geübtere Praktikanten*).

Safety considerations were crucial to the building's design, their priority
enhanced by the constraints on funding. Ladenburg himself oversaw the
installation of the institute's evaporation niches—but even his oversight
was not a guarantee of success. Having initially declared the overall building ventilation "not perfect but, given the means at hand, fairly successful," Ladenburg soon reported "the deficiency in these arrangements" to
the architects and the ministry.[132] The response to Ladenburg's repeated
complaints remains unclear, but the new institute certainly received additional funding during the early 1880s and was subsequently significantly
altered and extended.[133]

FIGURE 8.8

The exterior of Ladenburg's Kiel Chemical Institute embodied Gropius and Schmieden's style, especially in the pillars and arches surrounding the main door and ground floor windows. Steeply sloping land to the rear meant that the building was not modeled on the open courtyards found at Bonn and Berlin, instead having a much more compressed, linear layout.

Source: Kurt Feyerabend, *Die Universität Kiel: Ihre Anstalten, Institute und Kliniken* (Dusseldorf: Lindner, 1929), 43 (Image 32). Author collection.

Chemistry's more dangerous activities were excluded from the main laboratory spaces. Sometimes this was a matter of physical separation. A room devoted to the use of hydrogen sulfide (*Schwefelwasserstoff*), for example, was located in a corner of the ground floor, adjoining the lobby (*Offene Halle*) and main laboratory (*Saal für 24 Praktikanten*), while the room for "larger heating operations" (later designated *Physikal[ische] Zim[mer]*) was sandwiched between the balance room (*Waage Zim[mer]*) and storeroom (*Vorräte*)—neither of which were in constant use.

In other respects, however, the building itself fulfilled functions previously served by portable equipment. Ladenburg, as we know, favored Carius's method of analysis using sealed tubes. Although this may have

Chemisches Institut der Universität zu Kiel.

FIGURE 8.9

By the late 1880s, Ladenburg had designated one of the smaller laboratories ("Destillation") for distillation. This was one of the primary methods of purification used in his coniine synthesis, performed in specially designed glass distillation flasks (see chapter 9). Note that this later image indicates some changes in use from the plans as first published in "Preussische Staatsbauten, welche im Jahre 1877 in der Ausführung begriffen gewesen sind," *Atlas zur Zeitschrift für Bauwesen* 28 (1878): 52b.

Source: Josef W. Durm et al., eds. *Handbuch der Architektur,* 4. Teil, 6. Halbbd, Heft 2a (Darmstadt: Bergsträsser, 1905 [1889]), 338. Author collection.

been done in the laboratory for organic analysis (*Organische Analyse*), it is also possible that a specially isolated, ground floor room (later designated *Spectral Analyse*) was initially used for this purpose. Ladenburg's pioneering coniine synthesis would also, as chapter 9 explains, involve multiple sealed tube reactions. Before moving into the new institute, Ladenburg relied on an air bath fitted with a pressure regulator for performing sealed tube reactions. But he afterward reported that the "suitably improved state" of the new laboratory had rendered this piece of equipment superfluous, and that it had been sold to fund more useful apparatus.[134]

Reliance on sealed tubes and other glassware meant that glass and glassblowing were essential resources in Ladenburg's laboratory. At least two glassblowing tables fitted with torches were acquired in the early 1880s to serve this need.[135] In all probability, the same torches were used to sealed the hundreds of glass tubes that Ladenburg used in synthesizing coniine. No doubt they were also used by C. F. Roth to fashion the prototype melting point apparatus discussed in chapter 7 and that was transformed into a standard commodity item by Franz Müller of Bonn.[136]

Trained chemists were simultaneously a crucial resource and—as Germany's chemical industry flourished—one of the laboratory's most valuable outputs. Ladenburg's private laboratory accommodated some six chemists at any time, including at least two assistants, mainly working on commercially relevant chemistry. Several were stable presences over multiple years, acknowledged in publications as collaborators and sometimes coauthors— exposure no doubt intended to secure their future success.[137] At least three junior chemists helped synthesize coniine, including Wilhelm Friedrich Laun and Carl Stöhr as well as Roth. On leaving Ladenburg's laboratory, Roth set up as an apothecary, while Laun apparently joined Merck in Darmstadt. Stöhr's career spanned academy, industry, and the state—including a stint as director of Kiel's torpedo laboratory.[138]

Although significantly smaller and constructed on a much more limited budget than Hofmann's in Berlin, the Kiel institute nevertheless supplied the necessary resources for cutting-edge research. It is no surprise, given his family's connections to BASF, that Ladenburg's research was directed toward transforming coal tar into commercially useful products, including medicinally active natural alkaloids. As chapter 9 explains, this focus was a crucial factor in Ladenburg's successful coniine synthesis.

DISCIPLINE AND INSTITUTION

Developments in chemical practice and pedagogy explain why and how chemistry's laboratory landscape changed during the second half of the nineteenth century. The rise of analysis and the turn to synthesis shaped the laboratory's interior fittings and layout, and they dictated its increasing size. Growing numbers of chemists trained and worked in academic chemical laboratories built across Germany between about 1860 and 1890. Overwhelmingly, these chemists were doing organic synthesis.[139] In addition to its scientific value, much of their work was increasingly relevant to the manufacture of dyes and drugs.[140] In just over half a century, synthetic organic chemistry had become a powerhouse of German science, industry, and medicine.

Synthetic organic chemistry's economic significance was crucial to its institutionalization, bolstering its claim to increased funding and better facilities. But the nature of those facilities was driven by changes in the science. Organic synthesis, even more than analysis, placed new demands on the laboratory environment—especially in the academic context. Chemical laboratories were made larger to provide and accommodate the increasing number of trained practitioners needed to serve this new science. Innovations in practice, and especially the increasingly collaborative nature of research in organic synthesis, drove a transformation of the laboratory intended to foster the productivity and preserve the health of those working inside.

Organic synthesis was necessarily a hazardous venture—as chemists like Hofmann and Ladenburg well knew. Previously confined to the workshops of pharmacists, metallurgists, and other artisans, practical chemistry had struggled for legitimacy in universities. From the mid-century on, synthetic organic chemistry's dangers increasingly threatened chemistry's status as an acceptable academic discipline. As we have seen, chemists responded by deploying all the resources they could command to minimizing these risks.

This effort, however, entailed considerable circumspection. Tales of accidents, explosions, injuries, and fatalities were unlikely to attract state support and were unsuitable for friends and relations. Thus, chemistry's dangers were largely tacit and understated. Chemists learned safe practice through personal experience, discovering the risks of their calling in the process. Many nineteenth-century chemists suffered either from chronic exposure to toxic and carcinogenic substances, or from the immediate consequences of poisoning and explosion. Such afflictions, however, were recounted only occasionally, usually in memoirs and anecdotes intended for other practitioners. Seldom made explicit in either practical manuals or laboratory descriptions, the dangers of synthesis were nevertheless implicit in language and material culture. There can be little doubt, for example, of the hazards indicated by "heating sealed tubes (bombs) . . . in the so-called 'bomb furnace'."[141]

Discipline was crucial to laboratory safety and productivity. Experienced chemists instilled discipline in their juniors through systematic group training and individual oversight. But becoming a chemist entailed more than learning standard procedures and the proper use of instruments and glassware. Chemical training also introduced students to the rudiments of laboratory reasoning, helping them acquire an attitude toward experiment and reason intended to equip them for independent research. Thus, institutional chemical laboratories transcended their established role as places of experiment, becoming instead places where experiment and theory were brought together and made productive. Highlighting laboratory reasoning—and where and how this was learned—integrates aspects of chemistry that have hitherto been viewed as associated with textbooks, or classrooms, or lecture theaters in isolation. This discrete model of scientific training—which originated in historical studies of physics—therefore emerges as being of limited relevance to chemistry.

The chemical laboratory was a highly ordered social space that separated novices from initiates, while also facilitating the interactions between student and teacher, assistant and professor that were found most conducive to effective research. It marked all who worked within apart from wider society, whether by lingering smells, dye-stained hands, favored tavern, or badge of belonging—for example, the Kaliapparat-shaped brooch worn by Liebig's students in Giessen.[142] Even as chemists earned academic qualifications, established professional societies, and founded new industries, the ties that bound them were frequently more visceral.

Discipline and training implicated the laboratory's layout and construction as well as its interior fittings and equipment. A necessary part of chemistry's material culture, institutional laboratories helped manage risks so severe that they could no longer remain the responsibility of any individual. The expansion of teaching was not the only—or even the main—reason chemistry required purpose-built laboratories beginning in the mid-century.[143] As we have seen, the new methods of synthesis were powerful drivers of this change.

Elucidating the relationships between research, training, and laboratory infrastructure begins to explain why Hofmann, Ladenburg, and their peers devoted so much effort to building and equipping laboratories. It also reinforces the importance of distinguishing chemistry's growing scale, changing nature, and diversifying goals. Larger laboratories did not merely bolster the chemist's ego or affirm state identity. The laboratories built for Hofmann and Ladenburg certainly embodied chemistry's rising cultural status and economic value. But while Berlin's grandeur proclaimed chemistry's scientific credentials, in Kiel more pragmatic motives produced serviceable functionality that enabled successful synthesis.

Recognizing the institutional laboratory as the outcome of changes in chemistry explains why chemistry does not fit the revolutionary model provided by physics—a field far from its modern disciplinary form in the 1840s, when Liebig's laboratory was in its heyday.[144] By about 1870, organic synthesis made the institutional laboratory an essential scientific resource, its achievements helping secure the necessary funding. What had been accomplished with such effort in chemistry was then—as Hofmann predicted—transferred with relative ease to physics and physiology, making it seem inevitable that funds should be made available for building scientific institutes at this time. The concerted programs of institute building undertaken by a newly unified Germany, most notably in Strasbourg, thus appeared to be the product of political circumstances alone, obscuring chemistry's role in establishing the institutional laboratory as an essential disciplinary marker of academic science.

Chemists worked hard to establish and sustain the academic credentials of organic synthesis. Laboratories were fundamental to this effort, manifestations in brick and mortar of an integrated system of training and research that produced new knowledge and novel substances, as well as trained chemists—all against considerable odds. Fire, explosion, toxic fumes, broken glass, injury, chronic illness, and death itself were ever-present threats. Chemists responded by transforming the laboratory into a new material resource that embodied social and technical aspects of their discipline. Within its walls, chemists at every stage were encouraged to emulate the professor's laboratory reasoning, an approach to experiment and theory that was embedded in the pedagogical system. Chapter 9 shows what synthetic organic chemists could do, having mastered the manipulative techniques and reasoning skills such laboratories were built to enable.

9 THE SCIENCE OF SYNTHESIS

This chapter uses alkaloid chemistry—a field opened up by August Hofmann's development of the ammonia type around 1850—to explain how chemists established the science of synthesis.[1] Transformed by three decades of relentless work, by the 1880s, synthetic organic chemistry increasingly dominated the disciplinary landscape of the sciences. Of unparalleled significance to commerce and industry, it had created a rapidly expanding host of powerful methods and tools, reshaping the chemical laboratory, inside and out. Underpinned by chemists' laboratory reasoning, synthesis enhanced the connection between wet chemistry and abstract concepts. It produced new understandings of molecular nature and of what constituted an individual, pure substance. It was also bringing chemists closer to their long-held goal of making compounds identical to those found in nature.

The alkaloids remained attractive synthetic targets. From the outset, Hofmann had been committed to the possibility of making naturally occurring compounds such as quinine—an ambition shared by many chemists, including his mentor Justus Liebig. In 1867, Hofmann used a prize offered by the Prussian Academy of Sciences to promote the synthesis of medicinally important alkaloids—but without success.[2] Despite his unsurpassed command of alkaloid chemistry, it seems that Hofmann at this stage underestimated the difficulties of alkaloid synthesis. Indeed, he would never make any natural alkaloid in the laboratory—even though, as this chapter explains, he certainly tried.

Understanding why Hofmann—the pioneer of synthesis—could transform chemical understanding of natural alkaloids yet did not synthesize a single one of them takes us to the heart of what distinguished "synthetical experiments" from target synthesis.[3] It also focuses attention on coniine, the natural alkaloid that provides this chapter's illustrative and exemplary core.[4] Derived from the hemlock plant and considerably simpler than quinine, coniine was Hofmann's chosen synthetic target. It would, however, be the much less eminent chemist Albert Ladenburg who in 1886 claimed the "total synthesis of coniine," and with it the first synthesis of a natural alkaloid.[5]

Ladenburg's coniine synthesis was an important landmark in the history of chemistry, partly because it demonstrated chemistry's power to mimic nature, expressed in Ladenburg's use of the term *total synthesis*. But its primary significance—to chemistry and to this study—was to help establish the requirements, purpose, and productive possibilities of target synthesis. Target synthesis meant much more than producing nature-identical substances. In achieving this goal, Ladenburg and others were redefining molecular nature, building a new relationship between chemists and their object of study. Explaining how they did so is my purpose in this chapter.

ALKALOID CHEMISTRY, 30 YEARS ON

Before picking up the story around 1880, when coniine became the focus of Hofmann's ongoing research into natural alkaloids and the primary target of Ladenburg's synthetic ambition, let us pause to consider organic chemistry's development over the preceding 30 years. Chapters 7 and 8 showed how chemists used glass and glassblowing to gain improved control over substance and transformation, and how glassware-based techniques of separation, characterization, and reaction exposed them to new hazards, driving a transformation in the laboratory.

This period was one of rapid and pervasive change, in theory as well as experiment and material culture. In the field of alkaloid chemistry, Hofmann's 1850 introduction of the ammonia type refined the taxonomy of nitrogen-containing organic compounds. Hofmann produced the ammonia type by reasoning from experiments using alkyl halides—new reagents that acted on organic bases, usually at high temperatures and pressures in sealed glass tubes. The alkyl halides functioned as reliable classificatory tools and—as Hofmann's work with coniine illustrates—they proved to be discriminating methods of investigating the constitution of organic bases.

The combination of experiment and theory in Hofmann's laboratory reasoning was intimate and suggestive. The ammonia type explained many organic bases—both natural alkaloids and artificial laboratory products—as formed from ammonia by substitution. This mode of formation indicated that organic bases might be made in the laboratory by joining a suitable hydrocarbon with ammonia. As we saw in chapters 5 and 6, Hofmann's synthetical experiments using alkyl halides indeed produced many artificial bases, intensifying Hofmann's belief that natural alkaloids might also be made in this way.[6] This chapter explains why Hofmann's unsuccessful attempts to make coniine in the 1880s helped persuade him to abandon this view.

Hofmann's ammonia type was a significant advance in constitutional theory, with impact far beyond alkaloid chemistry. A key component of Charles Gerhardt's type theory, it provided inspiration for the English chemist Edward Frankland's notion of fixed atomic "combining power"—the starting point for emerging theories of valence and structure.[7] Learning that each element combined with others according to strict numerical rules—the foundation of valence theory—further constrained analytical outcomes. In Liebig's time, it mattered only that percentage composition determined by analysis translated into whole numbers of atoms (or equivalents) in the molecular formula. By the 1850s, the results of analysis also had to imply relationships between those numbers that were consistent with the rules of valence.

These regularities were an important impetus for the mid-1850s development of structural theory. As chemists' ability to distinguish and identify substances improved, they encountered ever more cases where different substances shared the same molecular formula. Structural theory explained such cases by formulating new kinds of isomerism based on relational concepts of connectivity and position. Thus, chemical structure summarized chemists' increasingly sophisticated taxonomy of substance, incorporating an expanding multiplicity of isomeric possibilities.

As this chapter illustrates, chemists began to use visual molecular representations as shorthand for the structural knowledge that they struggled to convey succinctly in words. Systematic names for even moderately complex structures were cumbersome and easy to confuse.[8] The shift from molecular and rational formulae to graphical formulae (similar to modern displayed or structural formulae) also initiated a change in how chemists reasoned. While the former expressed understanding of composition and sometimes constitution, the latter communicated knowledge of structure and reactivity in ways that made chemistry newly visual.

This shift makes it all the more important to appreciate that neither structural theory nor graphical formulae originally implied the distribution of atoms in three-dimensional space, instead facilitating chemists' work with relative arrangements of atoms (or equivalents) that were compatible with experiment and the rules of valence and structure. The next section explains how Hugo Schiff, believing he had made coniine, in 1870 inferred coniine's constitution from its method of preparation and illustrated this using a structural formula (see figure 9.1). But as Schiff's work makes clear, such formulae were interpretational and heuristic rather than being representational devices.[9]

Structural theory also did not explain optical isomerism, a phenomenon whereby structurally identical substances rotate plane-polarized light in opposite directions. In 1874, Joseph Achille le Bel and Jacobus Henricus van't Hoff independently introduced a theory of atoms in space that correlated optical rotation with molecular geometry. Most organic chemists, however, regarded this theory as highly speculative during the 1880s.[10] They knew many naturally derived organic substances rotated light to the right (dextrorotatory) or the left (laevorotatory), whereas laboratory products usually lacked this activity. Thus, chemists learned that separating right- and left-rotating isomers from optically inactive laboratory products was vital for making nature-identical compounds.[11] However, they did not yet interpret these isomers in three-dimensional, spatial terms.

Identifying structural isomers proved particularly fruitful when applied to what Hofmann in 1855 termed "aromatic compounds."[12] Originally identified by their characteristic odor, the simplest members of this group were benzene and pyridine. Benzene contained only carbon and hydrogen, but pyridine also contained nitrogen—thereby connecting aromatic chemistry to the chemistry of organic bases and alkaloids. Indeed, Hofmann's original model alkaloid, aniline—chosen because of its similarity to natural alkaloids, including coniine—was also aromatic, derived from benzene. Coniine was not aromatic. But as we shall see, Hofmann's demonstration during the 1880s of its relationship to pyridine would prove crucial in elucidating coniine's composition, constitution, and structure.

Analysis revealed another indicative property of aromatic compounds: They contain far less hydrogen than fully saturated organic compounds.

Benzene, for example, contains just six hydrogens (C_6H_6) compared with 14 in the related saturated hydrocarbon, hexane (C_6H_{14}). Low hydrogen content was frequently explicable in terms of unsaturation, in which valencies of carbon usually satisfied by hydrogen were instead fulfilled by other carbons. Aromatic compounds, however, contained so little hydrogen that it was unclear how their composition might be reconciled with carbon's established valency of 4. Nor did they react like unsaturated compounds, making it difficult to rationalize aromatic chemistry in structural terms.

This was the conundrum that August Kekulé sought to resolve by his 1865 proposal that benzene's six carbons were joined in a closed ring.[13] In structural terms, the ring closure explained how just six hydrogens (rather than the 14 in a saturated six-carbon hydrocarbon, or the eight expected in a fully unsaturated six-carbon hydrocarbon) could satisfy carbon's valency of 4. As later work would show, molecules with carbon rings contained fewer hydrogens than their saturated counterparts did—regardless of whether they were aromatic. For example, the saturated closed ring hydrocarbon cyclohexane (C_6H_{12})—produced by reducing benzene (C_6H_6)—contained two fewer hydrogens than its open-chain equivalent, hexane (C_6H_{14}).[14]

In hindsight, Kekulé's benzene ring is widely viewed as an immediate breakthrough in understanding aromatic compounds. In fact, aromatic structural theory in the 1860s, 1870s, and 1880s was far from providing a satisfactory explanation of aromatic chemistry, and its extension from benzene to pyridine around 1880 proved highly controversial. As we shall see, Hofmann's synthetical experiments helped establish the ring's plausibility, while synthesizing coniine persuaded Ladenburg of its explanatory usefulness.

Many alkaloids continued to defy constitutional and structural interpretation during the mid-century. At the same time, medical reliance on natural alkaloids was growing. Even coniine—famous as Socrates's poison of choice—found uses in treating a range of diseases, from bronchitis and whooping cough to rheumatism, syphilis, and neuralgia.[15] The alkaloids that Hofmann selected in 1867 as synthetic targets for the Prussian Academy's prize included quinine, cinchonine, strychnine, brucine, and morphine—essential drugs that were widely used to reduce pain, fever, and inflammation, and in managing heart disease and other ailments.[16]

This combination of medical potency, commercial potential, and scientific challenge continued to draw chemists into the burgeoning field of alkaloid chemistry. There was certainly plenty for them to do. Many new alkaloids were identified during the second half of the nineteenth century—from artificial as well as natural sources. To give just one particularly relevant example, piperidine—a close relative of coniine—was first produced in 1853 by Auguste Cahours, a protégé of Michel Chevreul and Jean Baptiste Dumas, while studying the black pepper alkaloid piperine.[17]

By about 1880, both Hofmann and Ladenburg were studying piperidine and coniine. Each chemist would make fundamental yet distinct contributions, reflecting their different goals and approaches. The remainder of this chapter explains what drew Hofmann and Ladenburg to coniine. It begins by highlighting the crucial new methods and the novel criteria of substance, purity, and identity that they would later deploy in its investigation. Together, as we shall see, Hofmann's synthetical experiments and Ladenburg's target synthesis constituted a new model for chemical mastery of organic nature.

MISTAKEN CLAIM

Standards for identifying organic compounds were altered by the turn to synthesis. Hugo Schiff—a German chemist working in his adopted Italian homeland—learned this when he claimed to have synthesized coniine in 1870. Schiff's claim did not convince other chemists, and it therefore reveals how disciplinary expectations for product identification were changing. Experts like Hofmann recognized that demonstrating the identity of synthetic products with naturally occurring targets required physical data, including melting and boiling points, quantitative analysis, and reagent tests. Schiff, however, evidently did not recognize these requirements.

Pressed to substantiate his claim, Schiff subjected his synthetic product to several of these methods. Hampered by limited sample size and purity, he repeated his synthesis three times, using successively larger amounts of starting material, before finally retracting his claim. After 2 years, reaction with one of Hofmann's reagents (ethyl iodide, by now a standard method of classifying organic bases) at last convinced Schiff that he had not, in fact, made coniine. By the 1870s, characterizing a substance required understanding its constitution, taxonomic information that was acquired by using standard reactions and reagents. Schiff's work also illustrates how increasingly stringent requirements for product identification magnified the difficulties and dangers of synthesis, exposing him to poisoning and explosions.

Schiff had not set out to make coniine, seeking instead to elucidate "the action of ammonia and amine bases on aldehydes." Heating butyraldehyde with ammonia inside sealed tubes, he isolated a "strong poison." Schiff lacked sufficient material for purification and analysis but nevertheless inferred his product's composition and constitution from its mode of production.[18] Thus, his original identification was almost entirely based on toxicity. Schiff believed that his synthetic product was coniine because "it shows poisoning phenomena which are completely characteristic of coniine."[19]

The commentator for London's Chemical Society was far from convinced. He agreed that Schiff had "succeeded in producing by synthesis a product which possesses the characteristic properties of the active principle of hemlock." But when it came to Schiff's product identification, he conceded only that "there is obtained amongst the other products, a final one which has the composition of the alkaloid in question."[20] In fact, even this was generous, since Schiff had provided no direct evidence for his assigned formula.

Schiff's background helps explain why he was prepared to base a grand claim on such limited evidence. Schiff had trained with Friedrich Wöhler, receiving his PhD in 1857, in a laboratory mainly engaged in inorganic chemistry. His exposure to organic chemistry, and especially to the new methods of synthesis, was therefore limited. By 1870, Schiff was in Florence, on the periphery of the German chemical community and evidently somewhat out of touch with currently accepted practice.[21]

Schiff's outsider status also sheds light on his speculative use of structural theory in proposing a constitution and structure for coniine (figure 9.1). Schiff's "constitutional formula" for coniine in turn illustrates what graphical formulae and chemical structures did, and did not, mean ca. 1870. Schiff's constitution was consistent with an earlier observation. In 1854, the Swiss chemist Adolf Planta and his assistant August Kekulé—who had been recommended to Planta by Liebig following Kekulé's Giessen training—reported that coniine contained only one replaceable hydrogen and was therefore to be classified as an imide base or secondary amine. Chemists today associate the $C=NH$ group contained in Schiff's formula with a distinct atomic grouping called a "primary imine," whose properties are quite distinct from those of a secondary amine. Schiff's constitutional formula, however, did not convey this meaning.[22] Nor did it imply that the coniine molecule was U-shaped. For Schiff, this constitution was merely intended to reflect coniine's established

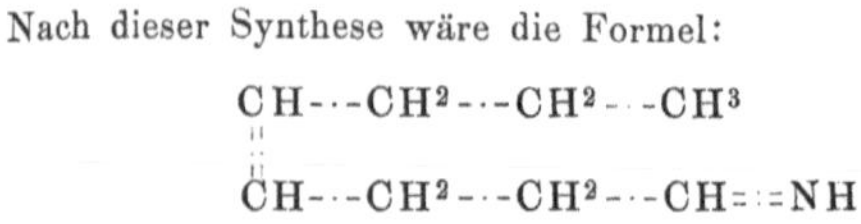

Nach dieser Synthese wäre die Formel:

$$CH-\cdot\cdot CH^2-\cdot\cdot CH^2-\cdot\cdot CH^3$$
$$CH-\cdot\cdot CH^2-\cdot\cdot CH^2-\cdot\cdot CH=\!=\!NH$$

FIGURE 9.1

Schiff's proposed constitution for coniine was based on his claimed synthesis. It sat-isfied basic rules of valence and indicated that coniine was a secondary amine—in agreement with Planta and Kekulé's 1854 report. Schiff's later determination that his product was a tertiary amine forced him to retract this claim.

Source: Hugo Schiff, "Erste Synthese eines Pflanzenalkaloids. (Synthese des Coniins)," *Berichte der deutschen chemischen Gesellschaft* 3 (1870): 947. Courtesy of the University of Wisconsin-Madison Libraries.

molecular formula ($C_8H_{15}N$) and classification, while complying with the rules of valence.

Despite other chemists' skepticism, Schiff remained convinced that he had made coniine. The following year, he repeated the serendipitous synthesis, recasting it as an intentional preparation of coniine. But he still failed to produce enough product for analysis, therefore performing only qualitative tests. As before, toxicity was his main evidence of identity. His product "possessed the smell of coniine to the highest degree," and poi-soned frogs in a way "which my brother [the physiologist, Moritz Schiff] regards as characteristic of poisoning with coniine."[23]

Pressed for further evidence, in 1872, Schiff finally made enough prod-uct to analyze. Fractional distillation enabled him to purify a substance that boiled "mainly about 168 °C," and "showed the composition of coniine." But again, Schiff reported no analytical data. He also admitted that—unlike natural coniine, which rotated light to the right—"the artificial alkaloid possesses no [optical] rotational ability," renaming it "paraconiine."[24] Schiff, however, believed that paraconiine might still contain natural coni-ine, together with its optical isomer. For Schiff, therefore, paraconiine's lack of optical activity was not a strong argument against his claim.

This situation prompted Schiff to attempt yet another large-scale prepa-ration, exposing him—and his assistant Icilio Guareschi—to considerable difficulties and dangers. A "violent explosion" destroyed a considerable portion of material. Eventually, the desired product "betrayed itself by [its] numbing smell," causing bouts of dizziness and high blood pressure that forced Schiff to stop work on several occasions. After purification, he obtained 24 g of product with formula $C_8H_{15}N$—an overall yield of less than 10%.

Schiff had learned that physical properties were required to identify a synthetic product—but he seems not to have appreciated how to make these values serve as useful characteristics. His "artificial coniine" boiled at 168–170 °C (at 759 mm mercury, corrected), not "significantly different" from reported values for natural coniine, which ranged between 163.5 and 171 °C. By the 1870s, as chapter 7 explained, only closely similar boiling points measured under identical conditions offered convincing proof of identity.

Schiff now took what would prove a decisive step, reacting his synthetic product with Hofmann's ethyl iodide. Introduced in 1851, ethyl iodide classified organic bases according to how many hydrogens were associated with nitrogen.[25] In fact, as mentioned above, Planta and Kekulé had already reacted natural coniine with ethyl iodide. They found coniine contained one replaceable hydrogen (i.e., was an imidogen base or secondary amine), and this was part of the evidence that Schiff's proposed structure sought to accommodate.[26] But Schiff now learned that his synthetic product contained no such hydrogen (i.e., was a nitrile base or tertiary amine). Here, at last, was a "clearly defined chemical difference" between "artificial coniine" and the natural alkaloid.[27]

Two years after his original claim, Schiff admitted that he had not made coniine after all.[28] His reluctant shift from qualitative description to quantitative evidence and indicative reagent tests exposes the barriers to disciplinary change. Schiff's claim foundered on his insufficiently discerning product identification and on his failure to manage the relationship between laboratory reasoning and evidence—struggles reflected in poorly supported structural speculation. Successful synthesis required careful integration of experiment and theory—as later work on coniine will show.

TAXONOMIC INCONSISTENCY

Earlier chapters explained how Hofmann used reaction with alkyl halides during his synthetical experiments on aniline. The alkyl group in these reagents replaced hydrogens associated with nitrogen in organic bases, making it possible to count the number of replaceable hydrogens. By 1851,

Hofmann developed a taxonomy based on this method, classifying organic bases into amidogen, imidogen, and nitrile bases (later primary, secondary, and tertiary amines, containing 2, 1, and 0 replaceable hydrogens, respectively).

Hofmann initially focused on artificial bases, but his method could also be applied to natural alkaloids. In 1853, Cahours used ethyl iodide to classify piperidine as an "imide" base (secondary amine), and as we have seen, Planta and Kekulé similarly assigned coniine to the same category a year later.[29] Hofmann continued to develop his new reagents, transforming them into the "most valuable agents of substitution in the hands of chemists," and powerful tools for investigating the constitution of alkaloids.[30]

In this chapter, we rejoin Hofmann in 1879 as he embarked on fresh experiments intended to clarify the constitution of more complex organic bases using these methods. His initial focus was piperidine, but as we shall see, practical difficulties soon prompted Hofmann to include coniine in his investigations. Ultimately, laboratory reasoning from experiments using methyl and ethyl iodide led Hofmann to a taxonomic inconsistency between piperidine and coniine—with important consequences for theoretical understanding of what coniine was.

Hofmann had learned a great deal more about methyl and ethyl iodide during the intervening 30 years.[31] Ethyl iodide was his preferred reagent for taxonomizing organic bases. But Hofmann had also developed an important method of constitutional analysis using methyl iodide that involved two sequential reactions: exhaustive methylation and elimination. Elimination expelled an unsaturated hydrocarbon (containing less hydrogen than saturated compounds) from the original base, usually without further constitutional change. Identifying the unsaturated hydrocarbon therefore offered valuable information about the starting material. Hofmann characterized numerous unsaturated hydrocarbons during this research. Thus, he was generally able to identify the expelled hydrocarbon, frequently by comparing its boiling point with known values.

Hofmann's decision to study piperidine was prompted by the apparent contradiction between its low hydrogen content—which suggested piperidine might be unsaturated—and its unexpected behavior when treated with standard inorganic reagents used to identify unsaturation. Hydrogen chloride caused no reaction, while bromine produced a substance Hofmann identified as a substituted pyridine. This result suggested that pyridine might be the initial reaction product—highlighting a potential relationship between piperidine and pyridine.[32]

This connection attracted the attention of Wilhelm Koenigs, a former student of Hofmann's who was now investigating Wilhelm Körner's 1869 extension of aromatic structural theory to pyridine.[33] Hofmann, meanwhile, recognized pyridine's relevance to alkaloid chemistry and was therefore monitoring Koenigs's and Körner's publications. In 1879, Koenigs claimed piperidine could be understood by analogy with Körner's pyridine ring. But even though Hofmann found Koenigs's proposal that piperidine also contained a ring structure "very tempting," he was not convinced.[34] Theoretical speculation that was empirically weakly founded had no place in Hofmann's laboratory reasoning.

Instead, Hofmann—assisted by Carl Ludwig Schotten[35]—reacted piperidine with methyl iodide, producing trimethylamine and a novel unsaturated hydrocarbon that Hofmann named "piperylene" (C_5H_8). This outcome was unusual, because it suggested that piperidine was a secondary amine in which two of nitrogen's three valencies were satisfied by a single hydrocarbon group. This peculiarity prompted Hofmann to perform the same experiment on coniine. Coniine indeed reacted similarly, producing trimethylamine and another new hydrocarbon that Hofmann named "conylene" (C_8H_{14}) (figure 9.2).[36]

The similarity in these results indicated that coniine was "a simple homologue of piperidine."[37] But for Hofmann, it also introduced a fundamental taxonomic inconsistency. Piperidine ($C_5H_{11}N$) contained three more hydrogens than piperylene (C_5H_8). Homology therefore indicated that

$$C_5H_8 + H_3N = C_5H_{11}N$$
Piperylen   Piperidin

$$C_8H_{14} + H_3N = C_8H_{17}N$$
Conylen   Coniin

FIGURE 9.2

Hofmann displayed the homology between coniine and piperidine, showing how these bases related to the hydrocarbons conylene and piperylene. He expressed no constitutional interpretation apart from the long-held conviction that alkaloids might be produced by the addition of ammonia (H_3N) to a suitable hydrocarbon.

Source: August Wilhelm Hofmann, "Einwirkung der Wärme auf die Ammoniumbasen," *Berichte der deutschen chemischen Gesellschaft* 14 (1881): 713. Courtesy of the University of Wisconsin-Madison Libraries.

coniine should contain 17 hydrogens (conylene's 14 plus 3). But Planta and Kekulé's formula included only 15 ($C_8H_{15}N$). Hofmann therefore directed Schotten to perform new analyses. These settled the matter, "speak[ing] unambiguously for the formula with 17 atoms of hydrogen" ($C_8H_{17}N$).[38] Thus, Hofmann's laboratory reasoning led him to an unexpected conclusion: Despite its relative simplicity, coniine's established formula had been wrong for almost 30 years.

RING OR NO RING?

Coniine's new formula explained why Schiff's synthesis had failed: No substance with formula $C_8H_{15}N$ could ever be coniine. It also refocused Hofmann's attention on the low hydrogen content observed in piperidine and coniine.[39] Neither showed the typical reactions of unsaturated compounds, which raised the possibility that they might contain rings. Thirty years before, Cahours had speculatively divided piperidine's hydrocarbon portion into two distinct groups. But piperidine's reaction with methyl iodide suggested to Hofmann that it contained a single "bivalent" hydrocarbon group related to piperylene, a constitution he represented as $(C_5H_{10})^{II}HN$.[40]

In 1881, this constitution incorporated an implicit question. Even as chemists debated Kekulé's ring structure for benzene and wondered whether related compounds including pyridine and piperidine might also contain rings, Hofmann's identification of a hydrocarbon group that satisfied two of nitrogen's three valencies implied that piperidine—and by extension, coniine—might contain a ring. Reasoning from experiment led Hofmann to this possibility, which he now set out to resolve.

Preliminary evidence suggested that the answer was "no." Piperylene, the hydrocarbon derived from piperidine, was "unsaturated," easily absorbing four bromine atoms. This was consistent with piperylene's hydrogen content, leading Hofmann to infer that the molecule contained "an elongated carbon chain."[41] For Hofmann, this countered Koenigs's recent speculation that pyridine and piperidine were rings. If piperidine were a ring, Hofmann reasoned, then piperylene should be, too. But, if this were the case, piperylene—whose putative constitution Hofmann also illustrated—should absorb only two bromine atoms (figure 9.3). Thus, experimental evidence from a standard reagent test for unsaturation appeared to refute the possibility that piperidine contained a ring.

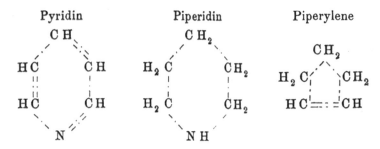

FIGURE 9.3

Koenigs's graphical formulae for pyridine and piperidine (left) as reproduced by Hofmann, alongside the constitution Hofmann argued they implied for piperylene (right, name added). When reacted with bromine, however, piperylene easily absorbed four bromine atoms, rather than the two that would have been consistent with a ring. Thus, Hofmann regarded the ring constitutions for piperidine and piperylene (and, by analogy, for coniine and conylene) as unfounded.

Source: August Wilhelm Hofmann, "Ueber die Einwirkung der Wärme auf die Ammoniumbasen," *Berichte der deutschen chemischen Gesellschaft* 14 (1881): 668; Wilhelm Koenigs, "Ueberführung von Piperidin in Pyridin," *Berichte der deutschen chemischen Gesellschaft* 12 (1879): 2343. Courtesy of the University of Wisconsin-Madison Libraries. Composite image courtesy of Andrew C. Warwick.

The homology between piperidine and coniine implied that if piperidine did not contain a ring, neither should coniine. But a disparity in physiological activity between piperidine and coniine meant that this reasoning remained uncertain. Experiments undertaken at Hofmann's request by the Berlin physiologist Hugo Kronecker had shown that piperidine was an anesthetic, while coniine was a paralytic. Thus, Hofmann came to no definite conclusion, inferring there might be "fundamental deviations in the atomic structure of the bivalent groups C_5H_{10} [in piperidine] and C_8H_{16} [in coniine]."[42]

Hofmann's study of piperidine and coniine nevertheless opened several future investigative avenues. First, identifying conylene as coniine's hydrocarbon precursor produced "a simpler formulation of the problem" of synthesizing coniine, so that "its solution appears to move a step closer."[43] Since the 1840s, Hofmann had believed that alkaloids might be made by reacting the appropriate hydrocarbon with ammonia.[44] He now had a perfect test case for this hypothesis. Converting conylene back to coniine would also close the analysis-synthesis loop that remained chemists' surest evidence for composition and constitution.[45] But Hofmann reported no success in developing this synthetic route to coniine—no doubt an instructive preliminary exposure to the difficulties of target synthesis.

Thus, Hofmann appears rapidly to have abandoned target synthesis in favor of the synthetical experiments suggested by his recent work. Some of these would apply "exhaustive treatment with methyl iodide" to "other more complicated nitrogen compounds, whose nature is still concealed." Nicotine, pyridine, and quinoline were therefore "already drawn into the circle of [Hofmann's] research."[46] Other compounds led him from coniine to the coniceïnes—complex substances whose constitution even the skillful and persistent Hofmann would struggle to elucidate.[47]

Within 2 years, Hofmann also returned to what he came to recognize as the crucial question of whether piperidine and coniine were related to the aromatic compound pyridine. In 1880, shortly before his premature death, the Russian chemist A. Vyschnegradsky claimed to have isolated pyridine from oxidation of coniine via a pyridine carboxylic acid, therefore assigning coniine to a family of alkaloids derived from pyridine.[48] But when Schotten—after leaving Hofmann's laboratory in 1882 for Berlin's nearby physiological institute—tried to replicate Vyschnegradsky's result, he failed. Thus, Hofmann continued to view such a relationship as unfounded.[49]

Things began to change in 1883, when Hofmann—now assisted by Franz Mylius[50]—produced "a not insignificant quantity of pyridine" from piperidine.[51] As we have seen, Hofmann had hitherto doubted Koenigs's 1879 proposal that piperidine was a "fully reduced pyridine."[52] His own experiments now suggested Koenigs might be right. But when Hofmann tried to confirm his findings by converting pyridine back into piperidine and so closing the analysis-synthesis loop, he could not.

This failure prompted a most productive shift: Hofmann switched to coniine instead. Within a year, he and Mylius showed that distillation in the presence of zinc dust converted coniine into conyrine, "an indubitable pyridine base." They also accomplished the reverse transformation.[53] Whereas piperidine offered only circumstantial evidence via its homology to coniine, direct evidence now established coniine's crucial link to aromatic chemistry.[54] Hofmann's results also indicated that coniine and piperidine shared a central core related to pyridine, confirming their close relationship to each other and to aromatic compounds.[55]

Hofmann's next step was to investigate how the additional atoms (C_3H_6) that differentiated coniine from piperidine were connected to its pyridine-derived core. Conyrine soon began to resolve this question. Oxidation with potassium permanganate produced crystalline picolinic acid (α-pyridine carboxylic acid, carboxylic acid substituent adjacent to nitrogen), identified by its melting point (134 °C). This key reaction indicated that conyrine contained a single three-carbon side chain adjacent to nitrogen—which in turn suggested that coniine was piperidine containing a propyl group in the same position relative to nitrogen.[56]

Hofmann's nomenclature now shifted in ways implying that he accepted coniine was a ring and that aromatic structural theory offered a valid explanation of conyrine. In 1884, he described conyrine and coniine as "ortho-propyl pyridine" and "ortho-propyl piperidine," respectively.[57] Introduced during investigations of substitution patterns in benzene, by the 1880s, ortho was displacing α as a way of describing substitution adjacent to the primary position on the ring.[58]

Hofmann's likely reasoning can be inferred by interpreting the chain of evidence laid out above in terms of structural theory. By the 1880s, structural theory incorporated only two explanations for low hydrogen content: unsaturation, reflected in typical reactivity established using test reagents, such as bromine; and the presence of a ring. In the case of aromatic compounds, rings explained lower hydrogen content than could be explained by unsaturation alone.

Piperidine and coniine contained two fewer hydrogens than usual for a fully saturated compound. They did not, however, show the behavior characteristic of unsaturated compounds. Nor were they aromatic. But Hofmann had now shown that piperidine and coniine were related to the aromatic compounds pyridine and conyrine. Not only was it becoming more likely that a ring constitution explained the low hydrogen content in piperidine and coniine, this result also in turn supported the idea that pyridine was a ring.

Hofmann's acceptance of rings for coniine and piperidine therefore went hand in hand with his increasing confidence that aromatic structural theory plausibly explained pyridine and conyrine. This view found expression the following year in images reminiscent of Koenigs's earlier depictions of pyridine and piperidine (figure 9.4; cf. figure 9.3). This was

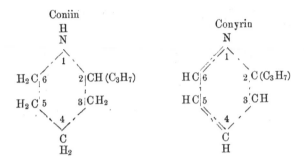

FIGURE 9.4

Hofmann used rings to display coniine's constitutional relationship to its aromatic counterpart conyrine. (Note that conyrine's graphical formula should contain a double bond between carbons 2 and 3.)

Source: August Wilhelm Hofmann, "Zur Kenntniss der Coniin-Gruppe," *Berichte der deutschen chemischen Gesellschaft* 18 (1885): 129. Courtesy of the University of Wisconsin-Madison Libraries.

an important step, pointing to a nascent role for visual representation in Hofmann's chemistry.

Hofmann had hitherto used synoptic tables to display the constitution of organic bases (see figures 5.1 and 6.9). Studying coniine now prompted him to incorporate rings in schematic diagrams that communicated at a glance Hofmann's experimentally derived knowledge of connectivity and relative arrangement. Moreover, his side-by-side depictions of coniine and conyrine made apparent the analogy between them, prompting the viewer to visualize their interconversion. Images like this initiated a change in how chemists, including Hofmann, understood and manipulated structure and reactivity, and they began to make chemistry into a visual science.

The power of visual structural reasoning was potentially enormous. But this did not mean every possibility that might be visualized could be realized. As we shall see—and as the cautious Hofmann certainly understood—visual speculations would only acquire validity when integrated into laboratory reasoning. In this instance, Hofmann's visual representation of coniine prompted an alteration in his view of alkaloid synthesis. As late as 1881, he remained convinced that alkaloids would be made by introducing nitrogen into an appropriate hydrocarbon. Now Hofmann suggested instead that coniine might be made by joining two organic molecules—one of which, pyridine, already contained coniine's essential nitrogen.

This idea was not entirely original, because Ladenburg was already attempting to alkylate pyridine, hoping to prepare a substituted pyridine that he could reduce to coniine. Hofmann, however, considered that the two required steps—alkylation (to add the propyl side chain) and reduction (to convert pyridine to piperidine)—could be carried out in either order. Thus, he proposed two alternative routes, seeming to regard these as of equivalent plausibility. It is important to note that Hofmann's decision to represent coniine using a ring did not imply the laboratory knowledge needed to make it and that is implicit in such a structure for chemists today. Ladenburg's approach might work, but Hofmann considered that "the synthesis of coniine will perhaps be even more simply accomplished through the action of propyl iodide on piperidine at high temperature."[59]

Believing his alkyl halide reagents might effect this crucial transformation, Hofmann embarked on his second attempt to synthesize coniine, assigning this project to his student Paul Ehestädt. Ehestädt, however, seems to have made little progress, and Hofmann published no results from this work. Within months, Hofmann learned that alkylating piperidine was a lot harder than he had anticipated, and he again responded to these difficulties by prioritizing synthetical experiments over target synthesis, redirecting Ehestädt toward investigation of coniine's curious derivatives, the coniceïnes.[60]

Coniine's new constitution incorporated a structural ambiguity that would be of vital importance to Ladenburg's synthesis. As figure 9.4 shows, it remained unclear whether coniine's side chain (C_3H_7) was attached by its terminal or central carbon. The side chain might therefore be either a propyl or an isopropyl group. Any successful target synthesis would have to resolve this question. But for Hofmann, this issue was now of apparently little consequence—suggesting that his desire to bring order to alkaloid chemistry once more trumped interest in synthesizing any individual alkaloid.

Hofmann's constitutional analysis of coniine was a remarkable feat of laboratory reasoning with far-reaching consequences for chemists' theoretical understanding of alkaloids. It demonstrated the power of his reagents and methods—while also exposing their limitations.[61] For Hofmann, establishing how the coniine molecule was constituted prompted a fresh approach to the synthesis of natural alkaloids. This strategy was more consistent with the alkaloids' complexity and diversity. Yet it did not enable Hofmann to synthesize coniine—a failure reflecting both the difficulty of target synthesis and the distinction between synthetical experiments and target synthesis.

AROMATICS TO ALKALOIDS

Albert Ladenburg's interest in coniine had very different origins from Hofmann's—and it would have a very different outcome. Natural alkaloids and artificial bases had been central to Hofmann's career since the 1840s. Ladenburg's background—as we saw in previous 8—was much more eclectic. Understanding how Ladenburg came to focus on synthesizing natural alkaloids, and why he selected coniine as his first target, helps explain how he acquired the tools and resources he would later deploy in synthesizing coniine.

Following initial training in Heidelberg with Robert Bunsen and Ludwig Carius, Ladenburg spent 6 months working with Kekulé in Ghent in 1865. This formative period brought Ladenburg to the heart of debates surrounding aromatic structural theory. During the mid-1870s, he was using aromatic amines to study benzene's positional isomers and hence its molecular symmetry.[62] This investigation contributed to Ladenburg's rejection of Kekulé's benzene ring—a stance that led to increasingly acrimonious disputes with the ring's early supporters.[63] It also drew Ladenburg onto Hofmann's territory, leading to occasional overlaps between their work.[64]

By 1879, Ladenburg was trying to synthesize the belladonna alkaloid, atropine—his first engagement with the unsolved "problem of alkaloid synthesis." Atropine's complexity meant that his results were limited—but Ladenburg believed the problem's significance made it "already of importance, when even a step in this direction succeeds."[65] Ladenburg's absolute focus on synthesizing alkaloids would prove a defining feature of his chemistry from this point on.

Atropine, like many alkaloids, had important medical uses—in treating heart disease, as well as in ophthalmology. In addition to atropine's academic interest, it is plausible that Ladenburg was motivated by more practical and commercial concerns. As chapter 8 explained, the Ladenburg family was a major investor in Germany's nascent chemical giant, the Badische Anilin- und Soda-Fabrik (BASF). His father had also provided considerable financial and other support during Ladenburg's early career—a family obligation that the son perhaps hoped his science might discharge.

As he struggled with atropine, Ladenburg explored Hofmann's reagents and methods. In 1883, he found a way of alkylating pyridine and soon began trying to reduce alkylated pyridines. These reactions were crucial in diverting Ladenburg's attention from atropine toward the natural alkaloids related to piperidine. Well before Hofmann published coniine's constitution and likely mode of synthesis, Ladenburg sought methods of making what he later called the "piperidine bases"—a group including coniine.[66]

At first, Ladenburg alkylated pyridine using Hofmann's ethyl iodide, initially describing this as "the application of a reaction, which led A. W. Hofmann to the preparation of aniline derivatives."[67] Indeed, Hofmann later claimed that his method of alkylating aniline was bound to work on pyridine.[68] Just as with his proposals concerning possible synthetic routes to coniine, the boundary between aromatic and aliphatic chemistry seems to have been highly permeable for Hofmann.

In fact, extending the reaction from aniline to pyridine was not straightforward. Ladenburg would do considerable work to make alkylating pyridine a reliable "synthetic method."[69] This was partly because there was no recognized analogy between aniline and pyridine—an argument that Ladenburg later used in reclaiming credit for his innovation.[70] But as Ladenburg discovered, alkylating pyridine also produced numerous isomeric products that were hard to differentiate and identify.

Problems of chemical identity were a serious obstacle to Ladenburg's efforts to make piperidine alkaloids, as highlighted in the account of his coniine synthesis below. Solving these problems relied on Ladenburg's expertise in aromatic chemistry. It is therefore worth emphasizing that at this time, Ladenburg was no more a fan of the pyridine ring proposed by Körner (and James Dewar) than of Kekulé's benzene ring.[71] Indeed, Ladenburg would not accept that "a ring formula is fitting for pyridine" until after synthesizing coniine.[72]

As with Hofmann, Ladenburg's theoretical views directed his choice of terminology. By the mid-1880s, Hofmann used ortho, meta, para—names indicating positions on a ring, and implicit acceptance of aromatic structural theory. In contrast, Ladenburg used α (alpha), β (beta), and γ (gamma) to indicate positions adjacent to, and successively farther away from, nitrogen—terms without any implication of a ring. Believing ethyl replaced hydrogen on the third carbon from nitrogen, for example, Ladenburg identified his first alkylated pyridine as "γ-ethyl pyridine."

Ladenburg's initial efforts to reduce pyridine involved repeating Koenigs's 1881 "hydrogenation" using zinc and hydrochloric acid. After numerous failures, Ladenburg's knowledge of aromatic chemistry came to the fore. Switching to sodium in alcohol—reagents previously used by Vyschnegradsky—produced more promising results. The yield remained very poor, but Ladenburg isolated a substance he identified by analysis and melting point as piperidine's platinum salt. Thus, Ladenburg converted pyridine into piperidine—which had yet to be isolated from a natural source, therefore continuing to be regarded as an artificial base.[73]

Within weeks, Ladenburg—aided by Carl Stöhr—successfully increased the efficiency of sodium in alcohol reduction, demonstrating its use on alkylated pyridines. His first starting material was commercial picoline—a mixture of isomeric methylpyridines derived from coal tar.[74] Thus, Ladenburg tested sodium reduction on a widely available, relatively cheap starting material—suggesting that converting coal tar into valuable products, including natural alkaloids, remained an attractive goal.

Reducing γ-ethyl pyridine—his first alkylated pyridine—Ladenburg isolated γ-ethyl piperidine, which smelled very like coniine and piperidine. Although smell, like toxicity, was no longer regarded as a reliable identifying characteristic, it remained a valuable marker of similarity. Hofmann's previous work had made a "homology" between coniine and piperidine "very probable." Ladenburg's observation that γ-ethyl piperidine, piperidine, and coniine smelled alike "strengthen[ed]" the likelihood Hofmann was right. In late February 1884, well before Hofmann's constitution for coniine appeared, Ladenburg published a method of making substituted piperidines that he hoped would "enable the synthesis of coniine."[75]

TARGET SYNTHESIS

In fact, it would take Ladenburg another 2 years to synthesize coniine—despite already being equipped with reliable reactions for alkylation and reduction. His work exemplifies how command of synthesis and chemical identity enabled chemists to elucidate structural information, even while they pursued specified molecular targets. Having initially adopted a synthetic strategy similar to Hofmann's, Ladenburg ultimately detected an unanticipated structural rearrangement that meant his approach was bound to fail. He therefore switched to a new starting material and different chemistry before finally making coniine—work that highlights the tenacity and skill that target synthesis required.

Target synthesis made chemical identity a matter of central concern—as Schiff's struggles during the 1870s showed. By the mid-1880s, identifying a single molecule among the many now made by synthesis was pushing chemists to develop ever more discriminating methods of characterization. This in turn required working at ever higher levels of purity. Ladenburg's synthesis involved solving a series of increasingly complex problems of isomerism, expanding the frontiers of chemical identity and redefining what coniine was.

It would be impractical to relate in detail every aspect of Ladenburg's coniine synthesis. Nor is this necessary to communicate the essence of his approach—something my account accomplishes by selecting four crucial phases of this work. These phases highlight how positional, structural, and optical isomerism challenged Ladenburg's ability to produce a pure substance he could prove was identical to natural coniine. Overcoming these barriers again relied on Ladenburg's expertise in aromatic chemistry. It also required crucial resources introduced in earlier chapters, as well as enhanced methods of purification, and techniques specific to the preparation and identification of single optical isomers.

POSITION AND STRUCTURE

In early 1884, Ladenburg began using his alkylation and reduction reactions to prepare propyl piperidines, searching for one identical to natural coniine. This work involved making and identifying numerous positional and structural isomers. Distinguishing isomers was difficult. They inevitably shared the same molecular formula, and their properties were frequently similar. Assigned position and structure were almost invariably tentative— an uncertain empirical base that could destabilize future experiments. Ladenburg discovered and corrected several such erroneous assignments on his route to coniine.

By the end of March, Ladenburg produced his first candidates, two piperidines that differed in the position of the propyl side chain. Potassium permanganate oxidation—used by Hofmann to show that coniine's side chain occupied the α-position—converted Ladenburg's major product to γ-picolinic acid, identified by its melting point (305 °C). Thus, Ladenburg inferred his major product was γ-propyl piperidine, reporting melting and boiling points, specific weight, smell, and toxicity—assessed by his colleague, the physiologist August Falck.[76]

Ladenburg could not isolate enough of the other, minor product to purify and fully characterize it—which meant that he was unable to determine whether it was α- or β-propyl piperidine. He observed its similarity to coniine, especially regarding toxicity. As with the γ-compound, however, Ladenburg also noted significant differences from natural coniine—including the observation that both synthetic products were optically inactive, whereas natural coniine was dextrorotatory.[77] Despite resembling the natural alkaloid, neither base was coniine.

Ladenburg therefore switched to the isopropyl compounds. His student Ludwig Schrader successfully prepared two isopropyl pyridines. Schrader identified the major product with lower boiling point as γ-isopropyl pyridine—again, by oxidation to γ-picolinic acid, identified by its melting point (303 °C, 2 °C lower than reported by Ladenburg). Despite lacking definitive evidence, Schrader suggested the minor, higher boiling product "probably belonged to the α-series"—that is, α-isopropyl pyridine. He also noted its similarity to conyrine, Hofmann's aromatic coniine derivative.[78]

Schrader's results modified Ladenburg's view of coniine. Up to this point, he had anticipated that coniine would be α-propyl piperidine. Now, it seemed likely that it was α-isopropyl piperidine instead, and that it might be made by reducing Schrader's minor product.[79] Thus Ladenburg—now assisted by Carl Stöhr—hoped "perhaps to prove the identity of α-isopropyl piperidine with coniine." He therefore set out to repeat Schrader's preparation and confirm his positional assignments.[80]

Oxidizing the minor, higher boiling-point isomer, which he anticipated was α-isopropyl piperidine, Ladenburg identified α-picolinic acid—but did not report its melting point, since it remained contaminated by the γ-isomer. He made repeated efforts to increase the purity of his material by improving the separation of the two isomers—both as pyridines and following their reduction to piperidines. Eventually, Ladenburg isolated what he believed was pure α-isopropyl piperidine, finding it to be very similar to natural coniine, especially where its effect on frogs was concerned. According to Falck, synthetic base and natural alkaloid showed "qualitatively" identical toxicity.[81]

Thus, Ladenburg remained persuaded that α-isopropyl piperidine might be coniine. In a move reminiscent of Schiff's earlier work, he dismissed discrepancies of almost 10 °C in the melting points of their hydrochloride and hydrobromide salts as "small differences" that could be explained by optical isomerism.[82] "It is possible," he speculated, "that if one were to succeed in separating the base into its two active components, the dextrorotatory part will prove completely identical to coniine."[83]

It did not. Like Schiff before him, Ladenburg now expended considerable effort—over almost a year—before admitting that he was wrong. During this period, he uncovered numerous errors in his previous work, including a conflation of α- and γ-isomers that seriously undermined his progress to date. Having previously agreed that Schrader's minor product was α-isopropyl pyridine, Ladenburg now showed that it was in fact γ-isopropyl pyridine—which in turn suggested that the major product was the α-isomer.[84]

From this point on, Ladenburg would exercise even greater care in purifying and identifying the many isomers involved in this work.[85] He also began entirely new experiments, repeating pyridine's reaction with propyl iodide and isopropyl iodide on a larger scale and making extensive use of oxidation to identify positional isomers among the products. By June 1885, an exhaustive comparison of melting points and other data showed that both reactions produced the same major product, which Ladenburg identified as α-isopropyl pyridine.[86]

This outcome had important structural consequences. It indicated that a molecular rearrangement occurred when pyridine reacted with propyl iodide, converting the reagent's propyl group into an isopropyl group in the product. Reagents and melting points—key tools of synthesis—enabled Ladenburg to determine structural changes taking place during a reaction. Synthesis promoted structural explanations of substance and reaction, establishing new links between laboratory chemistry and increasingly dynamic, visual notions of chemical structure. Thus, laboratory reasoning increasingly encompassed the visual, relating chemists' preparative capability to formulae that displayed molecules as rings and chains.

Ladenburg also showed that α-isopropyl pyridine was not the same as Hofmann's conyrine. This result implied that conyrine was α-propyl pyridine, and therefore, that coniine was the dextrorotatory isomer of α-propyl piperidine, resolving the intrinsic structural ambiguity in Hofmann's constitution. This resolution, however, introduced a major problem, leaving Ladenburg without a viable approach to making coniine.

NEW STRATEGY, NEW STARTING MATERIAL

Ladenburg once more used his expertise in aromatic chemistry to formulate a new strategy, switching starting material to α-picoline (α-methyl pyridine). This change offered many advantages, even though the new approach would require different chemistry. Most importantly, the side chain's position was already fixed—which ended Ladenburg's struggles with positional isomers. Ladenburg could also obtain pure α-picoline from commercial picoline using a procedure developed with Roth.[87]

The prospect of converting picoline, a low-cost constituent of coal tar, into a natural alkaloid was philosophically significant as well as commercially interesting. Ladenburg's student Otto Lange had recently obtained α-picoline from pyridine. Pyridine, in turn, had been produced by William Ramsay from acetylene, which Berthelot had previously formed by the direct combination of carbon and hydrogen. Any synthesis starting from α-picoline would therefore imply the possibility of a synthesis from the elements, then called a "total synthesis." As Ladenburg recognized, switching starting material to α-picoline considerably raised the stakes surrounding coniine's synthesis.[88]

By early 1886, Ladenburg—aided by Wilhelm Laun—reported "the synthesis of a base extremely similar to coniine" from α-picoline. Made cautious by recent experiences, Ladenburg described it as merely "probable" that his product was "chemically identical" to coniine.[89] He nevertheless published his preliminary results, no doubt wishing to secure priority while continuing work to verify his product's identity.

Because he knew from previous work how difficult it was to purify α-picoline, Ladenburg was concerned that remaining isomeric contaminants in the starting material might lead to similar impurities in the product. He also feared the kind of rearrangement that had thwarted his previous efforts.[90] Finally, Ladenburg's synthetic base—like most laboratory products—was optically inactive, requiring separation into its optically active components before any definite conclusions concerning its identity could be drawn.[91]

Ladenburg's concerns reinforced the importance of pure starting materials. Remaining impurities in the starting material frequently produced impurities in the product that were extremely difficult to remove. This problem was responsible for many instances where substances that nineteenth-century chemists believed were pure were later shown to be mixtures.[92] It also explains why Ladenburg, as he developed each of his key reactions, strove for ever-higher levels of purity. In some cases, this could be done by improving the reaction itself.[93] Elsewhere, as Ladenburg's work with α-picoline shows, obtaining adequately pure starting materials required innovative glassware and skillful manipulations.

PURITY IN PRACTICE

Ladenburg's next move—assisted by Lange, following Laun's tragically early death—was to repeat his most recent preparation "in larger scale and with chemically pure material." In doing so, Lange refined Roth's method for obtaining pure α-picoline from commercial picoline. Ladenburg later claimed that Lange's purification could be performed "with great ease and security"—even though their initial large-scale attempt produced just 380 g pure α-picoline from 1 kg of commercial picoline.[94]

Lange's procedure combined fractional distillation with fractional crystallization of α-picoline's mercury double salts. Glass and glassblowing were vital to both steps. The first step involved "very careful and multiple repeated fractionations" using a special distillation flask that Ladenburg had developed while purifying pyridine some years previously. His design incorporated an "extremely elongated neck . . . blown into many bulbs," a form that had "proved very practical, much better than the usual flasks with bulb-shaped attachments." Although it was now available "in various sizes" from Peter Desaga's Heidelberg scientific instrument firm, Ladenburg nevertheless illustrated his flask so that glassblowers could make it for themselves (figure 9.5).[95]

Ladenburg originally reported that purifying α-picoline by fractional crystallization required distinguishing the "large glittering prisms and leaves" of its mercury double salt from the "characteristic fine needles" of the analogous pyridine salt. This simple description belied the necessary manipulative skill. Only carefully controlled conditions caused one substance to crystallize while leaving the other in solution, and filtration at just the right moment was vital in achieving the separation. Above all, this method required recognizing particular crystals as they formed within glass. The difficulties Ladenburg experienced are reflected in his growing focus on crystal morphology and his later illustration of crystal forms alongside his distillation flask. Thus, Ladenburg used crystallographic data to highlight the necessity of accurate crystal identification in purifying α-picoline (figure 9.5).[96]

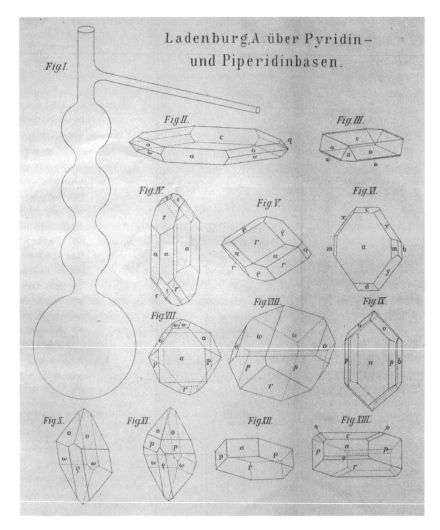

FIGURE 9.5

Ladenburg's later summary of his work with pyridine and piperidine bases illustrated his new distillation flask alongside diagrams of the crystal morphology of many substances that he had isolated. Ladenburg's flask combined distillation vessel, fractionating column, and still head in one. His diagram enabled chemists elsewhere to obtain similar apparatus and so to use his method. The crystal diagrams complemented interfacial angles provided in the text. Their juxtaposition with the flask reflected the very great difficulty of obtaining pure material using fractional crystallization, a highly skilled manipulation that relied on recognizing specific crystal forms.

Source: Albert Ladenburg, "Ueber Pyridin und Piperidinbasen," *Annalen der Chemie und Pharmacie* 247 (1888): 5–6. Courtesy of the University of Wisconsin-Madison Libraries.

Using these methods, Ladenburg obtained 380 g pure α-picoline, which he reacted with paraldehyde (acetaldehyde's cyclic trimer) to produce 45 g crude α-allyl pyridine (Ladenburg's term for pyridine with a 3-carbon unsaturated substituent at the α-position[97]), purified by fractional distillation. Before proceeding, he again confirmed substitution at the α-position: Potassium permanganate oxidation produced a crystalline acid, which Ladenburg identified as α-picolinic acid by its melting point (133 °C).[98]

Things took a turn for the better when sodium reduction of α-allyl pyridine proceeded in "almost quantitative yield" to give a synthetic base that showed "the greatest similarity with coniine and may be regarded as chemically identical to it."[99] Examination of its physical, chemical, and physiological properties indicated that Ladenburg had prepared optically inactive α-propyl piperidine, from which he anticipated natural coniine could be separated as the dextrorotatory isomer.

Ladenburg sought to confirm this crucial result without the complication of optical isomerism. He knew that Hofmann's method for converting natural coniine into conyrine (zinc dust distillation) destroyed coniine's optical activity. Ladenburg therefore subjected his synthetic base and natural coniine to this procedure, isolating the products as their platinum salts. Comparing their melting points, he found that these were identical to within a degree (158–160 °C for conyrine produced from his synthetic base cf. 159–160 °C for conyrine derived from natural coniine).[100]

This and other evidence convinced Ladenburg that his synthetic product was "physically isomeric" with natural coniine, differing only in its optical inactivity and the melting point of its hydrochloride salt.[101] Almost 3 years after first alkylating pyridine, Ladenburg had synthesized α-propyl piperidine as an optically inactive mixture of two optical isomers. It only remained to separate the dextrorotatory isomer from synthetic α-propyl piperidine.

TOTAL SYNTHESIS

The process by which Ladenburg accomplished this final step inserted optical activity into the relationship between melting point and chemical identity, so that synthetic α-propyl piperidine's optical inactivity became the explanation for the difference in melting point between its hydrochloride salt and that derived from natural coniine. Ladenburg also established the converse, showing that, when pure, both optical isomers melted at the same temperature, different from the melting point of an optically inactive mixture of the two. The isolation of α-propyl piperidine's dextrorotatory isomer—identified by optical rotation and melting point—thus became constitutive of the total synthesis of coniine. Thereafter, optical rotation was increasingly a requirement of characterizing individual molecules.

Ladenburg's first attempts to resolve synthetic α-propyl piperidine into its optical isomers involved fungi. Following repeated failures, he eventually accomplished the separation by making bi-tartrate salts—a method developed by Louis Pasteur.[102] Because their solubility differed, these salts could be separated from one another by crystallization. Having prepared a concentrated solution, Ladenburg introduced a tiny crystal of natural coniine bitartrate supplied by the Austrian chemist J. Schorm. Recovering the free base from the crystals that formed, Ladenburg found that they were dextrorotatory like natural coniine, with almost identical "angle of rotation" (13.87 degrees for the synthetic base cf. 13.79 degrees for natural coniine, both measured for the pure substance over a length of 1 dm).[103]

Ladenburg now used melting points measured in Roth's apparatus (see figure 7.5) to prove the identity of synthetic dextrorotatory α-propyl piperidine with natural coniine. Working once more with the hydrochloride salts, he measured 217.5 °C as the melting point of the salt derived from the synthetic dextrorotatory base and 217.5–218.5 °C for that derived from natural coniine. Ladenburg considered that this final evidence established the "complete identity" of dextrorotatory α-propyl piperidine and natural coniine.[104]

In October 1886, Ladenburg claimed the "first total synthesis of [a natural] alkaloid," and with it a major landmark in natural product synthesis. His peers were quick to agree.[105] This was a great tribute to his reputation—but it also reflected other chemists' agreement that Ladenburg had substantiated his claim. Obtaining this evidence required the solution of complex problems of chemical identity, including those introduced by optical isomerism. In making dextrorotatory α-propyl piperidine and showing this was identical to natural coniine, Ladenburg simultaneously synthesized coniine and re-defined it in terms of abstract understanding of molecular nature.

THE POWER AND MEANING OF SYNTHESIS

This chapter continues to explain how and why chemists created the science of synthesis. We have seen how Schiff's claim to have made coniine foundered on inadequate evidence, revealing analytical limitations that only synthesis could overcome. Synthesis was a powerful investigative method—illustrated here by Hofmann's skillful use of synthetical experiments, taxonomy, and laboratory reasoning to revise coniine's formula, rationalize its reactivity, and elucidate its constitution. Natural product synthesis was no exception. Ladenburg's target synthesis produced not only coniine but also a new definition of its molecular identity that linked melting point with optical rotation as well as chemical structure. Synthesis has always been about much more than making substances.

The comparison between Hofmann and Ladenburg clarifies the distinction between synthetical experiments and target synthesis and the ongoing, synergistic value of both approaches when they are linked to solid laboratory reasoning. Both chemists operated to stringent, labor-intensive standards of evidence and logic, and both reasoned from the results of experiments in synthesis to structural information. But Ladenburg's single-minded focus on product over reaction contrasts with Hofmann's preference for exploring novel chemistry using reagents—a commitment that diverted him from engaging the difficulties of target synthesis. Ladenburg strove to bridge the gap between synthetic products and his chosen target, developing reactions that he used in making coniine and other piperidine alkaloids. Hofmann's synthetical experiments established structurally informative reagent tests with wide application and laid the groundwork for broadly useful synthetic methods.[106]

To an even greater extent than did synthetical experiments, target synthesis made visible the gulf between what theory indicated should be possible and what it took to accomplish such an outcome in practice. Ladenburg's alkylation is a case in point. Having initially considered this merely an "application" of Hofmann's earlier alkylation of aniline, Ladenburg soon learned better.[107] His new alkylation might be "based upon" Hofmann's earlier reaction but it "by no means followed from this that its transfer [from anilines] to the pyridine group had been a matter of course, as Hofmann liked to present it." We have seen that making coniine ultimately required Ladenburg to use quite different chemistry. As he later recollected, "I worked for a long time in vain, until I found the method."[108]

Ensuring that reactions functioned reliably, producing identifiable products with reasonable efficiency, required working at unprecedented levels of purity. Poor yields frequently necessitated starting from kilogram quantities to produce grams or less of the final product. Chemists therefore operated on widely varying amounts of material, a situation that highlights the possibilities as well as the dangers of working in glass. Some methods could be adapted by blowing the same item of glassware in different sizes—as in the case of Ladenburg's distillation flask. But for sealed tube reactions—a staple for Ladenburg as well as Hofmann—this resizing was not possible. Instead, larger quantities of material necessitated multiple tubes—with extremely hazardous consequences. Ladenburg's largest reactions involved hundreds of sealed tubes.[109] Chemists' reliance on sealed tube reactions tested infrastructure as well as practical skill, leading to the incorporation of dedicated enclosures in the structure of laboratory buildings (discussed in chapter 8).

Practical expertise in aromatic chemistry was crucial at multiple junctures in Ladenburg's coniine synthesis. But—although Hofmann adopted ring structures to represent pyridine, piperidine, and coniine by 1885—Ladenburg made no use of rings during this work. Up to this point, Ladenburg remained unconvinced by Kekulé's benzene ring and the pyridine ring proposed by Körner and Dewar, because they could not explain experimentally established molecular symmetry. Instead, as we have seen, Ladenburg relied on experimental knowledge and laboratory reasoning to guide his work.

Synthesizing coniine and confirming Hofmann's constitution, however, convinced Ladenburg that the ring's explanatory advantages outweighed its limitations, leading him to offer a "schematic expression" of coniine that included a ring (figure 9.6).[110] Synthesis—whether synthetical experiments or target synthesis—thus prompted even such skeptics as Ladenburg to accept the validity of visual representation and structural reasoning.

This was not, however, because synthetic capability was always intrinsic to such visual representations, or because ideas about structure disconnected from laboratory reasoning were likely to be productive. On the contrary, Ladenburg's synthesis invested coniine's ring structure with new meaning, so that it signified something quite different for him than it had done for Hofmann. Hofmann, considering the transformation of pyridine into coniine, envisaged two alternative synthetic routes. His preferred method soon failed, while the other—despite initially seeming more promising—ultimately also proved incapable of making coniine. Ladenburg now saw in the same structure a single, quite different way of making coniine—somewhat in the way that a modern chemist might. Thus, by encompassing the visual, laboratory reasoning began to endow chemical

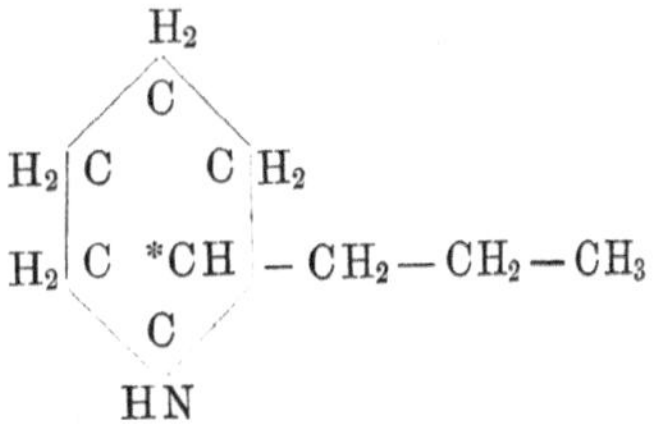

$$H_2$$
$$C$$
$$H_2\,C \qquad C\,H_2$$
$$H_2\,C \quad {}^*CH \; - CH_2 - CH_2 - CH_3$$
$$C$$
$$HN$$

FIGURE 9.6

Having separated optically inactive α-propyl piperidine into its optical isomers, Ladenburg used this result to test the le Bel–van't Hoff hypothesis that optical activity was caused by molecular asymmetry. His "schematic expression" of coniine's constitution identified the molecule's only asymmetric carbon (marked *. Note that the carbon atom shown at the ring's bottom apex is an error). Coniine's constitution was thus consistent with the theory of atoms in space, suggesting that every α-substituted piperidine should also be "separable into optically active isomers," which Ladenburg had already confirmed was the case for α-pipecoline (α-methyl piperidine) and α-ethyl piperidine. Note that Ladenburg also displayed coniine's side chain, reflecting his determination that this was an extended propyl (not a branched isopropyl) group.

Source: Albert Ladenburg, "Ueber das spezifische Drehungsvermögen der Piperidinbasen," *Berichte der deutschen chemischen Gesellschaft* 19 (1886): 2584. Courtesy of the University of Wisconsin-Madison Libraries.

structures with new possibilities, embedding within them the practical knowledge that gave life to structural theory.

Ladenburg's synthesis successfully integrated laboratory reasoning and aspects of visuality, with consequences for his understanding of the then-controversial connection between molecular structure and optical activity. We have seen that optical activity was central to Ladenburg's coniine synthesis. He now used his schematic expression (figure 9.6) to verify that his synthetically based understanding of coniine's constitution, and α-propyl piperidine's separation into two optical isomers, was consistent with the theory of atoms in space. As expected, the structure contained just one asymmetric carbon atom.

Ladenburg also used this diagram to make a prediction concerning the optical behavior of β- and γ-propyl piperidines. His prediction exemplifies how such illustrations prompted visual reasoning—but it also reveals the difficulty of adjusting to this novel way of thinking: Ladenburg mistakenly suggested that neither β- nor γ-propyl piperidine would be separable into optical isomers, when in fact this is only the case for the γ-compound.[111]

This work helped establish optical activity as an essential characteristic of many organic compounds and most natural products. During his coniine synthesis, Ladenburg compared the optical rotation of synthetic and natural coniine using an established instrument (Laurent's half-shade polarimeter). But even though these measurements were indicative, confirming the direction and approximate magnitude of coniine's rotational ability—they were not yet characteristic. In developing his inquiry into the optical activity of coniine and other piperidine bases, Ladenburg switched to "specific rotation."[112] Thus, the 1880s were a period of transition, during which Ladenburg helped standardize specific rotation alongside melting and boiling points as necessary characteristics of individual substances.[113]

This chapter has explained the power of synthesis by showing how synthetical experiments and target synthesis worked together. Synthetical experiments elucidated the nature of organic compounds and expanded the field of chemical enquiry. Yet, as Hofmann well knew, they remained an unreliable approach to making any particular substance. Ladenburg, meanwhile, belonged to a new generation of synthetic organic chemists for whom mastering organic nature meant overcoming the difficulties of target synthesis to match and ultimately transcend nature's constructive capability. Target synthesis produced natural products, such as coniine. It also drove the development of chemical theory, generating new productive possibilities. The science of synthesis gave chemists unprecedented power over organic nature—their most valuable source of starting materials and immediately useful products, and the very substance of life.

CONCLUSION: MAKING MODERN CHEMISTRY

This book promised a historical explanation of nineteenth-century organic chemistry, based on an account of what chemists did, and why and how they did it. Elucidating what chemists, including Justus Liebig, August Wilhelm Hofmann, and Albert Ladenburg knew, and—equally important—what they did not, has made visible the methods, tools, and resources that enabled their growing mastery over organic nature. As highlighted in the Introduction, these chemists—by virtue of their disciplinary impact—were instrumental in, and exemplary of, broader changes in and beyond chemistry. Understanding how Liebig, Hofmann, and Ladenburg worked and what they accomplished thus explains how, by about 1900, organic chemists transformed not only their own science but also the disciplinary landscape of the sciences. During the 1820s and 1830s, when this book began, the study of natural organic substances was a minor medical subfield. By the end of the century, organic chemistry was one of the world's most productive and industrially important sciences.

Nineteenth-century chemists established analysis and synthesis as inseparable, core methods of investigating organic compounds, critically dependent on novel human and material resources. Liebig pioneered the use of small, home-blown glassware that, combined with a new system of training, enabled students to contribute to research. His struggles to reshape the discipline around labor-intensive, glassware-based analysis met considerable resistance, especially because his methods were not guaranteed to produce superior results in any individual case. At the disciplinary level, however, these changes were transformative, driving an unprecedented increase in the number of chemists in training, and in what those chemists were able to accomplish once trained.

The turn to synthesis began in early-1840s Giessen as a direct consequence of Liebig's analytical program. The cause of this shift, however, was not success but deadlock. Liebig is widely celebrated for having solved the problem of analyzing organic compounds by his twin innovations of research school plus Kaliapparat. But, as we have seen, the more Liebig—aided, and sometimes provoked, by his students—learned about analysis, the more clearly he understood that analysis alone could never elucidate the chemistry of organic compounds.

Antoine Lavoisier is famous for defining chemistry as the science of analysis, an axiom that was fundamental to his introduction of chemical elements.[1] This study suggests that Liebig might warrant similar status for having promoted the necessary complementarity of analysis and synthesis in investigating organic nature.[2] Liebig's method and approach so inspired the young Hofmann that it stayed with him for the rest of his career. "Which of us," Hofmann asked in 1875, "returning to-morrow to his lonely post in the laboratory, and resuming his obstinate labour in penetrating Nature's stubborn depths, will not feel animated and cheered in his work by [Liebig's] example?" "For us chemists," he answered, "it will be . . . imperative and delightful . . . to work, not only with Liebig's instruments in our hands, but also with his noble spirit in our hearts."[3] Hofmann's tone was doubtless shaped by his eulogizing purpose. Yet there is no reason to doubt the fundamental sincerity of one who owed so much to his mentor.

Uncovering the temporal and geographical origins of organic synthesis helps elucidate its multiple meanings, contemporary as well as retrospective. For modern practitioners, *synthesis* and *synthetic* are frequently synonymous with molecular construction. But as we have seen, at least two generations of organic chemists encountered synthesis primarily as an investigative method. Developing synthesis into a reliable way of making natural products required the adoption of new methods of identifying substances to the exclusion of earlier factors, including origin. Despite the impossibility of isolating any substance—whether of natural or laboratory origin—that is absolutely pure, natural product synthesis queries the routine characterization of synthetic products as artificial. Although "synthetic" continues to be equated with "artificial" in general usage, frequently with negative connotations, the aim of natural product synthesis is to produce compounds identical to those found in nature. Synthesis, together with the novel criteria of identity and purity that it entailed, thus disrupted an established relationship between the artificial and the natural.[4]

Recognizing purity, identity, and the very notion of substance as inextricably linked historical concepts destabilizes a widespread tendency to essentialize all three. This book has provided numerous examples of substances whose identities were changed by the application of modified techniques of purification, most notably among the many substances that chemists encountered while attempting to synthesize the natural alkaloid, coniine. Rather than treating melting and boiling points as intrinsic properties of pure substances that awaited only adequately pure samples and sufficiently reliable measurement techniques to emerge as useful characteristics, this study has shown the interdependence of all these factors.[5]

Similarly, we have seen the pluralism in chemists' constitutional and structural interpretation of individual substances. Today, organic chemists confidently specify each organic substance by a single (usually skeletal) formula. Their nineteenth-century predecessors, however, generally worked without such a one-to-one correspondence between formula (of whatever kind) and substance. Learning how they could do so—and how synthesis in turn helped them establish a unitary link between substance and formula—contributes to understanding of the historical relationship between experiment and theory and thus to one of this study's main goals. Synthesis was not merely possible in the pre-structural era, it was essential in chemists' creation of a molecular world in which substances could be known by their structure.

That world was not created in any ivory tower. Quite the contrary. This book has shown that industry and imperialism motivated and provided essential resources for the turn to synthesis. Without indigo and coal tar, Hofmann could not have developed synthetical experiments on aniline. And without the global quinine market, there could not have been either raw material or demand for Liebig's amorphous quinine—and hence no compelling reason for Hofmann's move to London, with its consequences for chemistry in Britain and for chemical theory. Developments in academic chemistry thus emerge as more deeply intertwined with commerce and industry than is apparent in existing histories. I do not mean that previous studies have overlooked a relationship between science and industry—only that this connection was more fundamentally constitutive of nineteenth-century chemistry as a science, of its very methods and theories, than has hitherto been demonstrated.

This constitutive connection played out in material ways, most notably in the development of institutional chemical laboratories during the second half of the century. Existing histories explain the construction of such laboratories as either competition among states in the pre-1871 period or the expression of a research ideal by the new German empire (*Kaiserreich*).[6] The present study, however, shows that the material and human demands of organic synthesis were crucial in driving this change. It also points to an explanation of what drove the German state to fund such costly projects. Industry was central to Germany's response to Britain's empire. Just as Britain's prosperity relied on the agrarian space of its imperial possessions, so the laboratory became fundamental to Germany's competitive success.[7] Laboratories capable of this task had to be constructed according to chemists' requirements, which demanded government and—ultimately— industrial funding.[8]

Hofmann was a key player in these developments, overseeing the construction of great chemical laboratories and serving multiple terms as president of Germany's Chemical Society, founded in association with Adolf Baeyer in 1867. Hofmann's career spanned the crucial phase in the development of organic synthesis, from his early synthetical experiments on aniline to what his peers recognized as pathbreaking studies of coniine and the coniceïnes. In 1849, Hofmann envisaged the primary goal of synthesis as "not the host of new substances, which we are continually discovering, . . . but new methods of operation, by means of which we may imitate . . . the formative forces of nature."[9] Hofmann's words conjured into being a future in which chemists might overcome any distinction between the methods of laboratory chemistry and in vivo natural product formation.[10] By 1885, when his attempts to make coniine failed, Hofmann understood—perhaps better than any of his contemporaries—the distinction between synthetical experiments and target synthesis.

Over a period of 50 years, Hofmann had pioneered synthesis as a sophisticated tool of constitutional and structural elucidation. His approach transformed unwanted by-products of industrial production, notably coal tar, into commercially valuable products, including numerous artificial dyes. Synthetical experiments had the power to advance chemical theory and create manufacturing wealth. But as Hofmann had learned, they were not a viable method of preparing nature-identical compounds in the laboratory. Synthesis was the "key to Hofmann's thinking," but considerable further work would be necessary to fulfil the promise of target synthesis, especially of natural products.[11]

Successful target synthesis built on methods, tools, and resources introduced during the development of synthetical experiments. Ladenburg's coniine synthesis—this study's exemplary natural product synthesis—relied on skilled labor, chemical glassware, and the facilities of his Kiel laboratory. It highlighted issues of isomerism and identity, focusing renewed attention on purity, melting points, and optical properties. Ladenburg's synthesis used Hofmann's reagents, the alkyl halides—but in a new way. Where Hofmann's primary goal was investigative, Ladenburg developed the alkyl halides as one of many emerging "synthetic methods" capable of effecting predictable molecular change.[12]

Ladenburg's efforts to establish reliable methods of changing molecules reflected his experience of the difficulties of target synthesis. Throughout most of the nineteenth century, chemists' published descriptions of reactions varied enormously in the level of detail they provided—a situation exemplified in this study by the difference between Hofmann's publications concerning the amines and those of Adolphe Wurtz. In contrast, a synthetic method specified the reagents, conditions, and techniques required to accomplish a certain transformation. It typically itemized solvent, temperature, and duration; recommended how starting materials and products were to be purified; and indicated other factors to be controlled. Such information was vital in enabling other chemists to attain the desired result, and it was therefore essential in the development of target synthesis.

From about 1880, synthetic methods constituted a new and rapidly expanding kind of knowledge in organic chemistry, spawning a range of novel publications. Friedrich Konrad Beilstein's famous *Handbook*—first published in 1881—helped chemists maintain oversight of what they knew by listing known compounds in structurally defined families and describing their properties.[13] Compendia such as Karl Elbs's *Synthetic Methods of Preparing Carbon Compounds* served a different purpose, guiding chemists' approach to target synthesis by gathering experience of what their colleagues and predecessors had shown to be feasible, perhaps even reliable.[14] Thus, nineteenth-century chemists established approaches to knowledge management that provided a solid foundation for later developments driven by chemistry's continuing expansion and the advent of digital technologies.[15]

Despite its rapid development, nineteenth-century organic chemistry encompassed some important commonalities, as shown by this account of how Liebig, Hofmann, and Ladenburg worked. What I have called "laboratory reasoning" was crucial in each chemist's ability to make stable connections between systematic experimentation and theoretical advance. That reasoning built on established theory. For example, when Liebig sought to rationalize the nitrogen content of alkaloids such as morphine, his laboratory reasoning assumed the validity of atomic theory—at least insofar as this constrained the combining proportions of elements in the molecule. Yet in 1831, Liebig did not invoke Jöns Jacob Berzelius's electrochemical dualistic theory to interpret the outcome of his alkaloid analyses. Nor were Hofmann's synthetic investigations of the organic bases during the 1840s—experiments that produced the ammonia type—guided by then-contested theories of molecular constitution. Finally, we have seen that synthesizing coniine in 1886—far from being made possible by aromatic structural theory—was in fact crucial in Ladenburg's nascent acceptance of this as-yet-unstable theory.

This book's claim that practice and not theory was responsible for chemistry's development is therefore not equivalent to a claim that theory played no role in organic chemistry. Instead, the examples in this study explain how a combination of systematic experiments and established theory, interpreted through laboratory reasoning, drove the advancing front of theory. In moving beyond historical reliance on structural theory as the decisive enabler of synthetic success, this study calls for the reinterpretation of key episodes in the history of organic synthesis in light of laboratory reasoning.

ORGANIC SYNTHESIS AND LABORATORY REASONING

The developing science of synthesis was permeated by a tension between experiment and theory that was mediated by laboratory reasoning. This tension is illuminated by the gulf between chemists' increasingly sophisticated conception of molecules and their persistently limited ability to mimic nature's formative forces. Three landmark events in the history of synthesis, all concerned with aromatic dyes, illustrate complementary, developing aspects of this tension. The analysis that follows is not based on new historical evidence. Instead, it shows how different these well-known episodes appear when viewed in the context of the present study. Recognizing the role played by laboratory reasoning destabilizes accepted fixed points in an existing historiography of organic synthesis that routinely prioritizes theory over experiment.

The first landmark is William Perkin's 1856 attempt to prepare quinine, a synthetical experiment performed at the request of his teacher Hofmann. Perkin's approach was based on the arithmetical relationship between quinine's molecular formula and that of his starting material, allyl toluidine. Despite this rationale, he failed to make quinine, eventually producing the purple dye mauve instead. Mauve made Perkin rich, and it launched the synthetic dye industry. It has also been made to stand for the power of serendipity in the face of what later chemists, with the benefit of modern structural knowledge, have tended to dismiss as ludicrously flawed planning. Yet Perkin's attempt to make quinine was both entirely logical in the light of contemporary knowledge and contributed to chemists' understanding of constitution and reactivity. In relation to this study, Perkin's experiment was consistent with Hofmann's research program, a noteworthy outlier only because it proved so commercially significant.

The second landmark is the1869 preparation of the red madder dye alizarin by Carl Graebe and Carl Liebermann, then working in Baeyer's laboratory at Berlin's Trade Academy (*Gewerbeakademie*). Like Perkin, Graebe and Liebermann patented their discovery.[16] And as with mauve, synthetic alizarin became a lucrative industrial product. But, unlike Perkin's preparation of mauve, Graebe and Liebermann's achievement is widely regarded as one of the earliest structurally determined target syntheses and as a key example of aromatic structural theory's power to enable natural product synthesis. In fact, the case of alizarin has much more in common with Perkin's preparation of mauve than such narratives indicate.

In 1868, Graebe and Liebermann produced anthracene, an aromatic constituent of coal tar, from natural alizarin (using Baeyer's zinc dust reduction). This result confirmed that alizarin and anthracene were related, suggesting that it might be possible to make alizarin from anthracene (itself recently synthesized from benzoyl chloride).[17] Graebe and Liebermann therefore used a published procedure to oxidize anthracene to anthraquinone before subjecting anthraquinone to standard reactions they hoped might produce alizarin by introducing its two hydroxyl substituents (bromination followed by hydrolysis).[18]

Having successfully isolated alizarin, Graebe and Liebermann claimed the "first example of the artificial preparation of a vegetable dye." As they themselves recognized, however, they were fortunate to produce alizarin rather than any of its numerous isomers—an outcome that is attributable to alizarin's unusually high degree of molecular symmetry. Indeed, their proposed structure for alizarin—based on its relationship to anthracene—did not specify which positions on its constituent aromatic rings the hydroxyl groups occupied. Nor, having made alizarin, did they ascertain which isomer they had made—and which, therefore, was identical to natural alizarin.[19]

Just as with Perkin's mauve, Graebe and Liebermann's alizarin resulted from experiments that were both rationally planned and involved a hefty dose of serendipity. Moreover, immediate success meant that they were not forced to resolve issues of positional isomerism needed to assign alizarin's unique structure. Previously viewed as exemplifying the power of aromatic structural theory, theory in fact underdetermined Graebe and Liebermann's preparation of alizarin. In stark contrast to Hofmann's synthetical experiments and Ladenburg's coniine synthesis, its lucky outcome curtailed their application of laboratory reasoning, limiting the knowledge their experiment produced. Thus, alizarin assumes a more liminal role in the history of organic synthesis.

My final landmark continues this reshaping of the historical landscape of synthesis. The culmination of some 20 years of work, Baeyer's 1878 indigo synthesis is justly viewed as a foundational moment in the development of Germany's dye industry. Indigo, like mauve and alizarin, is an aromatic compound, and Baeyer—a former student of August Kekulé's—was an early adopter of the benzene ring. But even though indigo's aromatic nature was significant, this was not because aromatic structural theory either explained its chemistry or predicted its synthesis. Indeed, Baeyer's original publication made no reference to the ring. Instead, Baeyer's approach prioritized manipulative and experimental skills, celebrating what Baeyer called the ability "to think in phenomena"—a term akin to my "laboratory reasoning."[20] His synthesis relied on mastery of how carbonyl compounds reacted, requiring the selective reduction of a carbonyl group adjacent to nitrogen (rather than a carbonyl group adjacent to a phenyl group). Although he was certainly well versed in current theories, including the theory of aromatic structure, Baeyer's command of laboratory chemistry—rather than his theoretical convictions—enabled his achievement.[21]

In fact, Baeyer's synthesis exemplified many of this book's arguments. It mobilized the considerable resources at his disposal, including a well-equipped laboratory and numerous skilled assistants, as well as the financial backing of his industrial collaborator, Heinrich Caro. He nevertheless struggled to isolate a product identical to natural indigo and, having done so, spent several years before definitively establishing its structure.[22] This was the theoretically significant outcome of Baeyer's indigo synthesis—much as Ladenburg's synthesis would later confirm coniine's structure and molecular identity. But, as with Perkin's mauve, Baeyer's indigo also had global consequences. Once a commercially viable route from coal-tar-derived aniline was established, synthetic indigo drove a transition from plantation agriculture to industrial manufacture.[23] Thus, indigo's synthesis changed not only chemists' conceptualization of its molecular nature but also its related patterns of international trade, manufacture, labor, and consumption.

Mauve, alizarin, and indigo occupy a central position in the history of industrial chemistry and dye manufacture. Seen in the context of the present study, however, they are de-centered in favor of the methods and resources that made synthesis possible. Existing histories of organic chemistry might be characterized as juxtaposing a historiography built around taken-for-granted theories of atoms, valence, structure, and stereochemistry with a historiography of science-driven industry. This study's emphasis on laboratory reasoning produces a new historiography of organic chemistry founded on chemists' changing goals and capabilities, and whose major landmarks are the glassware revolution, the turn to synthesis, and the development of target synthesis.

The limited success of Liebig's analytical program, Hofmann's use of synthetical experiments to produce the ammonia type, and Ladenburg's coniine synthesis take center stage in this new framing, elucidating the origins of organic synthesis and establishing its power to drive theoretical advance. In particular, Ladenburg's coniine synthesis attains new prominence, as both the first synthesis of a natural alkaloid and a more illuminating example of planned natural product synthesis than alizarin. Where alizarin's unusual molecular symmetry was crucial to Graebe and Liebermann's rapid success, coniine's optical activity better typified the challenges of natural product synthesis. Thus, the new historiography both revises our view of what famous nineteenth-century chemists accomplished and focuses fresh attention on the achievements of previously neglected actors.

ORGANIC SYNTHESIS AND AROMATIC STRUCTURE

The syntheses of mauve, alizarin, and indigo—like Hofmann's use of aniline as model alkaloid and Ladenburg's coniine synthesis—capitalized on
the greater experimental tractability of aromatic compounds, highlighting
aromatic chemistry as fundamental to how nineteenth-century chemists
worked. They also illuminate the tension between the relative ease of manipulating aromatic compounds and the problems that chemists faced when
trying to develop theories to explain them. Synthetical experiments around
1850 convinced Hofmann that a latent simplicity lay beneath the apparent
complexity of aromatic organic bases. As Kekulé and others would discover,
explaining that simplicity in structural terms was very difficult indeed.[24]

Nineteenth-century chemists knew aromatic compounds by their dis-
tinctive properties and very low hydrogen content—increasingly referred to
as "unsaturation." Crucially, these included a propensity to form crystalline
derivatives that made aromatic compounds generally easier to work with
than their nonaromatic counterparts. Cheaply available from coal tar, aromatic compounds were also acquiring ever greater importance as the raw
materials of industrial production. Aromatic chemistry was an established
subfield of organic chemistry well before Kekulé's theory of aromatic structure began to gain traction.

Despite its central importance, aromatic chemistry presented a serious
challenge to developing constitutional and structural theory. By about 1860,
the search for structural concepts capable of reconciling the composition
and constitution of aromatic compounds with the rules of valence emerged
as one of organic chemistry's most pressing theoretical problems. It is therefore no surprise that several chemists, Kekulé among them, attempted to
theorize about aromatic compounds at around the same time.[25]

Why, then, did other chemists begin to find Kekulé's benzene ring persuasive? From the perspective of modern chemistry, a ring closure perhaps seems
to be the only plausible solution to the high level of unsaturation displayed by
aromatic compounds. At a time when not all chemists were convinced of the
virtues of structural theory, the notion of a hexagonal ring remained underdetermined by the experimental evidence, and it was not the only solution
championed during the 1860s, 1870s, and 1880s.[26] Although the majority
of Kekulé's competitors based their proposals on six-membered rings, there
were nevertheless significant differences between these structures—especially

with regard to the symmetry they implied. Indeed, for Ladenburg and others, Kekulé's hexagonal carbon ring was incompatible with evidence of molecular symmetry produced by counting benzene's substitution isomers—the very evidence on which Kekulé's theory was based.[27]

The present study indicates that the extension of Kekulé's theory from benzene to pyridine was highly problematic for Hofmann—even though, unlike Ladenburg, he never opposed Kekulé's benzene ring. Synthetic experience convinced Hofmann, and ultimately Ladenburg, of the ring's explanatory utility for coniine and piperidine, and hence for pyridine. Yet even after adopting a ring structure for pyridine, Ladenburg continued to query Kekulé's benzene ring.[28] Its later success has made acceptance of the ring seem inevitable. There remains much we do not understand about what persuaded chemists that Kekulé's benzene ring was useful despite its empirical inadequacies, and how they began to develop and exploit its power.

These issues go to the heart of how chemists made chemistry both visual and spatial. As chapter 9 showed, Hofmann's work with coniine prompted a shift from tabulating the relations between similar compounds to working with visual representations. Nevertheless, as we have seen, the manipulation of graphical formulae alone did not provide a reliable guide to synthesis. Such visuality became productive only when integrated with extensive, systematic, and practical knowledge of substance and reactivity, becoming embedded in a new form of laboratory reasoning that encompassed the visual.

Having synthesized coniine and accepted its representation by a ring structure, Ladenburg attempted to use the ring to predict the optical activity of isomeric propyl piperidines.[29] His prediction was only partly successful, exemplifying the difficulties that chemists encountered when learning to use ring structures. Only experience taught chemists how to interpret and internalize Kekulé's ring, molding it into a productive, even predictive, theory. The *Benzolfest*'s timing indicates that this process took several decades. Organized in 1890 around a plenary lecture by Kekulé's former student Baeyer, this international event stamped a heavily modified benzene ring with a disciplinary seal of approval some 25 years after its introduction.[30]

This study suggests that chemists' adoption of aromatic structural theory may be elucidated along similar lines to those followed here. It has shown how synthetical experiments underpinned developing constitutional theory, and how target synthesis prompted increasing engagement with structural theory. I have likewise previously explained how Emil Fischer's 1890s sugar syntheses validated and modified the stereochemical theory proposed in 1874 by le Bel and van't Hoff, helping establish organic molecules in three dimensions.[31] Extending this approach—something I intend to do in a sequel to the present volume—thus promises an account of how and why chemistry became fully visual, and how chemists ultimately stabilized a three-dimensional molecular world.

ORGANIC SYNTHESIS AND CHEMICAL THEORY

Synthetic organic chemists are frequently characterized as indifferent—or even hostile—to theory. This book has shown the inadequacy of this view. Revealing the practical and material basis of nineteenth-century chemists' knowledge explains why they prioritized experiment over speculative, as-yet-unstable theories, and how their practice connected experiment with theory through laboratory reasoning. Establishing the primacy of practice in synthetic organic chemistry in no way excludes theory—but it does require that we revise our conception of what theory meant to chemists.[32] As Ladenburg put it, chemists' theories might "be regarded as the temporary expression . . . of a certain series of facts, which are thereby . . . explained in the most practical way for us." This transience meant that—although accepted theory could sometimes provide a starting point for investigation—laboratory reasoning was chemists' vital bridge between experiment and conceptual advance. Because, as Ladenburg put it, "our natural laws are not incontrovertible truths or revelations," chemists held fast to empirical knowledge.[33]

My claim that chemistry was built on practice may seem banal. After all, other historians have written about famous chemists and chemical industry; and they have studied reactions and laboratories, apparatus and melting points. However, such studies almost invariably fall back on theory as chemistry's ultimate motivating power. This book shows that Ladenburg, Hofmann, and Liebig inhabited a very different world, in which speculative theory played no such role. Thus, any attempt to understand how the science of chemistry was built requires not merely greater emphasis on experiment but also a major shift in explanatory focus.

Elucidating how systematic experimentation enabled chemists—including Liebig, Hofmann, and Ladenburg—to make sense of an apparently disordered nature has begun to explain how their laboratory reasoning connected the behavior of liquids and crystals in glass with an abstract molecular world. It was therefore chemistry's experimental order—rather than the controversial leading edge of theory—that provided its driving force. And it was this order that enabled chemists to establish a reliable structural theory during the latter decades of the nineteenth century.

This book has offered a glimpse of that transformation. In doing so, it has shown that nineteenth-century chemical theory—whether it was expressed in words or images—relied on embedded practical knowledge for its meaning. Neither Hofmann's ammonia type, nor the structure of coniine as established by Ladenburg's synthesis, functioned without necessary connections to reactions and reagents, glassware, and melting points. One might go so far as to say that these theories could work only within the confines of the laboratories that nineteenth-century chemists built to enable their reasoning processes.

In other words, chemical theories of the kind examined in this book were unlike their mathematical counterparts in physics, which until recently have dominated the philosophy of knowledge in the physical sciences. Emergent theories of molecular constitution and structure were not subject to mathematization but remained tethered to messy reality, dependent for their validity on chemists' manipulative capabilities. Insofar as these theories could be distributed, their movement necessitated the simultaneous transfer of material resources and embodied skills. This book's account of the relationship between experiment and theory exemplifies the possibility of a fresh approach to chemical theory. In so doing, it both reinforces the desirability of positioning chemistry at the heart of any philosophical account of the sciences and illustrates the contribution that the history of chemistry might make to that endeavor.[34]

Today, chemists use structural theory to impose order on the molecular world—a change whose early stages were reflected in the distinction between Elbs's compendium of synthetic methods and Beilstein's catalog of substances, classified by formula and structure, and susceptible to chemical theory. Thus, chemists eventually created a world in which molecular structures and transformations reliably conformed to their theoretical suppositions, so that the material and the abstract interacted in ways that were practically and theoretically productive.

Theory's predictive power over organic nature is now so firmly established that it is hard for us to see that it was ever otherwise. Yet, as this book has begun to explain, this was what Liebig, Hofmann, and Ladenburg, and their peers and successors, worked so hard to accomplish. This was how they made modern chemistry, creating the molecular world as we know it today.

NOTES

INTRODUCTION

1. Marcellin Berthelot, *La Chimie Organique Fondée sur la Synthèse,* vol. 2 (Paris: Mallet-Bachelier, 1860), 811. Adolf Baeyer, *Ueber die chemische Synthese* (Munich: Akademie der Wissenschaften, 1878), 4, made a very similar claim, albeit using different language and metaphor.

2. K. C. Nicolaou, "The Art and Science of Total Synthesis at the Dawn of the Twenty-First Century," *Angewandte Chemie International Edition* 39 (2000): 54.

3. Robert B. Woodward, "Art and Science in the Synthesis of Organic Compounds: Retrospect and Prospect," in *Pointers and Pathways in Research: Six Lectures in the Fields of Organic Chemistry and Medicine,* ed. Maeve O'Connor (Bombay: CIBA of India, 1963), 21.

4. Otto Theodor Benfey and Peter J. T. Morris, *Robert Burns Woodward: Architect and Artist in the World of Molecules* (Philadelphia: Chemical Heritage Foundation, 2001); Robert B. Woodward, "The Total Synthesis of Vitamin B12," *Pure and Applied Chemistry* 33 (1973): 145–178; Robert B. Woodward and Roald Hoffmann, "Stereochemistry of Electrocyclic Reactions," *Journal of the American Chemical Society* 87 (1965): 395–397. One in his series of papers about Woodward, Jeffrey I. Seeman, "R. B. Woodward's Letters: Revealing, Elegant and Commanding," *Helvetica Chimica Acta* 100 (2017): e1700183, cast fresh light on Woodward's persuasive literary style.

5. Dieter Seebach, "Organic Synthesis: Where Now?" *Angewandte Chemie International Edition* 29 (1990): 1320–1367, 1321, was especially perturbed by a series of comments made by John Maddox, then editor of *Nature,* during a speech in Maastricht in 1988. Fears about funding are evident in a slew of more recent publications written by chemists. See, for example, K. C. Nicolaou, "The Emergence of the Structure of the Molecule and the Art of Its Synthesis," *Angewandte Chemie International Edition* 52 (2013): 131–146; K. C. Nicolaou, "The Emergence and Evolution of

Organic Synthesis and Why It Is Important to Sustain It as an Advancing Art and Science for Its Own Sake," *Israel Journal of Chemistry* 58 (2018): 104–113; Eugenio J. Llanos et al., "Exploration of the Chemical Space and Its Three Historical Regimes," *Proceedings of the National Academy of Sciences* 116 (2019): 12660–12665.

6. Nicolaou, "Emergence and Evolution," 104.

7. Seebach, "Organic Synthesis," 1352, citing Robert B. Woodward, "Synthesis," in *Perspectives in Organic Chemistry*, ed. Alexander R. Todd (New York: Interscience, 1956), 155–184.

8. Nicolaou, "Art and Science," 47. Neither Vancomycin nor Taxol® is manufactured by laboratory synthesis alone. Nicolaou's synthetic work nevertheless helped open the path to variant drugs with the potential to overcome problems of resistance and toxicity. Note that Robert A. Holton's group at Florida State University published an alternative total synthesis of Taxol® within a week of Nicolaou.

9. Nicolaou, "Art and Science," 48.

10. Jeffrey I. Seeman, "On the Relationship between Classical Structure Determination and Retrosynthetic Analysis/Total Synthesis," *Israel Journal of Chemistry* 58 (2018): 43. I sympathize with Jeff Seeman's contention that "[a]n overly positive view of organic synthesis . . . leads non-synthetic chemists to conclude that [synthetic accomplishments] are routinely simple."

11. That literature falls broadly into the following categories: biographies and institutional histories; accounts of theory change and development; history of textbooks, instruments, and laboratories; and studies of industry, application, and regulation.

12. Alan J. Rocke, "What Did 'Theory' Mean to Nineteenth-Century Chemists?" *Foundations of Chemistry* 15 (2013): 146, described organic chemistry as nineteenth-century chemistry's "crowning achievement."

13. Cf., for example, Robert E. Kohler, *From Medical Chemistry to Biochemistry* (Cambridge: Cambridge University Press, 1982). Robert E. Kohler, "Lab History: Reflections," *Isis* 99 (2008): 761–768, examined lab history as institutional history.

14. I use the term "material culture" to encapsulate how the developing science of chemistry shaped its material environment at every scale from instruments to architecture. Interpreting the laboratory as an essential component of chemistry's material culture enables this account to develop longstanding historical interest in the venues of scientific activity, contributing to a literature whose foundational studies include Owen Hannaway, "Laboratory Design and the Aim of Science: Andreas Libavius versus Tycho Brahe," *Isis* 77 (1986): 585–610; and Steven Shapin, "The House of Experiment in Seventeenth-Century England," *Isis* 79 (1988): 373–404. See also Antonio García-Belmar, "Sites of Chemistry in the Nineteenth Century," *Ambix* 61 (2014): 109–114, which introduced a special issue focused on the sites of nineteenth-century chemistry.

15. Catherine M. Jackson, "The 'Wonderful Properties of Glass': Liebig's Kaliapparat and the Practice of Chemistry in Glass," *Isis* 106 (2015): 43–69.

16. John H. Brooke, "Wöhler's Urea and Its Vital Force: A Verdict from the Chemists," *Ambix* 15 (1968): 84–114; Peter J. Ramberg, "The Death of Vitalism and the Birth of Organic Chemistry: Wöhler's Urea Synthesis in Textbooks of Organic Chemistry," *Ambix* 47 (2000): 170–195.

17. The view that synthesis followed without difficulty from knowledge of chemical structure originated in historical accounts written by late nineteenth-century chemists, such as Baeyer, *Ueber die chemische Synthese,* and it has stood almost unchallenged ever since. James R. Partington's highly influential 4-volume history of chemistry reinforced this message, dismissing organic synthesis in remarkably few pages. See James R. Partington, *A History of Chemistry,* vol. 4 (London: Macmillan, 1964), 259–260 (Wöhler's urea), 468–471 (on Berthelot), 504–505 (on Kolbe), and 799 (on Ladenburg). Subsequent generations of historians of chemistry largely followed Partington's lead, according organic synthesis a central role in the development of applied and industrial chemistry while giving the impression that this capability resulted from theoretical advances driven in other ways. For example, Bernadette Bensaude-Vincent and Isabelle Stengers, *A History of Chemistry* (Cambridge, MA: Harvard University Press, 1996), 144, perpetuated this understanding by concluding that "[o]nce certain ideas about structure . . . had been acquired, synthesis became a way of making new substances." Writing specifically about the synthetic dye industry, Alan J. Rocke, *The Quiet Revolution: Hermann Kolbe and the Science of Organic Chemistry* (Berkeley, CA: University of California Press, 1993), 294, similarly characterized the (1869) laboratory preparation of the madder dye alizarin as "the first dramatic payoff" of August Kekulé's (1865) theory of aromatic structure.

18. This practice likewise began with chemist historians but has since been widely adopted by historians of chemistry. See, for example, Carl Schorlemmer, *The Rise and Development of Organic Chemistry* (Manchester: Cornish and London, 1879); Carl Graebe, *Geschichte der organischen Chemie,* vol. 1 (Berlin: Springer, 1972 [1920]).

19. Colin A. Russell, "The Changing Role of Synthesis in Organic Chemistry," *Ambix* 34 (1987): 169–180, offered six meanings of synthesis—all of which related to the production of organic substances.

20. Friedrich Wöhler, "Ueber künstliche Bildung des Harnstoffs," *Annalen der Physik und Chemie* 88 (1828): 253–256. See also Brooke, "Wöhler's Urea."

21. John Blyth and August Wilhelm Hofmann, "On Styrole, and Some of the Products of Its Decomposition," *Memoirs and Proceedings of the Chemical Society* 2 (1843): 348; Ursula Klein and Wolfgang Lefèvre, *Materials in Eighteenth-Century Science: A Historical Ontology* (Cambridge, MA: MIT Press, 2007), examined the analogous and, by the early nineteenth century, well-established pairing of analysis and synthesis in mineral (inorganic) chemistry.

22. Henry Edward Armstrong, *Introduction to the Study of Organic Chemistry: The Chemistry of Carbon and Its Compounds* (London: Longmans, Green & Company, 1874), 2, found it necessary to define synthesis as "putting together," a definition highlighting the term's continuing unfamiliarity and limited precision.

23. Alan J. Rocke, *Nationalizing Science: Adolphe Wurtz and the Battle for French Chemistry* (Cambridge, MA: MIT Press, 2001), 252.

24. The classic study of Germany's efforts to emulate British industrial success through chemistry remains Peter Borscheid, *Naturwissenschaft, Staat und Industrie in Baden, 1848–1914* (Stuttgart: Klett, 1976).

25. Anthony S. Travis, "Perkin's Mauve: Ancestor of the Organic Chemical Industry," *Technology and Culture* 31 (1990): 51–82; Anthony S. Travis, *The Rainbow Makers: The Origins of the Synthetic Dyestuffs Industry in Western Europe* (Bethlehem, PA: Lehigh University Press, 1993); Carsten Reinhardt and Anthony S. Travis, *Heinrich Caro and the Creation of Modern Chemical Industry* (Berlin: Springer, 2000); Anthony S. Travis, *The Synthetic Nitrogen Industry in World War I: Its Emergence and Expansion* (Berlin: Springer, 2015).

26. That other European nations looked to Germany as the leader in this field is confirmed by a multitude of sources, perhaps most notably the surveys of German laboratories commissioned during the latter part of the century by Britain, France, and Italy. Edward Festing, *Report of Visits to Chemical Laboratories at Bonn, Berlin, Leipzig, etc.* (London: Eyre and Spottiswoode, 1871); Edward C. Robins, *Technical School and College Building: Being a Treatise on the Design and Construction of Applied Science and Art Buildings, and Their Suitable Fittings and Sanitation, with a Chapter on Technical Education* (London: Whittaker, 1887); Adolphe Wurtz, *Les Hautes Études Pratiques dans les Universités Allemandes* (Paris: Imprimerie Impériale, 1870); Adolphe Wurtz, *Les Hautes Études Pratiques dans les Universités d'Allemagne et d'Autriche-Hongue* (Paris: Masson, 1882); Giorgio Roster, *Delle Scienze Sperimentali e in Particolare della Chimica in Germania* (Milan: Civelli, 1872).

27. Jeffrey A. Johnson, "Academic Chemistry in Imperial Germany," *Isis* 76 (1985): 509.

28. Reinhardt and Travis, *Heinrich Caro*. See also Nadia Berenstein, "Making a Global Sensation: Vanilla Flavor, Synthetic Chemistry, and the Meanings of Purity," *History of Science* 54 (2016): 399–424.

29. Nicolaou, "Art and Science," 48.

30. Bernadette Bensaude-Vincent and Jonathan Simon, eds, *Chemistry: The Impure Science* (London: Imperial College Press, 2008).

31. Catherine M. Jackson, "Emil Fischer and the 'Art of Chemical Experimentation'," *History of Science* 55 (2017): 86–120, showed how Emil Fischer's study of the natural sugars established these as three-dimensional organic molecules. Fischer's work indicated both a necessary modification of stereochemical theory and the possibility of making numerous artificial sugars.

32. Peter J. T. Morris, ed., *From Classical to Modern Chemistry: The Instrumental Revolution* (Cambridge: Royal Society of Chemistry, 2002); Carsten Reinhardt, *Shifting and Rearranging: Physical Methods and the Transformation of Modern Chemistry* (Sagamore Beach, MA: Science History Publications, 2006).

33. For valuable contributions concerning the period up to about 1840, see Ursula Klein, "Techniques of Modelling and Paper-Tools in Classical Chemistry," in *Models as Mediators: Perspectives on Natural and Social Science*, ed. Mary S. Morgan and Margaret Morrison (Cambridge: Cambridge University Press, 1999), 146–167; and Ursula Klein, *Experiments, Models, Paper Tools: Cultures of Organic Chemistry in the Nineteenth Century* (Stanford, CA: Stanford University Press, 2003).

34. The relationship between experiment and theory has been extensively examined by philosophers and historians of physics, e.g., Peter Galison, *How Experiments End* (Chicago: University of Chicago Press, 1987). Yet studies of modern chemistry with this focus remain few. Exceptions include: Chang, *Inventing Temperature;* and Klein, *Experiments, Models, Paper Tools.*

35. The term "alkaloid" was coined in 1818 to describe organic bases of vegetable origin and is now used almost exclusively in relation to naturally occurring compounds. During the period of this study, however, it frequently denoted artificial organic bases, e.g., August Wilhelm Hofmann, "Researches Regarding the Molecular Constitution of the Volatile Organic Bases," *Philosophical Transactions of the Royal Society* 140 (1850): 97. I therefore use the term "natural alkaloids" to specify naturally occurring alkaloids (e.g., quinine and coniine), regardless of how they were obtained. In other words, this book considers both coniine extracted from hemlock and nature-identical coniine produced by laboratory synthesis to be natural alkaloids. John E. Lesch, "Conceptual Change in an Empirical Science: The Discovery of the First Alkaloids," *Historical Studies in the Physical Sciences* 2 (1981): 305–328, recounted the difficulties chemists experienced in recognizing and classifying alkaline principles derived from plants.

36. Sacha Tomic, *Aux Origines de la Chimie Organique: Méthodes et Pratiques de Pharmaciens et des Chimistes (1785–1835)* (Rennes: University of Rennes Press, 2010), examined the early phase of this development. The present book demonstrates the alkaloids' continuing importance in driving chemistry's development, as highlighted in numerous primary sources, including August Wilhelm Hofmann, *Die organische Chemie und die Heilmittellehre* (Berlin: Hirschwald, 1871).

37. C. M. Jackson, "Emil Fischer," introduced this term in relation to Fischer's sugar work. The present study shows how, half a century earlier, chemists such as Hofmann used analogous forms of reasoning to link experiment and theory.

38. Although neither comprehensive nor offered as a prescription, the following list identifies some of the studies that have been most influential on me: Chang, *Inventing Temperature;* Harry Collins, *Changing Order: Replication and Induction in Scientific Practice* (Chicago: University of Chicago Press, 1985); David Kaiser, ed., *Pedagogy and*

the Practice of Science: Historical and Contemporary Perspectives (Cambridge, MA: MIT Press, 2005); Klein, *Experiments, Models, Paper Tools;* Thomas S. Kuhn, *The Essential Tension: Selected Studies in Scientific Tradition and Change* (Chicago: University of Chicago Press, 1977); Bruno Latour, "Give me a Laboratory and I will Raise the World," in *Science Observed: Perspectives on Social Studies of Science,* ed. Karin Knorr Cetina and Michael Mulkay (London: SAGE Publications, 1983), 141–170; Alan J. Rocke, *Image and Reality: Kekulé, Kopp and the Scientific Imagination* (Chicago: University of Chicago Press, 2010); Steven Shapin, "History of Science and Its Sociological Reconstructions," *History of Science* 20 (1982): 157–211; Andrew C. Warwick, *Masters of Theory: Cambridge and the Rise of Mathematical Physics* (Chicago: University of Chicago Press, 2003).

39. C. M. Jackson, "Wonderful Properties," 43.

40. For example, William H. Brock, *Justus von Liebig: The Chemical Gatekeeper* (Cambridge: Cambridge University Press, 1997), 49, described Liebig's accomplishment as "Organic Analysis Perfected." Cf. the persistent difficulty of assigning alkaloid formulae, discussed in Emil T. Wolff, *Vollständige Uebersicht der elementar-analytischen Untersuchungen organischer Substanzen* (Halle: Eduard Anton, 1846).

41. The present work indicates that even Gerrylynn Roberts's classic studies underestimated the commercial interests surrounding the RCC's foundation. Gerrylynn K. Roberts, *The Royal College of Chemistry (1845–1853): A Social History of Chemistry in Early Victorian Britain* (PhD diss., Johns Hopkins University, 1973); Gerrylynn K. Roberts, "The Establishment of the Royal College of Chemistry: An Investigation of the Social Context of Early Victorian Chemistry," *Historical Studies in the Physical Sciences* 7 (1976): 437–485. Despite noting its imperial context, Hannah Gay's more recent institutional history similarly downplayed the importance of commerce in the RCC's early history in favor of emphasizing its multiple interconnected goals. Hannah Gay, *The History of Imperial College London, 1907–2007: Higher Education and Research in Science, Technology, and Medicine* (London: Imperial College Press, 2007).

42. Justus Liebig, "Amorphous Quinine: On Amorphous Quinine as It Exists in the Substance Known in Commerce as Quinoidine," *Lancet* 1 (1846): 585–587.

43. August Wilhelm Hofmann, "Chemische Untersuchung der organischen Basen in Steinkohlen-Teerol," *Annalen der Chemie und Pharmacie* 47 (1843): 37–87; August Wilhelm Hofmann, "Metamorphosen des Indigo's: Erzeugung organischer Basen, welche Chlor und Brom enthalten," *Annalen der Chemie und Pharmacie* 53 (1845): 1–57.

44. James S. Muspratt and August Wilhelm Hofmann, "On Toluidine, a New Organic Base," *Memoirs and Proceedings of the Chemical Society* 2 (1845): 367–383.

45. Such tropes are prevalent in popular historical material—though they also surface in professional historical literature. Four accounts of William Perkin's discovery of mauve are instructive in this regard: Philip Ball, "Perkin, the Mauve Maker," *Nature* 440 (2006): 429–429; Alan Dronsfield, "Mauveine," *RSC ChemHistory* (downloaded November 20, 2021, from https://edu.rsc.org/download?ac=15453); Simon Garfield, *Mauve: How One Man Invented a Colour that Changed the World* (London:

Faber and Faber, 2001); and Travis, "Perkin's Mauve." Despite their differences and distinct audiences, all stress the limited plausibility and theoretical basis of Perkin's attempt to make quinine while celebrating its serendipitous outcome.

46. Catherine M. Jackson, "Chemical Identity Crisis: Glass and Glassblowing in the Identification of Organic Compounds," *Annals of Science* 72 (2015): 187–205.

47. According to Ernst Homburg, "The Rise of Analytical Chemistry and Its Consequences for the Development of the German Chemical Profession (1780–1860)," *Ambix* 46 (1999): 1–32, an early nineteenth-century "revolution in the laboratory," in which portable apparatus, such as test tubes, displaced large items, including furnaces, promoted analysis as a professional activity. The present study furthers materially oriented understanding of chemistry's development. It explains why chemists, including Jöns Jacob Berzelius—who pioneered the test tube for its heat-resistant, stemless form and low cost—turned to lampworking; how working in glass transformed analysis, modifying chemical practice and training; and why those changes—especially in the context of organic synthesis—led to significant alterations in laboratory building and infrastructure during the second half of the nineteenth century.

CHAPTER 1

1. Accessed March 8, 2022, http://www.acs.org.

2. Quantitative organic analysis is the process by which chemists determine the composition of organic compounds, which frequently are substances derived from plant and animal sources, in terms of their constituent elements: carbon, hydrogen, oxygen, and (sometimes) nitrogen. The difficulties of early nineteenth-century organic analysis are examined later in this chapter under the header "The State of the Field."

3. Jacob Volhard, *Justus von Liebig,* vol. 1 (Leipzig: Barth, 1909). According to Volhard's pupil Daniel Vorländer, "Jacob Volhard," *Berichte der deutschen chemischen Gesellschaft* 45 (1912): 1856, Volhard lodged with Liebig, who was a close family friend, while studying in Giessen. See also James Campbell Brown, "Justus Liebig: An Autobiographical Sketch," in *Essays and Addresses,* 25–43 (London: Churchill, 1914).

4. Jack B. Morrell, "The Chemist Breeders: The Research Schools of Liebig and Thomas Thomson," *Ambix* 19 (1972): 1–46, began modern studies of Liebig. Morrell's study relied heavily on Volhard's biography, and his research school approach set the tone for much later work, including: Joseph S. Fruton, "The Liebig Research Group—A Reappraisal," *Proceedings of the American Philosophical Society* 132 (1988): 1–66; Joseph S. Fruton, *Contrasts in Scientific Style: Research Groups in the Chemical and Biochemical Sciences* (Philadelphia: American Philosophical Society, 1990); Frederic L. Holmes, "The Complementarity of Teaching and Research in Liebig's Laboratory," *Osiris,* 2nd series, 5 (1989): 121–164; and Frederic L. Holmes, "Liebig," in *Dictionary of Scientific Biography,* vol. 8, ed. Charles Gillespie, 329–350 (New York: Scribner, 1973). Liebig has since found his modern biographer in William H. Brock, *Justus von Liebig:*

The Chemical Gatekeeper (Cambridge: Cambridge University Press, 1997). Brock's book spans Liebig's diverse entrepreneurial and scientific career.

5. Alan J. Rocke, "Organic Analysis in Comparative Perspective: Liebig, Dumas, and Berzelius, 1811–1837," in *Instruments and Experimentation in the History of Chemistry*, ed. Frederic L. Holmes and Trevor Levere (Cambridge, MA: MIT Press, 2000), 275–310; and Melvyn C. Usselman, Christina Reinhart, Kelly Foulser, and Alan J. Rocke, "Restaging Liebig: A Study in the Replication of Experiments," *Annals of Science* 62 (2005): 1–55, helped focus my attention on Liebig's Kaliapparat.

6. Despite contributing much to our understanding of Liebig's life and work, previous studies have left the trajectory of Liebig's early independent career—including his focus on organic analysis and his crucial mobilization of glassblowing expertise—largely unexplained. On the final point, Marika Blondel-Mégrelis, "Liebig or How to Popularize Chemistry," *HYLE—International Journal for Philosophy of Chemistry* 13 (2007): 43–54, is rare in offering an explanation of Liebig's decision to abandon organic analysis that is not focused on theoretical disputes.

7. James R. Partington, *A History of Chemistry*, vol. 4, (London: Macmillan, 1964), 300.

8. Morrell, "Chemist Breeders," 8, commented in passing that synthesis was among Liebig's contributions, but did not expand on this claim.

9. Lampworking, sometimes called flameworking, is the practice of manipulating glass in the flame of a lamp or torch.

10. Wöhler to Liebig, 28 November 1830, in "Liebig-Wöhler Briefwechsel, 1829–1873," ed. Christoph Meinel and Thomas Steinhauser, in preparation.

11. Brock, *Gatekeeper*, 5–9, 27.

12. On Parisian scientific dominance during the revolutionary and Napoleonic periods, see: Maurice Crosland, *The Society of Arcueil: A View of French Science at the Time of Napoleon* (Cambridge, MA: Harvard University Press, 1967); Maurice Crosland, *Gay-Lussac: Scientist and Bourgeois* (Cambridge: Cambridge University Press, 1978); Robert Fox, "The Rise and Fall of Laplacian Physics," *Historical Studies in the Physical Sciences* 4 (1974): 89–136; John E. Lesch, *Science and Medicine in France: The Emergence of Experimental Physiology, 1790—1855* (Boston: Harvard University Press, 1984). Robert Fox, *The Savant and the State: Science and Cultural Politics in Nineteenth-Century France* (Baltimore: Johns Hopkins University Press, 2012), examined how Parisian dominance was shifted by changing relations between science and the French state following the 1814 Bourbon restoration.

13. Fulminates are dangerously unstable salts produced by reacting the metal with nitric acid in the presence of alcohol. Justus Liebig, "Sur l'Argent et le Mercure Fulminans," *Annales de Chimie et de Physique* 24 (1823): 294–317, described early investigations of silver and mercury fulminates, work that was brought to completion the following year in Justus Liebig and Joseph Louis Gay-Lussac, "Analyse du Fulminate d'Argent," *Annales de Chimie et de Physique* 25 (1824): 285–311.

14. Brock, *Gatekeeper*, 33–35, explained how this position was negotiated on Liebig's behalf.

15. Liebig to his parents, 30 March 1823 from Paris, in *Briefe von Justus Liebig: Nach neuen Funden*, ed. Ernst Berl (Giessen: Liebig-Museum-Gesellschaft, 1928), 52.

16. Liebig to August Walloth, 23 September 1824 from Giessen, in Berl, *Briefe*, 75.

17. Liebig to his parents, 28 April 1823 from Paris, in Berl, *Briefe*, 52–53.

18. Liebig to August Walloth, 23 September 1824 from Giessen, in Berl, *Briefe*, 75.

19. Cf. e.g., Holmes, "Complementarity," which took Liebig's focus on analysis completely for granted; and Rocke, "Comparative Perspective," which attributed Liebig's move into this field only to his awareness of the limitations of current analytical methods.

20. Alan J. Rocke, *Chemical Atomism in the Nineteenth Century: From Dalton to Cannizzaro* (Columbus, OH: Ohio State University Press, 1984), 173; Volhard, *Liebig*, 403. Jöns Jacob Berzelius, *Jahres-Bericht über die Fortschritte der physischen Wissenschaften: Eingericht an die Schwedische Akademie der Wissenschaften, den 31. März 1831*, trans. Friedrich Wöhler (Tübingen: Laupp, 1832), 44–48.

21. Liebig's quarrelsome behavior has become a common trope in the history of chemistry, e.g., Morrell, "Chemist Breeders," 32–33; Holmes, "Liebig," 337; and especially Partington, *History*, vol. 4, 299–300. On Liebig's vulnerability and his quest for social elevation, see Pat Munday, "Social Climbing through Chemistry: Justus Liebig's Rise from the *niederer Mittelstand* to the *Bildungsbürgertum*," *Ambix* 37 (1990): 1–19.

22. Heinrich Buff, "Ueber Indigsäure und Indigharz," *Jahrbuch der Chemie und Physik* 21 (1827): 38–59.

23. Liebig to his wife, Henriette "Jettchen" Moldenhauer, 2 Nov. 1828, in *Justus von Liebig in eigenen Zeugnissen und solchen seiner Zeitgenossen*, ed. Hertha Dechend (Weinheim: Chemie, 1953), 24, quoted in Volhard, *Liebig*, 124.

24. Holmes, "Liebig," 331, noted that Liebig's interest in uric acid drew him into mainstream organic chemistry.

25. Justus Liebig, "Ueber die Säure welche in dem Harn der grasfressenden vierfüssigen Thiere enthalten ist," *Annalen der Physik und Chemie* 17 (1829): 390–391. Liebig's paper was published in French translation as Justus Liebig, "Sur l'Acide Contenu dans l'Urine des Quadrupèdes Herbivores," *Annales de Chimie et de Physique* 43 (1830): 188–198.

26. Alkaloids are a family of medicinally important, nitrogen-containing compounds derived from plants.

27. Jean Baptiste Dumas and Pierre-Joseph Pelletier, "Recherches sur la Composition Élémentaire et quelques Propriétés Caractéristiques des Bases Salifiables Organiques," *Annales de Chimie et de Physique* 24 (1823): 163–191.

28. Paul Rossignol, "Les Travaux Scientifiques de Joseph Pelletier," *Revue d'Histoire de la Pharmacie* 77 (1989): 143. This essay was published to mark the bicentenary of Pelletier's (1788–1842) birth and the 75th anniversary of the foundation of the Société d'Histoire de la Pharmacie. In collaboration with Joseph Bienaimé Caventou, Pelletier isolated a series of medicinally important alkaloids including emetine (1817), strychnine (1818), and brucine (1819), as well as quinine (1820) and caffeine (1821). All alkaloids contained nitrogen, frequently in tiny amounts, and they were classified as a family by their ability to react with acids, a somewhat unusual property for organic compounds.

29. Rocke, *Nationalizing Science*, 55.

30. Liebig, "Ueber die Säure," 391.

31. Wöhler to Liebig, 8 June 1829, in "Liebig-Wöhler Briefwechsel," ed. Meinel and Steinhauser.

32. Liebig to Wöhler, 29 July 1829, in "Liebig-Wöhler Briefwechsel," ed. Meinel and Steinhauser.

33. Wöhler to Liebig, 22 November 1829, in "Liebig-Wöhler Briefwechsel," ed. Meinel and Steinhauser. Justus Liebig and Friedrich Wöhler, "Ueber die Zusammensetzung der Honigsteinsäure," *Annalen der Physik und Chemie* 18 (1830): 161–164.

34. Justus Liebig and Friedrich Wöhler, "Untersuchungen über die Cyansäuren," *Annalen der Physik und Chemie* 20 (1830): 369–400.

35. Liebig to Wöhler, 28 January 1830, and the response, Wöhler to Liebig, 10 February 1830, in "Liebig-Wöhler Briefwechsel," ed. Meinel and Steinhauser.

36. Liebig to Wöhler, 8 March 1830, in "Liebig-Wöhler Briefwechsel," ed. Meinel and Steinhauser.

37. Wöhler to Liebig, 21 March 1830, in "Liebig-Wöhler Briefwechsel," ed. Meinel and Steinhauser.

38. Wöhler to Berzelius, 9 July 1830, in Otto Wallach, *Briefwechsel zwischen J. Berzelius und F. Wöhler*, vol. 1. (Leipzig: Engelmann, 1901), 304. See also Rocke, "Comparative Perspective," 279, on Berzelius and Wöhler's disapprobation of the style of investigation that Liebig had acquired in Paris.

39. Jean Baptiste Dumas, "Sur l'Oxamide, Matière qui se Rapproche de Quelques Substances Animales," *Annales de Chimie et de Physique* 44 (1830): 129–143, 133; translated as Jean Baptiste Dumas, "Ueber das Oxamid, eine gewissen Thierstoffen verwandte Substanz," *Annalen der Physik und Chemie* 19 (1830): 478.

40. Rocke, "Comparative Perspective," 277.

41. Rocke, *Nationalizing Science*, 55.

42. Liebig to Wöhler, [4] August 1830, in "Liebig-Wöhler Briefwechsel," ed. Meinel and Steinhauser.

43. Berzelius to Wöhler, 14 October 1830, in Wallach, *Briefwechsel*, 315, retrospectively thanked Wöhler for arranging this meeting.

44. Wöhler to Liebig, 26 July 1830, in "Liebig-Wöhler Briefwechsel," ed. Meinel and Steinhauser.

45. Liebig to Wöhler, [4] August 1830, in "Liebig-Wöhler Briefwechsel," ed. Meinel and Steinhauser. This encounter, which took place during that year's Conference of German Doctors and Natural Scientists, was the only meeting between Liebig and Berzelius. Cf. Liebig to Wöhler, 2 September 1830, in "Liebig-Wöhler Briefwechsel," declining Wöhler's invitation to travel to Hamburg via Berlin as beyond the resources of a Giessen professor.

46. Liebig to Wöhler, 12 October 1830, in "Liebig-Wöhler Briefwechsel," ed. Meinel and Steinhauser.

47. See, for example, Berzelius to Wöhler, 9 July 1830, in *Briefwechsel*, ed. Wallach, 304.

48. Berzelius to Berthollet, 23 September 1811, in *Correspondance entre Berzelius et C. L. Berthollet (1810–1822)*, ed. H. G. Söderbaum (Uppsala: Académie Royale des Sciences de Suède, 1912), 116, attempted to initiate a correspondence about organic analysis with Gay-Lussac. Gay-Lussac took 4 years to respond.

49. Louis-Jacques Thenard, *Traité de Chimie élémentaire, théorique et pratique*, vol. 4 (Paris: Crochard, 1816), 200–202.

50. Jöns Jacob Berzelius, "Experiments to Determine the Definite Proportions in Which the Elements of Organic Nature Are Combined," Part 2 of 5, *Annals of Philosophy* 4 (1814): 402, 406, separated the determinations of hydrogen and carbon as follows: Berzelius obtained a direct measure of hydrogen by collecting condensed water in a receiver and trapping water vapor in a small tube filled with calcium chloride. According to Rocke, *Nationalizing Science*, 22, this approach was soon taken up by Gay-Lussac. In adopting this separation of hydrogen from carbon, however, Liebig, "Ueber die Säure," 392, credited Gay-Lussac rather than Berzelius.

51. Berzelius to Liebig, 13 December 1831, in *Berzelius und Liebig: Ihre Briefe 1831–1845 mit gleichzeitigen Briefen von Liebig und Wöhler*, 3rd edition, ed. Wilhelm Lewicki (Göttingen: Cromm, 1991), 19, expressed pleasure that Wöhler, thanks to Liebig, had overcome his previous "distaste" for organic analysis.

52. Justus Liebig, "Ueber die Analyse organischer Substanzen," *Annalen der Physik und Chemie* 18 (1830): 357, criticized William Prout, "On the Ultimate Composition of Simple Alimentary Substances; with some Preliminary Remarks on the Analysis of Organized Bodies in General," *Philosophical Transactions of the Royal Society* 117 (1827): 355–388, for failing to address the central problem of nitrogen determination.

53. Chemists will be aware that oxygen (unlike carbon, hydrogen, and nitrogen) is not now experimentally determined during analysis. Rather, oxygen is calculated so

as to bring the total percentage composition up to 100. This shift in analytical practice was taking place during the period discussed in this chapter. Whereas Dumas
and Pelletier attempted to determine oxygen experimentally in 1823, Liebig in 1830
calculated oxygen by difference.

54. It is important not to confuse this issue with the existence of isomers (different compounds with identical composition). Although chemists confronted similar
problems in identifying isomeric substances, the matter at issue here concerned
distinct compounds whose compositions were very close but not identical.

55. Dumas and Pelletier, "Recherches," 179 cf. 176.

56. Dumas and Pelletier, "Recherches," 185.

57. Liebig would soon identify problems of exactly this kind. Justus Liebig, "Ueber
einen neuen Apparat zur Analyse organischer Körper, und über die Zusammensetzung einiger organischen Substanzen," *Annalen der Physik und Chemie* 97 (1831):
24–26, reported a difference of only about 2% in the carbon content of the closely
related cinchona alkaloids, quinine and cinchonine: 75.76% and 77.81%, respectively, cf. theoretically required values of 74.39% and 78.67%, respectively. The formulae Liebig assigned differed by just 2 atoms of hydrogen and 1 of oxygen between
the two cinchona alkaloids, quinine ($C_{20}H_{24}N_2O_2$) and cinchonine ($C_{20}H_{22}N_2O$).

58. Jöns Jacob Berzelius, "Experiments to Determine the Definite Proportions
in Which the Elements of Organic Nature Are Combined," *Annals of Philosophy*
4 (1814): 323; and Jöns Jacob Berzelius, "Essay on the Cause of Chemical Proportions, and on Some Circumstances Relating to Them: Together with a Short and
Easy Method of Expressing Them," *Annals of Philosophy* 3 (1814): 51–62, argued for
the validity and usefulness of molecular formulae. Ursula Klein, *Experiments, Models,
Paper Tools: Cultures of Organic Chemistry in the Nineteenth Century* (Stanford, CA:
Stanford University Press, 2003), is the classic study of the introduction of Berzelian
formulae and their use in guiding early nineteenth-century chemists' reasoning.

59. Ursula Klein, "Shifting Ontologies, Changing Classifications: Plant Materials
from 1700 to 1830," *Studies in History and Philosophy of Science* 36 (2005): 261, highlighted this important shift in chemists' conception of the composition of organic
matter. As skepticism about the ontological significance of atoms grew during the
nineteenth century, many chemists (including Liebig) substituted the term "equivalent" for "atom." For more on this, see Rocke, *Chemical Atomism*.

60. Dumas and Pelletier, "Recherches," 189–191. Formulae, for Dumas and Pelletier,
were subservient to percentage composition, serving as heuristic devices that facilitated comparisons among the alkaloids.

61. I have adjusted Dumas and Pelletier's formula to bring its basis into line with
that used by Liebig in 1830.

62. Melvyn Usselman, "Liebig's Alkaloid Analyses: The Uncertain Route from
Elemental Content to Molecular Formulas," *Ambix* 50 (2003): 71–89, explained the

relationship between analytical data and combining weight in fixing the number of atoms of each kind.

63. Liebig, "neuen Apparat," 13; Robiquet's response appeared in Pierre Jean Robiquet, "Note sur la Narcotine," *Journal de Pharmacie* 17 (1831): 637–643.

64. Dumas and Pelletier, "Recherches," 188–190, dismissed this possibility without naming Robiquet.

65. As late as 1828, Joseph Louis Gay-Lussac, *Cours de Chimie* (Paris: Pichon et Didiers, 1828), 26, maintained that "[i]n the analysis of plants, it is necessary to try to obtain gaseous products rather than solid products, because with gaseous products, one attains a greater degree of precision."

66. Dumas and Pelletier, "Recherches," 184–185.

67. Liebig, "neuen Apparat," 13.

68. Liebig, "Ueber die Säure," 391.

69. Jöns Jacob Berzelius, "An Address to Those Chemists Who Wish to Examine the Laws of Chemical Proportions, and the Theory of Chemistry in General," *Annals of Philosophy* 5 (1815): 129.

70. Berzelius, "Experiments," Part 2, 406, measured the mass of water that condensed in a glass receiver and was trapped by a glass tube containing calcium chloride. Although he measured carbon in the same experiment, this was trapped in a quite different part of the apparatus. Berzelius determined each element only from measurements that related directly to that element and were independent of the other element.

71. Dumas and Pelletier had also analyzed veratrine, emetine, and caffeine.

72. Liebig to Wöhler, [4] August 1830, in "Liebig-Wöhler Briefwechsel," ed. Meinel and Steinhauser, shows that Liebig was contemplating a new apparatus for nitrogen determination at this time. Analyzing morphine also entailed finding out how much hydrogen and oxygen were present. Because these elements were not the focus of Liebig's dispute with Dumas, this chapter does not address these aspects of Liebig's analytical approach.

73. The Bavarian State Library in Munich, Germany, holds 12 of Liebig's laboratory notebooks, the earliest of which (Liebigiana I. C. 1, consisting of 89 [+2] leaves) bears the date "1830" on its cover.

74. Frederick L. Holmes, "Justus Liebig and the Construction of Organic Chemistry," in *Essays on the History of Organic Chemistry*, ed. James G. Traynham (Baton Rouge: Louisiana State University Press, 1987): 132.
I began work using a microfilm copy of Liebig's 1830 notebook held in the library of the Max Planck Institute for the History of Science, Berlin. In order to complete this study, however, I also worked with the original notebook held in the Bavarian State Library, Munich, Germany.

75. In-text citations from the notebook follow standard practice: manuscript page number followed by r(ecto) or v(erso), as required.

76. Usselman et al., "Restaging Liebig," outlined a timeline that was very helpful in the early stages of my study of Liebig's notebook. As Usselman et al. reported, Liebig did not date his notebook entries. Like them, my basic assumption is that Liebig filled the pages of his notebook in chronological sequence. Once able to access the original manuscript, however, I also observed that Liebig frequently revisited earlier pages to make amendments. Close examination of the original was thus vital in enabling me to link work done at the same time (and recorded using the same pen and ink), even where this work appeared on different pages of the notebook. In this way, I was able to elucidate further the most probable order of Liebig's work with morphine, and hence the process by which he developed the Kaliapparat.

77. Liebig, "neuen Apparat," 15. With help from Wöhler and others, Liebig acquired morphine from two pharmacists, Emanuel Merck in Berlin and Carl Wittstock in Darmstadt, which he analyzed without further purification.

78. Liebig weighed the Kaliapparat before and after analysis, whereas Berzelius had measured the gain in weight of a receiver in which water condensed and a calcium chloride tube that trapped water.

79. Liebig demonstrated that the presence of nitrogen did not affect the Kaliapparat's reliability as a device for measuring carbon by analyzing two nitrogen-rich substances, urea (62v) and cyanic acid (64v). These analyses were published in Liebig, "neuen Apparat," 8.

80. Liebig to Wöhler, 12 October 1830, in "Liebig-Wöhler Briefwechsel," ed. Meinel and Steinhauser, tends to confirm the provisional date suggested by Usselman et al., "Restaging Liebig," 8, for Liebig's invention of the Kaliapparat.

81. Dumas and Pelletier, "Recherches," 185, calculated 72.02% as the average of two readings (71.36% and 72.68%) obtained by analyzing morphine from different sources.

82. Liebig, like Dumas and Pelletier, was working with morphine supplied by two different pharmacists, Merck and Wittstock. The Kaliapparat gave 71.4% carbon for morphine from Merck, but no more than 69.1% carbon for morphine from Wittstock. Liebig used Wittstock's morphine for the volumetric analysis, which gave the carbon content as 72.0%. A discrepancy of almost 3% between these two values was well outside the acceptable range.

83. "After a series of futile experiments," Liebig ("neuen Apparat," 4) "felt himself obliged entirely to separate the determination of nitrogen from that of carbon." Since Berzelius had no personal experience of nitrogen determination, he was presumably unable to provide Liebig with specific, practical guidance concerning how to do this.

84. Justus Liebig, *Anleitung zur Analyse organischer Körper* (Braunschweig: Vieweg, 1837), 48. See also Justus Liebig, *Instructions for the Chemical Analysis of Organic Bodies*, trans. William Gregory (Glasgow: Griffin, 1839).

85. Liebig, "neuen Apparat," 9–10, and table 1, figure 3.

86. Cf. Liebig, "neuen Apparat," 17.

87. Liebig, "neuen Apparat," 11–12, and table 1, figure 4. Liebig recognized the cause of this problem as what chemists today would call the nonideal behavior of ammonia. In Liebig's experiments, this led to considerable variation in the final volume of nitrogen.

88. This problem had long been recognized by Parisian analysts: see Joseph Louis Gay-Lussac, "Recherches sur l'Acide Prussique," *Annales de Chimie et de Physique* 95 (1815): fn. on 183.

89. Nitrogen tended to oxidize under the conditions of combustion analysis, producing acidic nitrogen oxides.

90. In fact, the problem was worse for the Parisian method: Because side reactions doubled the apparent volume of nitrogen, they elevated the combined volume of carbon dioxide plus nitrogen; but since this volume was then removed from the mixture, the final, residual volume of nitrogen was too low. Overall, therefore, side reactions made the ratio of nitrogen to carbon dioxide much too small.

91. Since nitrogen and carbon were linked in the standard volumetric analysis, this would also tend to produce values for carbon that were too high.

92. Liebig, "neuen Apparat," 10.

93. Liebig, "neuen Apparat," 9–10.

94. Cf. Liebig, "neuen Apparat," 17.

95. Liebig, "neuen Apparat," 10.

96. Liebig to Berzelius, 8 January 1831, in Lewicki, *Berzelius und Liebig,* 5.

97. This differential effect on nitrogen and carbon also explains how Liebig could simultaneously calibrate his Kaliapparat against Dumas and Pelletier's carbon determinations and use the Kaliapparat to argue that their nitrogen determinations were flawed.

98. Liebig, "neuen Apparat," 2–4.

99. Liebig did not explain the source of this average result for nitrogen. Although it was similar to nitrogen determinations recorded in his notebook, I was unable to produce this exact figure by averaging any of those reported values.

100. Liebig used at least some of his own analytical data as the basis for this calculation, including a volumetric analysis recorded at 68v.

101. Liebig, "neuen Apparat," 9.

102. Liebig, "neuen Apparat," 13, adopted Berzelius's atomic weights during this work. Liebig's published paper used 76.437 as the atomic weight of carbon (on a scale where oxygen is 100; equivalent to 12.23 on a scale where oxygen is 16), whereas he had previously relied on the "usual abbreviated" atomic weights. See

Jöns Jacob Berzelius, "Ueber die Bestimmung der relativen Anzahl von einfachen Atomen in chemischen Verbindungen," *Annalen der Physik und Chemie* 7 (1826): 397–416. Jöns Jacob Berzelius, "Ueber das Atomgewicht des Kohlenstoffs," *Annalen der Pharmacie* 30 (1839): 241–249, discussed continuing debates concerning carbon's atomic weight. By 1845, Jöns Jacob Berzelius, *Atomgewichts-Tabellen* (Braunschweig: Vieweg, 1845), 5, reported carbon's atomic weight as 75.12 (equivalent to 12.01).

103. For example, Liebig converted the measured mass of carbon dioxide to the mass of carbon using the percentage by mass of carbon in carbon dioxide. According to Jöns Jacob Berzelius and Pierre Louis Dulong, "Nouvelles Déterminations des Proportions de l'Eau et de la Densité de Quelques Fluides Élastiques," *Annales de Chimie et de Physique* 15 (1820): 395, this value was 27.65%. But Liebig's notebook indicates that he also used other values in this calculation, e.g., 27.64% on 53v.

104. It is likely that this failure reinforced Liebig's view that the Kaliapparat had yet to produce adequately accurate values for morphine's carbon content. Equally, assigning a satisfactory formula for morphine was probably significant in convincing Liebig that the Kaliapparat had begun working properly.

105. Liebig, "neuen Apparat," 18. Liebig proposed formulae for morphine containing 34 and 36 carbon atoms during this work. Adjusted to modern values, this means that Liebig's analyses left him in doubt as to whether morphine contained 17 or 18 atoms of carbon.

106. As Usselman, "Liebig's Alkaloid Analyses," 74, noted, Liebig believed (for reasons he did not explain) that molecular mass was far more stable than percentage composition derived from analysis—even though both were experimentally determined.

107. Usselman, "Liebig's Alkaloid Analyses," also examined Liebig's data selection but with distinct goals and a different outcome.

108. Liebig, "neuen Apparat," 17.

109. Liebig, "neuen Apparat," 10.

110. Liebig, "neuen Apparat," 18–19.

111. Usselman et al., "Restaging Liebig," 32, concluded that Liebig's "implicit rationale" for selecting this value as the basis for his new formula for morphine "was that the largest sample of morphine was likely to give the best values." By undermining the equivalence of his results, this study suggests that Liebig's choice was driven by theoretical rather than empirical concerns and therefore that—contra Usselman et al.—the motive was to improve his analytical results.

112. Several classic studies have demonstrated that data selection is fundamental to science: John Earman and Clark Glymour, "Relativity and Eclipses: the British Eclipse Expeditions of 1919 and Their Predecessors," *Historical Studies in the Physical and Biological Sciences* 11 (1980): 85; Gerald Holton, "Subelectrons, Presuppositions, and the Millikan–Ehrenhaft Dispute," in *The Scientific Imagination: Case Studies*, 25–83

(Cambridge: Cambridge University Press, 1978), 52–54 and 71–72. Yet we remain without agreed-on boundaries concerning which kinds of selection are permissible.

113. Liebig, "neuen Apparat," 13, claimed his paper provided "numerical details," and this assertion led Melvyn C. Usselman, "Liebig's Alkaloid Analyses," 78, to infer that Liebig reported all his experimental data. Similarly, Usselman et al., "Restaging Liebig," 32, believed that Liebig did not "make any attempt to conceal the [data selection] process from his readers—quite the contrary." However we interpret Liebig's motives, the present reading of his notebook undermines both those conclusions.

114. Liebig, "neuen Apparat," 28. See also Liebig to Berzelius, 8 January 1831, in Lewicki, *Berzelius und Liebig*, 4.

115. Liebig to Berzelius, 8 January 1831, in Lewicki, *Berzelius und Liebig*, 5, stressed the extreme difficulty of this work.

116. Liebig, "neuen Apparat," 18, admitted the difference between them was "not significant."

117. Cf., for example, Brock, *Gatekeeper*, 48–51, which outlined Liebig's analytical contributions under the title "Organic Analysis Perfected."

118. Emil T. Wolff, *Vollständige Uebersicht der Elementar-Analytischen Untersuchungen organischer Substanzen,* (Halle: Eduard Anton, 1846), 491–492, reported no less than seven distinct analyses of morphine.

119. Usselman, "Liebig's Alkaloid Analyses," 87.

120. Berzelius to Liebig, 22 April 1831, in Lewicki, *Berzelius und Liebig*, 7–8; Berzelius to Liebig, 13 December 1831, in Lewicki, *Berzelius und Liebig*, 21.

121. Liebig to Berzelius, 8 May 1831, in Lewicki, *Berzelius und Liebig*, 9–10.

122. Liebig to Wöhler, 3 April 1831, in "Liebig-Wöhler Briefwechsel," ed. Meinel and Steinhauser.

123. Although Liebig's lack of success with nitrogen has previously been noted, e.g., by Usselman, "Liebig's Alkaloid Analyses," and Usselman et al., "Restaging Liebig," 47, historians have not pursued its significance for either Liebig's accomplishments and career or for chemistry's future development.

124. Rocke, "Comparative Perspective," 290.

125. Cf. Ferenc Szabadváry, *History of Analytical Chemistry* (Chemin de la Sallaz, Switzerland: Gordon and Breach, 1992 [1966]), 285. Szabadváry's book remains the only dedicated history of analysis. In fact, both Antoine L. Lavoisier, *Oeuvres de Lavoisier,* vol. 1 (Paris: Imprimerie Impériale, 1864), 347–351, and Berzelius, "Experiments," part 2, 406, measured carbon by mass. This misleading idea is nevertheless persistent, as Usselman et al., "Restaging Liebig," 5, shows.

126. Usselman, "Liebig's Alkaloid Analyses," 74, and Usselman et al., "Restaging Liebig," 4, 33, used reconstructed alkaloid analyses to claim that Liebig's Kaliapparat improved the accuracy of carbon determination. Despite the many useful insights

328 NOTES TO CHAPTER 2

these studies produced, it is therefore worth noting that Usselman and his collaborators reconstructed Liebig's method and apparatus as described in 1837. A similarly
later viewpoint perhaps explains why Brock, *Gatekeeper*, 50, also claimed improved
accuracy for Liebig's carbon determinations. As we shall see in chapter 2, Liebig and
his students did a good deal of work to improve and stabilize carbon determination
using the Kaliapparat between 1830 and 1837.

127. Liebig, "neuen Apparat," 4–5.

128. Liebig, "neuen Apparat," 4–5, 7.

129. Berzelius to Liebig, 22 April 1831, in Lewicki, *Berzelius und Liebig*, 6.

130. Wöhler to Berzelius, 24 November 1831, in Wallach, *Briefwechsel*, 380–381.

131. Jean Baptiste Dumas, "Lettre de M. Dumas à M. Gay-Lussac, sur les Procédés de
l'Analyse Organique," *Annales de Chimie et de Physique* 47 (1831): 198.

132. Jean Baptiste Dumas, "Sur l'Esprit Pyroacétique," *Annales de Chimie et de Physique* 49 (1832): 209.

133. Dumas, "Lettre," 199, 204. See also Jean Baptiste Dumas, "Recherches de
Chimie Organique," *Annales de Chimie et de Physique* 53 (1833): 164–181.

CHAPTER 2

1. Morrell, "Chemist Breeders," 5, introduced the factory analogy.

2. Holmes, "Complementarity;" Alan J. Rocke, "Origins and Spread of the 'Giessen
Model' in University Science," *Ambix* 50 (2003): 90–115.

3. Brock, *Gatekeeper*, 48.

4. Justus Liebig, "Ueber die Zusammensetzung der Gerbesäure (Gerbestoff) und der
Gallussäuren," *Annalen der Pharmacie* 10 (1834): 173.

5. Catherine M. Jackson, "Visible Work: The Role of Students in the Creation of
Liebig's Giessen Research School," *Notes and Records of the Royal Society* 62 (2008): 39 ff.

6. Myles W. Jackson, *Spectrum of Belief: Josef von Fraunhofer and the Craft of Precision
Optics* (Cambridge, MA: MIT Press, 2000) and Mario Biagioli, *Galileo's Instruments of
Credit: Telescopes, Images, Secrecy* (Chicago: Chicago University Press, 2006) similarly
noted Fraunhofer's strategies for controlling his optical discoveries and Galileo's for
controlling the telescope.

7. Liebig to Berzelius, January 8, 1831, Lewicki, *Berzelius und Liebig*, 3–5.

8. Friedrich Wöhler and Justus Liebig, "Untersuchungen über das Radical der Benzoësäure," *Annalen der Physik und Chemie* 26 (1832): 325–343.

9. Rocke, *Nationalizing Science*, 49–50.

10. Liebig was well aware of this distinction. In 1832, responding to his rival Dumas,
Liebig condemned Dumas's attempt to build a research group supported by personal

patronage. See, Justus Liebig, "Bemerkungen zur vorhergehenden Abhandlung," *Annalen der Pharmacie* 2 (1832): 19–30. As Rocke (*Nationalizing Science*, 115) noted, whereas Dumas could afford to bear the cost of his students' research, Liebig's very different circumstances ensured his reliance on the labor of students who paid him fees.

11. Dumas, "Lettre," 198.

12. Rocke, "Comparative Perspective," 286–287, suggested Dumas's uptake of the Kaliapparat was "influenced" by Jules Gay-Lussac. In fact, as Rocke (*Nationalizing Science*, 46) explained, it was Oppermann and not Jules Gay-Lussac who brought the first Kaliapparat to Paris in December 1831, along with a recommendation from Liebig highlighting Oppermann's skill with the new device.

13. Jules Gay-Lussac and Théophile-Jules Pelouze, "Sur la Salicine," *Annales de Chimie et de Physique* 44 (1831): 220–221.

14. Jules Gay-Lussac and Théophile-Jules Pelouze, "Sur la Composition de la Salicine," *Annales de Chimie et de Physique* 48 (1831): 111. Cf. Rocke, "Comparative Perspective," 287.

15. Crosland, *Gay-Lussac*, 254.

16. Rocke, "Comparative Perspective," 286–287; Rocke, "Giessen Model," 98; and Rocke, *Nationalizing Science*, 46, began tracing the early spread of the Kaliapparat. Usselman et al., "Restaging Liebig," 4, nevertheless concluded that Liebig's apparatus "was rapidly adopted across European centers of organic-chemical research."

17. Surveying chemical journals during this period reveals remarkably few references to Liebig's new apparatus, almost all appearing in articles whose authors were connected to Liebig.

18. Berzelius to Liebig, January 15, 1833, in Lewicki, *Berzelius und Liebig*, 50–51.

19. Liebig, "neuen Apparat," 4–5.

20. As Usselmann et al., "Restaging Liebig," 9, noted, Liebig analyzed morphine by both methods but did not publish the comparison.

21. Justus Liebig, "Ueber die Zusammensetzung der Camphersäure und des Camphers," *Annalen der Physik und Chemie* 20 (1830): 41–47.

22. Liebig, "Camphersäure," 45.

23. Charles Oppermann, "Ueber die Zusammensetzung des Terpenthinöls und einiger von Demselben entstehenden Producte," *Annalen der Physik und Chemie* 22 (1831): 194.

24. Oppermann, "Terpenthinöls," 194.

25. Percy A. Houseman, "Camphor, Natural and Synthetic," *Science Progress in the Twentieth Century (1906–1916)* 3 (1908): 60–68, is a useful early twentieth-century review.

26. Oppermann, "Terpenthinöls," 200.

27. Oppermann, "Terpenthinöls," 207. Jean Baptiste Dumas, "Mémoire sur les Substances Végétales qui se Rapprochent du Camphre, et sur Quelques Huiles Essentielles," *Annales de Chemie et de Physique* 50 (1832): 226, 231, highlighted this cycle as a rare example in the organic realm of the "analyses and syntheses" that were typical of inorganic chemistry. While Dumas did not pursue this notion, Liebig may well have taken note.

28. Holmes, "Liebig," 333 ff.

29. Dumas, "Substances Végétales," 226, 230–231.

30. Brock, *Gatekeeper*, 57. According to Brock, *Gatekeeper*, 40, Liebig at this time still depended on financial support from his father.

31. As Brock, *Gatekeeper*, 45–46, explained, laboratory instruction in Giessen began with a series of graded exercises in inorganic analysis known as "the alphabet."

32. Justus Liebig, "Nachricht, das Chemisch-Pharmaceutische Institut zu Giessen Betreffend," *Magazin für Pharmacie* 20 (1827): 98–99; and Liebig, "Nachricht, das Chemisch-Pharmaceutische Institut zu Giessen Betreffend," *Jahrbuch der Chemie und Physik* 21 (1827): 376; cited in translation from Holmes, "Complementarity," 127–128.

33. Buff, "Indigsäure und Indigharz;" Friedrich Kodweiss, "Ueber die Zusammensetzung der Harnsäure und über die Producte, welche durch ihre Zersetzung mit Salpetersäure erzeugt werden," *Annalen der Physik und der Chemie* 19 (1830): 1–25.

34. Fruton, "Reappraisal," Appendix 1.

35. Several existing accounts assume or imply that this was the case, e.g., Brock, *Gatekeeper*, 50.

36. According to Usselman et al., "Restaging Liebig," 12, Ettling made the Kaliapparat smaller sometime during 1832. Lafond, "Ueber die Kunst Glas zu blasen, mit Verbesserungen von Danger," *Annalen der Pharmacie* 7 (1833): 298–313, provided the first published instructions for how to make a standard Kaliapparat, including specifying its dimensions.

37. Rocke, *Nationalizing Science*, 53. Rodolphe Blanchet and Ernst Sell, "Ueber die Zusammensetzung einiger organischer Substanzen," *Annalen der Pharmacie* 6 (1833): 304, 305.

38. Blanchet and Sell, "Zusammensetzung," 260.

39. Combining proportions refers to one of the fundamental tenets of Daltonian atomism, the law of multiple proportions. This law states that if two elements form more than one compound, then the ratios of the masses of the second element which combine with a fixed mass of the first element will always be reducible to ratios of small whole numbers. Many mineral substances had been shown to adhere to this requirement, but the difficulties of organic analysis meant that it was much harder to demonstrate for organic compounds.

40. Blanchet and Sell, "Zusammensetzung," 259–260.

41. Blanchet and Sell's paper therefore provides evidence for the notion put forward by Usselman et al., "Restaging Liebig," 14, that "analytical results acquired validity only when there was group consensus."

42. Numerous studies—beginning with Collins's classic account of the TEA laser (*Changing Order,* chapter 3)—have demonstrated the problems of replication at distant sites. Giessen analysis contributes to our understanding of the difficulties of replication by showing that—in the case of organic analysis in the 1830s—even those with similar training working side-by-side did not as a matter of course produce identical numerical results. In tackling this problem, Liebig required only that two analysts produced results consistent with the same molecular formula—a solution based on the newly established atomistic understanding of molecular composition.

43. For example, Blanchet and Sell, "Zusammentsetzung," 265–266, set out this comparison for commercial oil of turpentine.

44. Blanchet and Sell, "Zusammentsetzung," 279–280.

45. Rocke, "Comparative Perspective," 299, described how emerging commitment to atomism meant formula displaced composition as the primary outcome of analysis.

46. Fruton, "Reappraisal," Appendix 1.

47. Cf. Morrell, "Chemist Breeders;" Fruton, "Reappraisal."

48. Liebig to Wöhler, 1 May 1832, in "Liebig-Wöhler Briefwechsel," ed. Meinel and Steinhauser, contained a particularly emotional outpouring. See also, Liebig to Berzelius, 30 May 1832, in Lewicki, *Berzelius und Liebig,* 29, 31.

49. For example, Justus Liebig, "Bemerkungen der Redaktion zu der Abhandlung der Herren Pelletier und Couerbe," *Annalen der Pharmacie* 10 (1834): 203–210.

50. Wöhler and Liebig, "Radical der Benzoësäure."

51. Justus Liebig, "Ueber einige Stickstoff-Verbindungen," *Annalen der Pharmacie* 10 (1834): 2–3.

52. Liebig, "neuen Apparat," 8–9, applied this control to analyses of uric and cyanic acids.

53. Justus Liebig, "Analyse der Harnsäure," *Annalen der Pharmacie* 10 (1834): 48.

54. Liebig, "Harnsäure," 47. A "control" was Liebig's generic term for any procedure in which the results of multiple distinct experiments were used to validate each other.

55. Justus Liebig, "Ueber die Zusammensetzung der Hippursäure," *Annalen der Physik und Chemie* 32 (1834): 573.

56. For most compounds, he considered "Gay-Lussac's method perfectly adequate" (Liebig, "Stickstoff-Verbindungen," 13). But when revisiting the analysis of hippuric acid, which contained very little nitrogen, Liebig preferred a "direct" method of nitrogen determination (probably one recently developed by Dumas, though Liebig

did not specify). See Justus Liebig, "Ueber die Zusammensetzung der Hippursäure," *Annalen der Pharmacie* 12 (1834): 20.

57. Liebig, "Gerbesäure," 172–174.

58. Liebig, "Gerbesäure," 173.

59. Brock, *Gatekeeper*, 55–56; Rocke, *Quiet Revolution*, 68–71. See also Justus Liebig, "Analyse, organische," in *Handwörterbuch der reinen und angewandten Chemie*, vol. 1, ed. Justus Liebig, Johann C. Poggendorff, and Friedrich Wöhler (Braunschweig: Vieweg, 1842), 357–400; Liebig, *Anleitung*.

60. Cf. Rocke, *Nationalizing Science*, 53–54, which claimed that both Liebig's apparatus and his method remained essentially unchanged between 1831 and 1837.

61. Liebig, *Anleitung*, 22–23.

62. Liebig, *Anleitung*, 53–55, referring to the small molecules Liebig believed were most suitable for inexperienced analysts.

63. Cf. Liebig, *Anleitung*, 61.

64. Liebig, *Anleitung*, 48–49, described an adapted version of Dumas's 1833 method and proposed a numerical adjustment of 1% to the experimentally determined nitrogen content.

65. Liebig, *Anleitung*, 40–53. The section on nitrogen determination was followed in both cases by a description of the controls to be used in organic analysis.

66. Liebig, *Anleitung*, 31 (*untadelhaften*), 47 (*falsch*), cited in William Gregory's translation from Liebig, *Instructions*, 24, 37.

67. Liebig, *Anleitung*, 28, cited in William Gregory's translation from Liebig, *Instructions*, 23.

68. Liebig, *Instructions*, 28 (William Gregory's addition).

69. Liebig, *Anleitung*, 38–39. For more on the disagreement between Liebig and Berzelius, see Rocke, "Comparative Perspective," 294–295. Jöns Jacob Berzelius, "Nachtrag zum VI. Band: Ueber die Analyse organischer Körper durch Verbrennung," in *Lehrbuch der Chemie*, 3rd edition, vol. 7 (Dresden: Arnold, 1838), 610–630, contained Berzelius's refutation of Liebig's criticisms. Holmes, "Liebig," 338–344, explained how Liebig's investigation of organic acids led his theoretical views to diverge increasingly from those of his former mentor.

70. Liebig, *Anleitung*, 37.

71. Hermann Hess, "Ueber die Wasserstoffbestimmung bei der Analyse organischer Substanzen," *Annalen der Pharmacie* 26 (1838): 189–194.

72. Liebig to Berzelius, March 7, 1838, in Lewicki, *Berzelius und Liebig*, 148–151, defended the judgment expressed in Liebig's textbook in almost the same terms that he subsequently deployed in print against Hess.

73. Liebig note to Hess, "Wasserstoffbestimmung," 192–194; cited in William Gregory's translation from Liebig, *Instructions*, 28–30.

74. Fruton, "Reappraisal," 58, reported that Voskresensky matriculated in 1837.

75. Hermann Hess, "Ueber die Zusammensetzung der Zuckersäure," *Annalen der Pharmacie* 26 (1838): 1–9.

76. M. C. J. Thaulow, "Ueber die Zuckersaure," *Annalen der Physik und Chemie* 44 (1838): 497–513; Hermann Hess, "Ueber die Constitution der Zuckersäure," *Annalen der Pharmacie* 30 (1839): 302–313.

77. Justus Liebig, "Ueber die vorstehende Notiz des Hrn. Akademikers Hess in Petersburg," *Annalen der Pharmacie* 30 (1839): 313–319.

78. Hermann Hess, "Berichtigung zu meinem Aufsatz über die Constitution der Zuckersäure," *Annalen der Chemie und Pharmacie* 33 (1840): 116–117, also presented slightly amended analytical results.

79. Justus Liebig, "Bemerkungen zu vorstehenden Berichtigung," *Annalen der Chemie und Pharmacie* 33 (1840): 117–125.

80. See Klein, *Experiments, Models, Paper Tools*, 180–182, on Liebig's response to Zeise's first analyses.

81. W. C. Zeise, "Neue Untersuchungen über das entzündliche Platinchlorüs," *Annalen der Pharmacie* 23 (1837): 1–11. Zeise's results led him to criticize Berzelius's views concerning ethers (a category that included many compounds now called "esters"). Liebig to Berzelius, 21 December 1837, Lewicki, *Berzelius und Liebig*, 140, explained Liebig's desire to back Berzelius.

82. Justus Liebig, "Ueber die Aethertheorie," *Annalen der Pharmacie* 23 (1837): 13.

83. Friedrich Wöhler and Justus Liebig, "Ueber die Bildung des Bittermandelöls," *Annalen der Pharmacie* 22 (1837): 13.

84. Liebig, *Anleitung*, 35–36.

85. Liebig, "Aethertheorie," 16. Liebig attributed this excess of hydrogen to the presence of moisture, introduced by the oxidizing agent used in the analysis.

86. Liebig, "Aethertheorie," 18–19.

87. Zeise, "Neue Untersuchungen;" Liebig, "Aethertheorie," 19.

88. Klein, *Experiments, Models, Paper Tools*, 180–182.

89. Victor Regnault, "Neue Untersuchungen über die Zusammensetzung der organischen Basen," *Annalen der Pharmacie* 26 (1838): 10–41. Justus Liebig, "Bemerkungen zu vorstehender Abhandlung," *Annalen der Pharmacie* 26 (1838): 41–60, appeared immediately after Regnault's article. Regnault responded in a letter to Liebig, which Liebig published together with his further editorial commentary as Victor Regnault, "Weiterer Beitrag über die Zusammensetzung der organischen

Basen," *Annalen der Pharmacie* 29 (1839): 58–63. According to Fruton, "Reappraisal," 56, Regnault was in Giessen in 1835 and subsequently performed "meticulous" analyses of organic compounds, including the alkaloids.

90. Liebig, "Bemerkungen zu vorstehender Abhandlung," (1838), 56–57. For example, Liebig viewed Regnault's formula for morphine ($C_{35}H_{40}N_2O_6$) as preferable to his own 1831 formula ($C_{34}H_{36}N_2O_6$) but dismissed Regnault's new formula for narcotine.

91. Liebig, "Bemerkungen zu vorstehender Abhandlung," (1838), 42–43.

92. Liebig, "Bemerkungen zu vorstehender Abhandlung," (1838), 51, 59.

93. Liebig, "Bemerkungen zu vorstehender Abhandlung," (1838), 58–59.

94. Liebig, "Bemerkungen zu vorstehender Abhandlung," (1838), 42, set out the inconsistencies in current views concerning the relationship between the alkaloids' nitrogen content and their basicity.

95. Jean-Baptiste Dumas, *Eloge Historique de Henri-Victor Regnault* (Paris: Académie des Sciences, 1881). See also Robert Fox, *The Caloric Theory of Gases: From Lavoisier to Regnault* (Oxford: Oxford University Press, 1971); Chang, *Inventing Temperature*. Regnault's work in thermometry is examined further in chapter 7.

96. Wöhler to Liebig, 30 October 1840, in "Liebig-Wöhler Briefwechsel," ed. Meinel and Steinhauser.

97. Brock, *Gatekeeper*, 88–89. Based mainly on Wöhler's experimental work, this study contributed to Liebig's growing interest in physiological chemistry.

98. Wöhler dissolved alkaline ammonia in acidic hydrochloric acid. Converting ammonia to its chloroplatinate salt produced a solid that, when dry, could be weighed, enabling calculation of nitrogen's original mass.

99. Liebig to Wöhler, 28 June 1841, in "Liebig-Wöhler Briefwechsel," ed. Meinel and Steinhauser.

100. Franz Varrentrapp and Heinrich Will, "Neue Methode zur Bestimmung des Stickstoffs in organischen Verbindungen," *Annalen der Chemie und Pharmacie* 39 (1841): 257–296.

101. Varrentrapp and Will, "Neue Methode," fn. on 257.

102. Varrentrapp and Will, "Neue Methode," 264–265.

103. Holmes, "Complementarity;" Fruton, "Reappraisal," observed (but did not explain) this apparent delay.

104. Fruton, "Reappraisal," concluded with an appendix of those who had spent time working in Liebig's Giessen laboratory.

105. This account of Liebig's use of the *Annals* thus contributes to our understanding of how and why scientific journals became central to nineteenth-century science, questions posed just over a decade ago by James Secord, "Science, Technology

and Mathematics," in *The Cambridge History of the Book in Britain*, ed. D. McKitterick (Cambridge University Press, 2009), 443–474, and addressed by scholars including Melinda Baldwin, *Making "Nature": The History of a Scientific Journal* (Chicago: University of Chicago Press, 2015); and Alex Csiszar, *The Scientific Journal: Authorship and the Politics of Knowledge in the Nineteenth Century* (Chicago: University of Chicago Press, 2018). This study suggests that discipline building was as fundamental as the scientist's professional identity, or the social position and structure of science, in the rise of the scientific journal.

106. Cf. Brock, *Gatekeeper*, 48–49; Usselman, "Liebig's Alkaloid Analyses," 73; Rocke, "Comparative Perspective," 289; and Usselman et al., "Restaging Liebig," 13.

107. Holmes, "Liebig," 337.

108. Justus Liebig, "Basen, Organische," in *Handwörterbuch der reinen und angewandten Chemie*, vol. 1, ed. Justus Liebig, Johann C. Poggendorff, and Friedrich Wöhler (Braunschweig: Vieweg, 1842), 697–699, introduced the "amidogen" radical as a partial description of how basic nitrogen existed in the alkaloids. Rocke, *Nationalizing Science*, 162, fn. 13, explained the publication of this essay in 1840, despite the nominal 1842 publication date. See also Rocke, *Quiet Revolution*, 403, fn. 43.

109. The classic study of Wöhler remains Robin Keen, *The Life and Work of Friedrich Wöhler (1800–1882)* (Nordhausen, Germany: Bautz, 2011).

110. Wöhler to Liebig, May 9, 1842, in *Aus Justus Liebig und Friedrich Wöhler's Briefwechsel in den Jahren 1829–1873*, ed. August Wilhelm Hofmann (Braunschweig: Vieweg, 1888), 192–193, opened with the sentence, "I am very sorry that you do not wish to take part in the study of nicotinic acid." This began Liebig's gradual withdrawal from a joint investigation of the opium alkaloid narcotine with Wöhler. Subsequent letters detailed their ongoing investigations, which Liebig ultimately abandoned in Liebig to Wöhler, January 13, 1844, 237–238, passing his portion of the work on to John Blyth.
Readers may notice that these letters are cited from Hofmann's edition of the Liebig-Wöhler correspondence, rather than Meinel and Steinhauser's more recent transcriptions. Although both letters referenced above can be found in Meinel and Steinhauser's version, the sentence quoted from the earlier letter does not appear there. Hofmann's version of the Liebig-Wöhler correspondence is known to be heavily edited. Yet, there is no reason to doubt that Hofmann's insertion of such an uncontroversial statement regarding nicotinic acid was based on the original documentary record. I infer that Hofmann incorporated this text from another letter that he otherwise did not wish to reproduce and which appears not to have survived.

111. Justus Liebig, "Bemerkungen zu vorstehender Abhandlung," *Annalen der Chemie und Pharmacie* 38 (1841): 203 ff., was a particularly salient expression of Liebig's disillusionment, targeting Dumas's research program in substitution.
Brock, *Gatekeeper*, 92, followed Holmes, "Liebig," 343–344, in attributing Liebig's abandonment of organic chemistry to theoretical disputes, most notably those

involving Berzelius and Dumas. See also Blondel-Mégrelis, "Liebig," 44–45, on Liebig's switch to become a popularizer of chemistry.

112. Brock's term derived from Thomas Pynchon's (1973) novel *Gravity's Rainbow*, which described Liebig as having "occupied the role of a gate."

CHAPTER 3

1. C. M. Jackson, "Wonderful Properties."

2. Justus Liebig, *Familiar Letters on Chemistry in Its Relations to Physiology, Dietetics, Agriculture, Commerce and Political Economy*, 4th edition, ed. John Blyth (London: Walton and Maberly, 1859), 124.

3. Glass tubing was produced by furnace work, as Ernest Child, *The Tools of the Chemist: Their Ancestry and American Evolution* (New York: Reinhold, 1940), described and illustrated.

4. Marco Beretta, "Between the Workshop and the Laboratory: Lavoisier's Network of Instrument Makers," *Osiris* 29 (2014): 197–214, examined Lavoisier's extensive collaborations with instrument makers while seeking to develop precision methods in the study of gas chemistry. As Beretta explained, metal rather than glass was the material of choice because of its ability to withstand the conditions required for such experiments.

5. Joseph-Louis Gay-Lussac and Louis-Jacques Thenard, *Recherches Physico-Chimiques*, vol. 2 (Paris: Deterville, 1811), 269. According to Crosland, *Gay-Lussac*, 120, Fortin had previously made a similarly specialized tap for Gay-Lussac. See also Crosland, *Gay-Lussac*, 193, on Collardeau's involvement with Gay-Lussac's analytical work. Making taps and graduated vessels involves some of the most technically challenging glassblowing techniques.

6. J. Erik Jorpes, *Jac Berzelius: His Life and Work*, trans. Barbara Steele (Stockholm: Almquist and Wiksells, 1966), 49.

7. Jöns Jacob Berzelius, *Anwendung des Löthrohrs in der Chemie und Mineralogie*, trans. Heinrich Rose (Nuremberg: Schrag, 1821), 7, described learning mineralogical analysis using the blowpipe from Johann Gottlieb Gahn. Jorpes, *Jac Berzelius*, 19, reported that Berzelius learned how to blow glass from the Italian Jusuo Vaccano during a month-long stint as a pharmacist's apprentice in Vadstena in the summer of 1799.

8. Berzelius, "Address," 129.

9. Liebig to his wife, 2 November 1828, in *Justus von Liebig*, ed. Dechend, 24.

10. Wolfgang Götz, "Johann Baptist Batka, 'Arznei-Waarenhändler in Prag,'" *Geschichte der Pharmazie* 46 (1994): 2.
According to Götz, "Batka," 5, Liebig wrote to Johann Bartholomäus Trommsdorff on May 21, 1828, reporting the gift from Batka of a selenium medal showing a likeness of Trommsdorff—part of a new range Batka had commissioned commemorating

eminent chemists and pharmacists. In these circumstances, Liebig was under an obligation to make himself known to Batka when they were both in Hamburg. Philipp Lorenz Geiger, "Pharmacognostische Notizen," *Magazin für Pharmacie* 33 (1831): 134–136, described Batka's talk in Hamburg as concerning drugs extracted from bark, including the alkaloid strychnine. Given his interest in alkaloids, Liebig— who would shortly take over from Geiger as the *Magazin's* editor—surely attended.

11. Wenzel Batka, *Verzeichniss der neuesten chemischen und pharmaceutischen Geraeth- schaften: mit Abbildungen* (Nuremberg: Schrag, 1829).

12. Liebig, "neuen Apparat," fn. on 12, indicated that his analytical apparatus could be obtained from Batka. Made ambiguous by its position in the text (immediately following his failed approach to nitrogen determination), the Kaliapparat's inclu- sion in Wenzel Batka, *Verzeichniss der neuesten chemischen und pharmaceutischen Gerätschaften mit Abbildungen: Herausgegeben bei Gelegenheit der Versammlung deutscher Naturforscher in Wien* (Leipzig: Barth, 1832), seems to confirm that Liebig's intended meaning was indeed that Batka could supply all components of his apparatus.

13. Götz, "Batka," 4–5, detailed customers' dissatisfaction with Batka's pharmaceuti- cal products and reagent bottles.

14. Batka's 1832 catalog illustrated a poorly formed Kaliapparat (figure 78) that would be unlikely to produce results of sufficient accuracy and reliability; nor was his 1857 version much better. See Wenzel Batka, *Verzeichniss der neuesten chemischen, physikalischen und pharmaceutischen Apparate, Geräthschaften und Instrumente der Han- dlung* (Leipzig: Barth, 1857), 1.
The Kaliapparat nevertheless contributed to a significant expansion in Batka's offering of chemical glassware. In 1829, glass apparatus occupied just over a page of Batka's catalog, alongside a similar quantity of porcelain and a far greater range of equipment made from diverse metals, including platinum, tin, iron, and silver. By 1832, glassware extended to eight pages, roughly 100 items. This trend accelerated: by 1857, lampworked glassware dominated, comprising some 20% of Batka's much- expanded catalog (26 out of 144 pages).

15. The deficiencies of Batka's rendition are not surprising. Even consulting "On a New Apparatus" would have been of limited help, since it provided no details about the Kaliapparat. It is nevertheless possible that Batka supplied the Kaliapparat to Liebig's laboratory during 1831 and 1832. According to William H. Brock, "Liebig's Laboratory Accounts," *Ambix* 19 (1972): 47–58, Liebig made sizeable purchases from Batka in 1831 and 1832, with a final, much smaller purchase in 1833.

16. As Jackson, *Spectrum of Belief,* and Biagioli, *Galileo's Instruments of Credit,* show, Liebig was not alone in adopting such a strategy.

17. Blanchet and Sell, "Zusammensetzung."

18. The notable exceptions to this are Berzelius and the legendary British experi- menter Michael Faraday, both of whom were expert glassblowers—though rather

little is known about how they acquired this skill. Consider also that Liebig, having trained in Paris during the mid-1820s, had to return to the city in 1828 to learn basic glassblowing.

19. As far as I have been able to ascertain, Liebig's standard Kaliapparat did not become available from reputable suppliers of chemical glassware until about 1840.

20. Lafond, "Ueber die Kunst."

21. Just two texts were available in German: Friedrich Körner, *Anleitung zur Bearbeitung des Glases an der Lampe, und zur vollständigen Verfertigung der, durch das Lampenfeuer darstellbaren, physikalischen und chemischen Instrumente und Apparate* (Jena: Schmidt, 1831); and Jöns Jacob Berzelius, *Chemische Operationen und Gerätschaften, nebst Erklärung chemischer Kunstworter, in alphabetischer Ordnung*, trans. Friedrich Wöhler (Dresden: Arnold, 1831).

It is also significant that German-language glassblowing manuals directed at chemists began to appear in greater numbers from 1833, e.g., Ferdinand P. Danger, *Die Kunst der Glasbläserei vor dem Lothrohre und an der Lampe oder Darstellung eines neuen Verfahrens, um alle physikalische und chemische Instrumente, welche in den Bereich dieser Kunst gehören, als Barometer, Thermometer, Ureometer, Heber u.s.w. mit dem geringsten Kosten-Aufwande und auf die leichteste Art zu verfertigen* (Quedlinburg, Leipzig: Basse, 1833); Heinrich Rockstroh, *Die Glasblasekunst im Kleinen oder mittelst der Docht- oder der Strahlflamme, oder Anweisung, wie aus Glas mittelst der Docht- oder der Strahlflamme mancherlei Gegenstände im Kleinen zu gestalten: nebst einer Anweisung, wie Mikroskope, Barometer, Thermometer und Aräometer, Mikrometer und noch manche andere Gegenstände, bei welchen Glas das vornehmliche Material ist, verfertigt oder bewerkstelliget werden; auch einem Anhange von Glaskunststückchen und Glaskünsteleien* (Leipzig: Günther, 1833).

22. Michael Faraday, *Chemical Manipulation; Being Instructions to Students in Chemistry, on the Methods of Performing Experiments of Demonstration or of Research, with Accuracy and Success* (London: Phillips, 1827), vi–vii, ix.

23. Lafond, "Mémoire sur l'Art de Souffler le Verre," *Journal des Connaissances Usueles et Practiques* 16 (1832): 175–194; Ferdinand Danger, "Supplément à l'Art de Souffler le Verre," *Journal des Connaissances Usueles et Practiques* 17 (1833): 33–37. Liebig reprinted Lafond's article as: Lafond, "Ueber die Kunst."

24. This article advertised glassblowing classes offered in Paris by Lafond. It also provoked controversy: According to Danger, Lafond had only begun learning to blow glass some 15 months earlier. Danger claimed Lafond's article presented significant components of Danger's glassblowing course as his own work, also including numerous errors of fact and technique that Danger corrected in a lengthy response to Lafond. By the time Lafond's article appeared in German translation, it incorporated Danger's "corrections." See Lafond, with Danger's corrections, "Ueber die Kunst Glas zu Blasen," *Polytechnisches Journal* 48 (1833): 121–140.

25. The other was a wash bottle designed by Berzelius that helped chemists purify substances obtained as solid deposits (e.g., by precipitation) and was therefore

becoming an indispensable tool of mineral (inorganic) analysis. Liebig's reprint also dropped the section of Lafond's original article devoted to enameling.

26. Ferdinand P. Danger, *The Art of Glass-Blowing, or Plain Instructions for Making the Chemical and Philosophical Instruments Which Are Formed of Glass; Such as Barometers, Thermometers, Hydrometers, Hour-Glasses, Funnels, Syphons, Tube Vessels for Chemical Experiments, Toys for Recreative Philosophy, & c* (London: Bumpus & Griffin, 1831), 46, explained the difficulty of forming glassware containing multiple bulbs.

27. According to previous studies, the first description and illustration of a standard Kaliapparat appeared in Liebig's (1837) *Anleitung*. In fact, many accounts assume that Liebig's device was essentially unchanged from the start—despite solid historical evidence to the contrary. Rocke, "Comparative Perspective," 289, is almost alone in recognizing this crucial period of change.

28. A survey of Liebig's publications from 1831 on reveals that he ceased referring to the Kaliapparat in 1833. Poggendorff, commenting on Justus Liebig, "Ueber die Zusammensetzung der Hippursäure," *Annalen der Physik und Chemie* 32 (1834): fn. on 573, referred to Liebig's "now perfected device."
This revised timing does not undermine previous historians' inference that Liebig's original Kaliapparat was transformed into a standard device by his assistant and skilled glassblower Karl Ettling.

29. When constructing this train of three bulbs, master glassblower Tracy Drier uses a tool of German origin called an "Einschneider" (from *einschneiden*, to cut in) to create the constrictions between each bulb.

30. Varrentrapp and Will, "Neue Methode," 263.

31. Wöhler to Liebig, 30 October 1840, in "Liebig-Wöhler Briefwechsel," ed. Meinel and Steinhauser. Wöhler's innovation was inspired by earlier work with Liebig on uric acid: Justus Liebig and Friedrich Wöhler, "Ueber die Natur der Harnsäure," *Annalen der Physik und Chemie* 117 (1837): 561–569; Justus Liebig and Friedrich Wöhler, "Untersuchungen über die Natur der Harnsäure," *Annalen der Pharmacie* 26 (1838): 241–336.

32. This view of Wöhler derives from Rocke, *Quiet Revolution*, 16–20. On Wöhler's attitude to physiological chemistry—to which nitrogen determination was most relevant—see Joseph R. Fruton, *Proteins, Enzymes, Genes: The Interplay of Chemistry and Biology* (New Haven, CT: Yale University Press, 1999), 55.

33. Varrentrapp and Will, "Neue Methode," 266–267.

34. It seems likely that the skilled glassblower Will took the lead in this aspect of their joint project, while Varrentrapp's experience in quantitative inorganic analysis was surely helpful in perfecting the procedure for weighing ammonia in the form of an insoluble salt.

35. Varrentrapp and Will, "Neue Methode," 268, 273. Although "splash back" perhaps offers a more literal translation, I have chosen to use "suck back" (*Zurücksaugen*),

because it better describes this process. In contemporary use, *Zurücksaugen* displaced *Zurückspritzen* in chemists' German lexicon from about 1850 on. See, for example Liebig et al., eds., *Handwörterbuch,* vol. 7, 853. The term "suck back" remains in widespread use today.

36. Hydrogen released as organic carbon was converted to carbon dioxide; the presence of additional organic matter therefore made excess hydrogen available during the analysis.

37. Varrentrapp and Will, "Neue Methode," 272–273.

38. Varrentrapp and Will, "Neue Methode," 268.

39. As explained in chapter 2, Will and Varrentrapp's method measured nitrogen in the form of solid ammonium chloroplatinate, which was precipitated from this acidic solution by the addition of excess platinum chloride solution.

40. Varrentrapp and Will, "Neue Methode," 275.

41. Liebig to Wöhler, 28 June 1841, in "Liebig-Wöhler Briefwechsel," ed. Meinel and Steinhauser, mentioned in passing that Wöhler had "once occupied himself" with this method. There is no evidence that Wöhler was put out by this remark.

42. Varrentrapp and Will, "Neue Methode," 268–269.

43. See, for example, Holmes, "Liebig," 337, and James R. Partington, *A History of Chemistry,* vol. 3 (London: Macmillan, 1962), 299.

44. Liebig's correspondence, gathered and administered prior to his death by his grandson Justus Carrière, is cited from a range of mainly published sources. Carrière published Liebig's correspondence with Berzelius in 1892/1893, more widely available in its second edition: Justus Carrière, ed., *Berzelius und Liebig: Ihre Briefe von 1831–1845 mit erläuternden Einschaltungen aus gleichzeitigen Briefen von Liebig und Wöhler sowie wissenschaftlichen Nachweisen* (Munich: Lehmann, 1898). His closeness to Liebig perhaps accounts for Carrière's access to, and decision to publish, letters containing such raw emotions and extraordinary language.
The first published edition of the Liebig-Wöhler correspondence, meanwhile, was edited by Liebig's protégé August Wilhelm Hofmann. Published in 1888, Hofmann's edition was both selective and rearranged: Hofmann, ed., *Briefwechsel.* Hofmann's debt to Liebig plausibly resulted in unwillingness to see his mentor slighted, perhaps leading Hofmann to suppress certain aspects of this correspondence. Wherever possible, the Liebig-Wöhler correspondence has been cited from the new edition in preparation from extant manuscript sources under the direction of Professor Christoph Meinel in Regensburg.

45. Eilhard Mitscherlich, *Lehrbuch der Chemie,* 2nd edition, vol. 1 (Berlin: Mittler, 1834), iii–x ("Preface to the First Edition"). Liebig to Wöhler, 28 January 1830, in "Liebig-Wöhler Briefwechsel," ed. Meinel and Steinhauser, called Mitscherlich's textbook the best he had ever seen.

See also Justus Liebig, "Der Zustand der Chemie in Oestereich," *Annalen der Pharmacie* 25 (1838): 339–347; Justus Liebig, "Der Zustand der Chemie in Preussen," *Annalen der Chemie und Pharmacie* 34 (1840): 97–136.

46. Liebig to Wöhler, 1 May 1832, in "Liebig-Wöhler Briefwechsel," ed. Meinel and Steinhauser, described Mitscherlich's first visit to Giessen; Liebig to Wöhler, 14 September 1832, in "Liebig-Wöhler Briefwechsel," ed. Meinel and Steinhauser, the second visit, during which Mitscherlich took a course in organic analysis with Liebig; Liebig to Berzelius, 6 November 1832, in *Berzelius und Liebig*, ed. Carrière, 41–42; Eilhard Mitscherlich to Berzelius, 1 October 1832, in *Gesammelte Schriften von Eilhard Mitscherlich*, ed. Alexander Mitscherlich (Berlin: Mittler, 1896), 104. Eilhard Mitscherlich and Justus Liebig, "Ueber die Zusammensetzung der Milchsäure," *Annalen der Pharmacie* 7 (1833): 47–48.

47. Mitscherlich to Berzelius, late November 1831, in *Gesammelte Schriften*, ed. Alexander Mitscherlich, 98, expressed a wish to diversify from an increasingly unproductive research program in crystallography; Berzelius to Mitscherlich, 8 January 1832, in *Gesammelte Schriften*, ed. Alexander Mitscherlich, 100, approved this plan. Indeed, it seems likely—given the growing importance of organic analysis during this period, and Berzelius's relationship to Liebig—that Berzelius suggested to Mitscherlich that he become acquainted with Liebig and his work.

48. Mitscherlich to Berzelius, 18 December 1832, in *Gesammelte Schriften*, ed. Alexander Mitscherlich, 104.

49. Mitscherlich to Berzelius, 3 October 1833, in *Gesammelte Schriften*, ed. Alexander Mitscherlich, 106.

50. E. Mitscherlich, *Lehrbuch*, 205–210. Rocke, "Comparative Perspective," 308, fn. 81, noted that, according to its Preface, the book went to press in November 1832. The Preface also indicated the book's second section (which began at page 416) was complete by the end of March. When read alongside Mitscherlich's correspondence, however, it seems more likely that the new edition of his textbook appeared in late 1833 through the summer of 1834—as is also more consistent with the fact that the Preface itself is dated July 1, 1834. Mitscherlich to Berzelius, 3 October 1833, in *Gesammelte Schriften*, ed. Alexander Mitscherlich, 106, described work on the new edition as that summer's "almost exclusive" work, making it likely that the section on organic analysis was written during the summer of 1833, perhaps in Giessen. Mitscherlich's further investigations of benzene derivatives support this revised timing for the appearance of his textbook. Mitscherlich's correspondence with Berzelius dates this work to spring 1834; while Berzelius to Mitscherlich, 8 July 1834, in *Gesammelte Schriften*, ed. Alexander Mitscherlich, 108, cautioned Mitscherlich against first publishing research in his textbook. Mitscherlich ignored this advice and included his most recent findings as an appendix to the new edition (Eilhard Mitscherlich, *Lehrbuch*, 661–668). Berzelius to Mitscherlich, 31 October 1834, in *Gesammelte Schriften*, ed. Alexander

Mitscherlich, 109, conveyed Berzelius's mixed response to Mitscherlich's results and theoretical interpretation.

51. Although the Preface to the first edition of Eilhard Mitscherlich's textbook (reproduced in *Lehrbuch*, 2nd edition, vol. 1, x) stated the author's decision not to name "the discoverers of individual facts," Liebig's behavior following the appearance of Mitscherlich's description of his method of organic analysis shows how inadequate he found this disclaimer.

52. It is a measure of Mitscherlich's unrealistic self-assessment that Mitscherlich to Berzelius, 6 August 1841, in *Gesammelte Schriften*, ed. Alexander Mitscherlich, 116, could describe himself as having "served Germany" by developing Berzelius's approach into a widely used method of organic analysis, while Liebig—"because he came too late" to the field—deserved no such honor.

53. Liebig to Berzelius, 17 February 1834, in *Berzelius und Liebig*, ed. Carrière, 81.

54. Hans-Werner Schütt, *Eilhard Mitscherlich: Prince of Prussian Chemistry* (Philadelphia: Chemical Heritage Foundation, 1997) remains the definitive biography of Mitscherlich, adapted from the (1992) German original *Eilhard Mitscherlich: Baumeister am Fundament der Chemie*; while Kurt August Schierenberg, "Eilhard Mitscherlich and Justus von Liebig," *Giessener Universitaet* 2 (1977): 106–115, focused on the relationship between Mitscherlich and Liebig.

55. According to Liebig's editorial response to Eilhard Mitscherlich, "Ueber das Benzol und die Säuren der Oel- und Talgarten," *Annalen der Pharmacie* 9 (1834): 48, Péligot had described the decomposition of benzoic acid in an address to the Parisian Academy of Sciences on October 23, 1833, and therefore deserved priority.

56. Carl Löwig, *Der Chemiker Dr. Justus Liebig in Giessen vor das Gericht der öffentlichen Meinung* (Zurich: Orell, Fössli, and Co., 1833).

57. E. Mitscherlich, "Ueber das Benzol." Liebig's response appeared in extensive footnotes and appended commentary (48–56). Liebig to Berzelius, 17 February 1834, in *Berzelius und Liebig*, ed. Carrière, 81; Liebig to Berzelius, 25 March 1834, in *Berzelius und Liebig*, ed. Carrière, 84.

58. Berzelius to Liebig, 4 March 1834, in *Berzelius und Liebig*, ed. Carrière, 81. Berzelius evidently felt he owed allegiance to his former student over his newly adopted mentee.

59. Wöhler to Liebig, 3 March 1834, in "Liebig-Wöhler Briefwechsel," ed. Meinel and Steinhauser.

60. Liebig to Wöhler, 1 May 1832, in "Liebig-Wöhler Briefwechsel," ed. Meinel and Steinhauser; and Liebig to Berzelius, 30 May 1832, in *Berzelius und Liebig*, ed. Carrière, 29, document the physical and mental consequences of Liebig's punishing work schedule. Liebig to Wöhler, 29 February 1832, in "Liebig-Wöhler Briefwechsel," ed. Meinel and Steinhauser, described his state of health as "passable" following the "morbid excitement" occasioned by recent work.

61. Liebig to Berzelius, 25 March 1834, in *Berzelius und Liebig,* ed. Carrière, 84.

62. The classic study of the development of constitutional theory remains Colin A. Russell, *The History of Valency* (Leicester: Leicester University Press, 1971). On chemists' debates concerning the proper relationship between constitution and reactivity, see John H. Brooke, "Laurent, Gerhardt and the Philosophy of Chemistry," *Historical Studies in the Physical Sciences* 6 (1975): 405–429.

63. Liebig's commentary on Mitscherlich, "Ueber das Benzol," 56. Liebig to Berzelius, 25 March 1834, in *Berzelius und Liebig,* ed. Carrière, 84. This episode was apparently resolved following Liebig's admission ("Erklärung," *Annalen der Pharmacie* 9 (1834): 363) that Mitscherlich had prepared benzene before Péligot, just as Berzelius maintained.

64. Jöns Jacob Berzelius, *Lehrbuch der Chemie,* 3rd edition, vol. 6 (Dresden: Arnold, 1837).

65. Rocke, "Comparative Perspective," 292, established this timing.

66. Berzelius to Liebig, 20 February 1838; and Liebig to Berzelius, 7 March 1838, in *Berzelius und Liebig,* ed. Carrière, 146–147; and 148–149. Emphasis added but is implicit in the original.

67. Mitscherlich to Berzelius, 1834, in *Gesammelte Schriften,* ed. Alexander Mitscherlich, 108.

68. Carl Brunner, "Versuche über Stärkmehl und Stärkmehlzucker," *Annalen der Physik und Chemie* 4 (1835): 319–338, and Tafel III.

69. Liebig, *Instructions,* 3, described Brunner's apparatus and others as "employed by their inventors alone." This evaluation of other analytical methods did not appear in the original German edition.

70. Richard Felix Marchand, "Bemerkungen über die organische Analyse," *Journal für praktische Chemie* 13 (1838): 513–514.

71. Carl Brunner, "Beiträge zur organischen Analyse," *Annalen der Physik und Chemie* 44 (1838): 134–155.

72. Frederic Lawrence Holmes, *Antoine Lavoisier: The Next Crucial Year: Or, the Sources of His Quantitative Method in Chemistry* (Princeton, NJ: Princeton University Press, 1997), 89, concluded from his study of Lavoisier's notebooks that Lavoisier "sought reliability but not great precision." But neither Holmes's study nor others demonstrating the effective use of quantitative measurement by Lavoisier's contemporaries, including Joseph Priestley, have altered popular understanding of Lavoisier's pioneering role.

73. For example, see Antoine L. Lavoisier, "Mémoire sur la Combinaison du Principe Oxygine, avec l'Esprit-de-vin, l'Huile et Différents Corps Combustibles," *Histoire et Mémoires de l'Académie Royal des Sciences* 98 (1784): 593–608.

74. Rocke, *Quiet Revolution,* is a notable exception.

75. August Wilhelm Hofmann, *The Life-Work of Liebig. The Faraday Lecture for 1875* (London: Macmillan, 1876), 8. Hofmann's desire to secure his mentor's priority may also have influenced his decision to publish this portion of Liebig's correspondence with Wöhler largely unedited (as noted under the header "Glassblowing as Threat").

76. Partington, *History*, vol. 4, 300.

CHAPTER 4

1. For Hofmann's biographical details, see: Jonathan Bentley, "The Chemical Department of the Royal School of Mines: Its Origins and Development under A. W. Hofmann," *Ambix* 17 (1970): 153–181; Roberts, "Royal College of Chemistry;" Roberts, "Establishment;" Robert Bud and Gerrylynn K. Roberts, *Science versus Practice: Chemistry in Victorian Britain* (Manchester: Manchester University Press, 1984); Michael N. Keas, "The Structure and Philosophy of Group Research: August Wilhelm Hofmann's Research Program in London (1845–1865)" (PhD diss., University of Oklahoma, 1992); Christoph Meinel and Hartmut Scholz, eds., *Die Allianz von Wissenschaft und Industrie August Wilhelm Hofmann (1818–1892): Zeit, Werk, Wirkung* (Weinheim: VCH, 1992); Michael N. Keas, "The Nature of Organic Bases and the Ammonia Type," in Meinel and Scholz, eds., *Allianz;* Christoph Meinel, "August Wilhelm Hofmann: 'Regierender Oberchemiker,'" in Meinel and Scholz, *Allianz;* Colin A. Russell, "August Wilhelm Hofmann—Cosmopolitan Chemist," in Meinel and Scholz, *Allianz*.

2. J. Paul Hofmann, *Das chemische Laboratorium der Ludwigs-Universität zu Giessen* (Heidelberg: Winter, 1842).

3. Justus Liebig, "Ueber die Constitution der organischen Säuren," *Annalen der Pharmacie* 26 (1838): 113–190, acknowledged the limits of analysis. Disillusionment with Dumas's theoretically driven research program in substitution probably contributed to Liebig's decision to advocate a new approach to organic chemistry. See Liebig, "Bemerkungen zu vorstehender Abhandlung," (1841).

4. Liebig to Hofmann, 21 June 1845, in *Liebig und Hofmann in Ihren Briefen (1841–1873)* (Weinheim: VCH, 1984), ed. William H. Brock, 29, requested Hofmann to return some pages referring to ongoing research in Giessen.

5. Liebig, "Basen, Organische," argued for the alkaloids' significance to both chemistry and medicine.

6. As Klein, *Experiments, Models, Paper Tools,* 88–89, noted in the case of ethers, the relation between academic chemistry and commercial pharmacy during the early nineteenth century was one of "entanglement."

7. Muspratt and Hofmann, "On Toluidine," 367. This account substantiates Morrell's in passing claim ("Chemist Breeders," 8) that Giessen was the cradle of organic synthesis.

8. Amorphous quinine was a noncrystalline alkaloidal substance produced from the residue of quinine extraction.

9. The essays in Meinel and Scholz, *Allianz*, are especially noteworthy. Hofmann's academic and institutional accomplishments featured heavily in contemporary accounts of his life and work: Lyon Playfair, "Hofmann Memorial Lecture," *Journal of the Chemical Society* 69 (1896): 575–579; Frederick A. Abel, "Hofmann Memorial Lecture," *Journal of the Chemical Society* 69 (1896): 580–596; William H. Perkin, "Hofmann Memorial Lecture," *Journal of the Chemical Society* 69 (1896): 596–637; Henry E. Armstrong, "Hofmann Memorial Lecture," *Journal of the Chemical Society* 69 (1896): 637–732; Ferdinand Tiemann, "Gedächtnisrede August Wilhelm Hofmann," *Berichte der deutschen chemischen Gesellschaft* 25 (1892): 3377–3398.
For more on synthetic dyes, see Travis, *Rainbow Makers;* Reinhardt and Travis, *Heinrich Caro.*

10. The only relatively recent book with synthetic organic chemistry as its explicit focus is John Buckingham, *Chasing the Molecule* (Stroud: Sutton, 2004).
On the philosophical consequences of organic synthesis, see, for example: John H. Brooke, "Organic Synthesis and the Unification of Chemistry: A Reappraisal," *British Journal for the History of Science* 5 (1971): 363–392; Brooke, "Wöhler's Urea;" and Ramberg, "Death of Vitalism."

11. This integration is in contrast to the perceived divisions that allowed Meinel and Scholz to write of an "alliance of science and industry."

12. Tiemann, "Gedächtnisrede," 3382.

13. Prakash Kumar, *Indigo Plantations and Science in Colonial India* (Cambridge: Cambridge University Press, 2012).

14. Buff, "Indigsäure und Indigharz;" Justus Liebig, "Ueber die bittere Substanz, welche durch Behandlung des Indigs, der Seide und der Aloë mit Saltpetersäure Erzeugt wird," *Annalen der Physik und Chemie* 13 (1828): 191–208.

15. Julius Fritzsche, "Ueber die Produkte der Einwirkung von Kali auf Indigblau," *Annalen der Chemie und Pharmacie* 39 (1841): 76–91; Fred E. Sheibley, "Carl Julius Fritzsche and the Discovery of Anthranilic Acid, 1841," *Journal of Chemical Education* 20 (1943): 115–117.

16. Justus Liebig, "Ueber die Darstellung und Zusammensetzung der Anthranilsaure," *Annalen der Chemie und Pharmacie* 39 (1841): 91–96.

17. Nikolai Zinin, "Beschreibung einiger neuer organischer Basen, dargestellt durch die Einwirkung des Schwefelwasserstoffes auf Verbindungen der Kohlenwasserstoffe mit Untersaltpetersäure," *Journal für praktische Chemie* 27 (1842): 140–153; Nathan M. Brooks, "Nikolai Zinin and Synthetic Dyes: The Road Not Taken," *Bulletin for the History of Chemistry* 27 (2002): 26–36.

18. Julius Fritzsche, "Bemerkung zu vorstehender Abhandlung des Herrn. Zinin," *Journal für praktische Chemie* 27 (1842): 153.

19. Fruton, *Contrasts,* 289. The coal-gas distillery was either set up with Karl Gottlieb Oehler, another former Liebig student, or was soon bought by Oehler, later

becoming Teerfarbenwerk Oehler. According to Reinhardt and Travis, *Heinrich Caro*, 129, Oehler's was the first synthetic dye plant in Germany.

20. F. F. Runge, "Ueber einige Produkte der Steinkohlendestillation," *Annalen der Physik und Chemie* 31 (1834): 65–78.

21. Charles M. Gerhardt, "Untersuchungen über die organischen Basen," *Annalen der Chemie und Pharmacie* 42 (1842): 310–313, had recently produced quinoline by destructive distillation of alkaloids including quinine.

22. Hofmann, "Steinkohlen-Teerol."

23. Rocke, *Quiet Revolution*, 86–93, described Gerhardt's stormy career and difficult relationship with Liebig. After training with Liebig, Gerhardt had moved to Paris and from there to Montpellier, adopting an experimental approach of which his former teacher strongly disapproved and beginning to propound theoretical views concerning type theory (a challenger to the established electrochemical theory of chemical constitution), whose dogmatism and insecure empirical basis Liebig absolutely rejected.

24. Justus Liebig, "Chinolein oder Chinolin," *Annalen der Chemie und Pharmacie* 44 (1842): 279–280.

25. Hofmann, "Steinkohlen-Teerol," 78, reported differences in how leukol and quinoline reacted with chromic acid.

26. Justus Liebig, "Vorläufige Notiz über die Identität des Leucols mit Chinolin nach A. W. Hoffmann [sic]," *Annalen der Chemie und Pharmacie* 53 (1845): 427–428. This notice was the final publication in this volume, issued March 15, 1845.

27. Rocke, *Quiet Revolution*, 87.

28. Charles M. Gerhardt, "Quinoléine, Produit de Décomposition de la Quinine et de la Cinchonine," *Comptes Rendus* 1 (1845): fn. on 31; Gerhardt had abstracted Johann Conrad Bromeis, "Ueber das Chinolin," *Annalen der Chemie und Pharmacie* 51 (1845): 130–140.

29. Liebig to Hofmann, 19 October 1845, in *Liebig und Hofmann*, ed. Brock, 36.

30. Liebig to Hofmann, 6 February 1846, in *Liebig und Hofmann*, ed. Brock, 44–46.

31. Justus Liebig, "Herr Gerhardt und die organische Chemie," *Annalen der Chemie und Pharmacie* 57 (1846): 93–118.

32. Hofmann to Liebig, 15 January 1846, in *Liebig und Hofmann*, ed. Brock, 42–44. Liebig, "Herr Gerhardt," 115, described Gerhardt's failure to cite Hofmann's prior work.

33. Liebig to Hofmann 29 November 1845, in *Liebig und Hofmann*, ed. Brock, 41; August Wilhelm Hofmann, "New Researches upon Aniline," *Memoirs and Proceedings of the Chemical Society* 3 (1845): 26–28.

34. Rocke, *Quiet Revolution*, 88.

35. Hofmann's papers are somewhat ambiguous concerning how he sourced aniline; Hofmann, "Steinkohlen-Teerol," 40–43, referred to a large-scale distillation performed with Sell, suggesting coal tar as the starting material rather than indigo; otherwise, later accounts of Hofmann's work at the RCC indicate that indigo was the major source.

36. Hofmann, "Steinkohlen-Teerol," 40–43, 57.

37. In addition, as chapter 7 explains, the development of melting points as stable criteria of identity and purity from the mid-century on would make crystalline solids vastly easier to characterize.

38. Blanchet and Sell, "Zusammensetzung."

39. Liebig and Wöhler, "Natur der Harnsaüre," 241, asserted chemists' future acquisition of this ability. A decade later, August Wilhelm Hofmann, *Reports of the Royal College of Chemistry, and Researches Conducted in the Laboratories in the Years 1845-6-7* (London: Schulze, 1849), lx, expressed similar confidence.

40. Liebig, "organischen Säuren;" see also Holmes, "Liebig;" and Rocke, *Image and Reality*, 9–10.

41. Liebig's key example of the spectrum from basic to acidic behavior in otherwise similarly constituted substances was the artificial base melamine and its sequential decomposition products ammeline, ammelide, and cyanuric acid. See Liebig, "Stickstoff-Verbindungen."

42. Liebig, "organischen Säuren," 185–186.

43. Liebig, "Basen, Organische," introduced Liebig's "amidogen" theory of organic bases, discussed further in chapter 5.

44. August Wilhelm Hofmann, "On the Metamorphoses of Indigo: Production of Organic Bases Which Contain Chlorine and Bromine," *Memoirs and Proceedings of the Chemical Society* 2 (1843): 266–300. Also published as Hofmann, "Metamorphosen des Indigo's."

45. August Wilhelm Hofmann, "Uebersicht der in der letzten Zeit unternommenen Forschungen über den Indigo und seine Metamorphosen," *Annalen der Chemie und Pharmacie* 48 (1843): 241–343. This paper was dated 1843 but in fact appeared in two sections beginning in March 1844.

46. The classic study of the development of constitutional and structural theory remains Russell, *History of Valency*. For more on the origins of type theory, see Klein, "Techniques of Modelling." On Gerhardt and Laurent, see Rocke, *Quiet Revolution*, chapter 4; Marya E. Novitski, *Auguste Laurent and the Prehistory of Valence* (Chur, Switzerland: Harwood Academic, 1992).

47. Tiemann, "Gedächtnisrede," 3383.

48. Auguste Laurent, "Recherches sur l'Indigo," *Annales de Chimie et de Physique* 3 (1840): 393–434. On Laurent's contributions to constitutional theory, see Novitski,

Auguste Laurent. On Laurent's visit to Giessen, see Clara de Milt, "Auguste Laurent, Founder of Modern Organic Chemistry," *Chymia* 4 (1953): 85–114, esp. 103–104. Hofmann learned of this work during Laurent's visit to Giessen in the summer of 1843 and included it in his review of indigo chemistry.

49. Hofmann, "Metamorphoses of Indigo," 298.

50. Liebig, "Basen, Organische."

51. Hofmann, "Metamorphoses of Indigo," fn. on 268.

52. Muspratt and Hofmann, "On Toluidine," 367, explained the difference between synthesis and analysis in just such terms.

53. John Blyth and August Wilhelm Hofmann, "On Styrole, and Some of the Products of Its Decomposition," *Memoirs and Proceedings of the Chemical Society* 2 (1843): 348. Cf. Graebe, *Geschichte*, 148–149, which attributed this important terminological innovation to Hermann Kolbe. Because later scholars—notably chemistry's great chronicler James Riddick Partington—followed Graebe's lead, his misattribution has become established in the history of chemistry. Catherine M. Jackson, "Synthetical Experiments and Alkaloid Analogues: Liebig, Hofmann and the Origins of Organic Synthesis," *Historical Studies in the Natural Sciences* 44 (2014): 331–338, further analyzed this point.
I also differentiate Hofmann's use of "synthesis" and "synthetical" in the context of a novel investigative approach to organic chemistry from occasional prior uses with no such programmatic intent. For example, Dumas, "Substances Végétales," 230, referred to the formation of artificial camphor by reacting oil of turpentine with hydrochloric acid and the removal of the same acid using a base. These reactions were far more complex than Dumas realized.

54. Metastyrole was not identified as a polymer of styrene until the latter part of the nineteenth century and did not become a viable industrial product for roughly another half-century.

55. Blyth and Hofmann, "On Styrole," 348.

56. See Klein and Lefèvre, *Materials,* 115–116, on analysis and synthesis as complementary components of the investigative method in mineral chemistry; and Klein and Lefèvre, *Materials,* 230, on the importance of reversible chemical transformations.

57. These linked aspects of synthesis were similarly ordered in Kolbe's first uses of the term, some months after Blyth and Hofmann.

58. Muspratt and Hofmann, "On Toluidine," 367.

59. For example, Muspratt and Hofmann's "synthetical experiments" might usefully be distinguished from Thenard's (1809) use of the term "combinations" to describe the products (then called "ethers") of reacting organic acids with alcohols in the presence of inorganic acids. Whereas synthetical experiments sought to elucidate the constitution (internal molecular arrangement) of organic compounds, Louis-Jacques

Thenard, "De l'Action des Acides Végétaux sur l'Alcool, sans l'Intermède et avec l'Intermède des Acides Minéraux," *Mémoires de Physique et de Chimie de la Société d'Arcueil* 2 (1809): 5–22, examined the combination of acid and alcohol in a reaction he viewed as somewhat analogous to inorganic salt formation. As Klein, *Experiments, Models, Paper Tools*, 103, persuasively argued, Thenard's focus was the unity of inorganic and organic chemistry, instantiated by the joining and separation of organic components whose internal makeup was not his subject of enquiry.

60. According to Rocke, *Quiet Revolution*, 290, Hofmann in 1855 coined the term "aromatic" to describe substances related to benzene.

61. Muspratt and Hofmann, "On Toluidine," 369, 381. Readers may note that Muspratt and Hofmann's formula contains twice as many carbon atoms as aniline's modern molecular formula (C_6H_7N), a discrepancy attributable to their use of a relative atomic mass for carbon that was roughly half of its present-day value.

62. As Brooke, "Laurent, Gerhardt," explained, not all nineteenth-century chemists agreed that it was possible to infer constitution from reactivity. In the absence of other reliable methods of accessing molecular constitution, neither Liebig nor Hofmann ever doubted this approach—which Hofmann, as this study makes clear, applied with extreme caution and great skill.

63. Other chemists who had previously expressed belief in the ultimate feasibility of making natural products in the laboratory include Gay-Lussac and Liebig.

64. See, James S. Muspratt and August Wilhelm Hofmann, "On Certain Processes in Which Anilene Is Formed," *Memoirs and Proceedings of the Chemical Society* 2 (1845): 249; and Muspratt and Hofmann, "On Toluidine," 367–368.

65. Muspratt and Hofmann, "On Toluidine," 367.

66. John Gardner, "An Address Delivered in the Royal College of Chemistry, Hanover Square, on Wednesday Evening, June 3," *Lancet* 1 (1846): 637–641.

67. Hofmann, *Reports*, lx.

68. Gardner, "An Address," 639–640, provided an early account of "natural productions . . . produced artificially," including urea—an early step in the reformulation of Wöhler's work as a synthesis. Indeed, Zinin's laboratory preparation of aniline—which Hofmann showed was a component of coal tar—exemplified exactly this point.

69. Muspratt and Hofmann, "On Toluidine," 369–370.

70. Meinel, "Regierender Oberchemiker," 32, cited Liebig's somewhat anguished letter to Friedrich Mohr, dated November 9, 1844, explaining why he was unable to promote Mohr for the Bonn position.

71. Ultimately published almost a year later as Liebig, "Amorphous Quinine;" and Justus Liebig, "Ueber die Zusammensetzung und die medicinische Wirksamkeit des Chinoidins," *Annalen der Chemie und Pharmacie* 58 (1846): 348–356, this work was in fact complete by June 1845. See Liebig to Hofmann, June 21 1845, Letter 4, in

Liebig und Hofmann, ed. Brock, 27, which sent a draft of this paper to Hofmann for approval.

There is considerable circumstantial evidence for Hofmann's involvement in the practical work underpinning these publications. Most persuasive is that right from the start, Hofmann was intended to profit from this scheme equally with Liebig. But Liebig's later request that Hofmann perform a further nitrogen determination on amorphous quinine is also significant. Further corroboration is found in Hofmann's reply (Hofmann to Liebig, June 24 1845, Letter 5, in *Liebig und Hofmann,* ed. Brock, 29–32). Hofmann declined to do this work in Bonn but did tell Liebig that material for such an analysis was available "in the cupboard in your laboratory, where the platinum vessels are stored." In other words, Hofmann—and not Liebig—knew exactly where to find the Giessen laboratory's supply of amorphous quinine.

72. Gerhardt, "organischen Basen."

73. For example, Muspratt and Hofmann, "Anilene," 251–253, identified aniline as the common result of decomposing the isomeric substances, anthranilic acid and salicylamide.

74. Liebig, "On Amorphous Quinine," 585–586, explained how he had developed this test from the chemistry reported in Gerhardt, "organischen Basen."

75. For more on this contemporaneous scheme, see Brock, *Gatekeeper,* 120–129.

76. Both Roberts, "Royal College of Chemistry," and Brock, *Gatekeeper,* claimed that Bullock had studied with Liebig in Giessen during 1839—but his omission from Fruton's appendix may indicate he spent time in Giessen for different reasons.

77. Hofmann to Liebig, June 4, 1845, Letter 3, in *Liebig und Hofmann,* ed. Brock, 27–28. As discussed below, Bullock's proposed name appears to have been a translation from German.

78. Hofmann to Liebig, June 24, 1845, Letter 5, in *Liebig und Hofmann,* ed. Brock, 29–32.

79. Liebig to Linde, June 27, 1845, Letter 164, in *Universität und Ministerium im Vormärz: Justus Liebigs Briefwechsel mit Justin von Linde,* ed. Eva-Marie Felschow and Emil Heuser, Studia Giessensia, vol. 3 (Giessen: Giessen University Press, 1992), 223.

80. Linde to Liebig, August 28, 1845, Letter 171, in *Universität und Ministerium,* ed. Felschow and Heuser, 232.

81. Hofmann to Liebig, June 4 and 24, 1845, Letters 3 and 5, in *Liebig und Hofmann,* ed. Brock, 28–32.

82. This venture was brought to scholarly notice by Roberts, "Royal College of Chemistry;" Roberts, "Establishment;" and *Liebig und Hofmann,* ed. Brock, 14–17; and it was termed the "Quinidine-Quinine Conspiracy" by Brock, *Gatekeeper,* 129–136.

83. In addition to previously cited sources, this account of the founding of the RCC is based on my study of materials held in the Archives Imperial College, London.

84. Sir James Clark to Liebig, August 23, 1845, Letter 8, in *Liebig und Hofmann,* ed. Brock, 33–34.

85. Liebig to Linde, July 14, 1845, Letter 166, postscript, in *Universität und Ministerium,* ed. Felschow and Heuser, 225–226. Liebig also noted that he had been offered an annual salary of GBP 2,000–3,000.

86. Hofmann to Liebig, June 4, 1845, Letter 3, in *Liebig und Hofmann,* ed. Brock, 27–28. Hofmann's letter also showed he was actively developing lectures to deliver in Bonn, and for publication—following translation by Gardner—in Thomas Wakley's *Lancet.*

87. Hofmann to Liebig, June 24, 1845, Letter 5, in *Liebig und Hofmann,* ed. Brock, 29–32. In fact, Liebig would have recognized Linde as the third party.

88. Hofmann to Liebig, June 24, 1845, Letter 5, in *Liebig und Hofmann,* ed. Brock, 29–32. Hofmann—whose initial salary at the RCC (according to Roberts, "Royal College of Chemistry," 271) was to be GBP 400 per year—expected to clear a 10,000 thalers profit.

89. Liebig to Hofmann, July 8, 1845, Letter 6, in *Liebig und Hofmann,* ed. Brock, 32.

90. Liebig to Hofmann, July 8, and July 26, 1845, Letters 6 and 7, in *Liebig und Hofmann,* ed. Brock, 32–33.

91. Hofmann to Liebig, late 1845, in *Liebig und Hofmann,* ed. Brock, 39. Rocke, *Nationalizing Science,* 148, noted that by 1851, Hofmann's London research school was recognized by Gerhardt as exemplary and on a par with Giessen.

92. This concern was raised as early as June 1845 and reemerged at intervals throughout the year that followed. See Hofmann to Liebig, June 24, 1845, in *Liebig und Hofmann,* ed. Brock, 29–32.

93. Liebig to Linde, September 17 and 21, 1845, Letters 173 and 174, in *Universität und Ministerium,* ed. Felschow and Heuser, 233–235, were written from Newton[-le-Willows], the Cheshire home of the Muspratt family, and London.

94. Liebig to Hofmann, November 29, 1845, Letter 12, in *Liebig und Hofmann,* ed. Brock, 40–42.

95. Hofmann to Liebig, June 24, 1845, Letter 5, in *Liebig und Hofmann,* ed. Brock, 29–32; Liebig to Linde, October 6, 1845, Letter 178, *Universität und Ministerium,* ed. Felschow and Heuser, 238–239.

96. Liebig to Linde, September 21, 1845, Letter 174, in *Universität und Ministerium,* ed. Felschow and Heuser, 234–235, referred to a total of 2,000 f.

97. Hofmann to Liebig, January 15, 1846, Letter 13, in *Liebig und Hofmann,* ed. Brock, 42–44.

98. Liebig to Linde, January 4, 1846, Letter 187, in *Universität und Ministerium,* ed. Felschow and Heuser, 248–249.

99. Liebig to Hofmann, February 6, 1846, Letter 14, in *Liebig und Hofmann,* ed. Brock, 44–46.

100. Liebig to Linde, February 5, 1846, Letter 190, in *Universität und Ministerium,* ed. Felschow and Heuser, 252–253.

101. Liebig to Hofmann, February 6, 1846, Letter 14, in *Liebig und Hofmann,* ed. Brock, 44–46. In fact, given Hofmann's residence in London, it seems highly unlikely that Bullock was correct on this legal technicality. Although the English patent process prior to the 1852 reforms was complex and arcane, there seems to have been no stricture against foreign patentees. In 1804, for example, the German national Frederick Albert Winsor (Friedrich Albrecht Winzer) was granted a patent (number 2764) for an "improved oven, stove, or apparatus for the purpose of extracting inflammable air, oil, pitch, tar, and acids, and reducing into coke and charcoal all kinds of fuel" while residing at Cheapside, London. See Thomas Seccombe, "Frederick Albert Winsor," *Dictionary of National Biography, 1885–1900,* accessed November 18, 2021, https://en.wikisource.org/w/index.php?title=Winsor,_Frederick_Albert_(DNB00)&oldid=4491582.

102. Liebig to Hofmann, April 29, 1846, Letter 19, in *Liebig und Hofmann,* ed. Brock, 51, provides the only evidence that Hofmann complied with this request. Evidently pursuing the question of possible challenges to Bullock's patent, Hofmann had asked Liebig to supply copies of both Sertürner's and Winkler's [sic] original publications on quinoidine. The pharmacist Friedrich Sertürner is best known for his (1804) isolation of morphine and identification of its basicity.

103. It seems reasonable to infer that these terms were substantially those contained in the final agreement, reproduced in *Liebig und Hofmann,* ed. Brock, 55–56, as Document 22a. *Liebig und Hofmann,* ed. Brock, 56, fn. 1, identified Linde as the "capitalist" referred to in this contract. W. J. Hornix, "Tales of Hofmann," *Annals of Science* 44 (1987): 519–524, 520, correctly noted in his review of Brock's edition that the Liebig-Hofmann correspondence clearly identified the capitalist as Robert Barclay, a London dealer in patent medicines with whom—as the contract itself states—Bullock and Gardner had "already made" arrangements. Liebig's correspondence with Linde confirms this identification.
It is worth noting that, while several of Hornix's criticisms are borne out by my study of the relevant sources, some are not—as, for example, when he incorrectly claimed that Roberts, "Royal College of Chemistry," 265–273, gave the order of candidacy for the RCC directorship as "Will, Hofmann, Fresenius." In fact, this list order (on 267) described their appearance (with Bullock and Gardner) in a photograph reproduced in B. Lespius, *Festschrift zur feier des 50 jährigen Bestehens der deutschen chemischen Gesellschaft* (Berlin: Friedländer, 1919), opp. 8 (see my figure 4.1).

104. Liebig to Linde, March 26, 1846, Letter 196, in *Universität und Ministerium,* ed. Felschow and Heuser, 260; and Liebig to Hofmann, March 19 and 28, 1846, Letters 16 and 17, in *Liebig und Hofmann,* ed. Brock, 47–49.

105. Linde to Liebig, April 27, 1846, Beilage zu Brock-Brief 18, in *Justus von Liebig und August Wilhelm Hofmann in Ihren Briefen. Nachträge 1845–1869,* edited by Regine Zott and Emil Heuser (Mannheim: Bionomica-Verlag, 1988), 15–16, was Linde's refutation of Bullock's accusations; Liebig to Hofmann, April 30, 1846, Letter 19a, in *Liebig und Hofmann,* ed. Zott and Heuser, 16, meanwhile, expressed Liebig's outrage that Hofmann believed that Liebig might have been complicit.

106. Liebig to Knapp, May 1, 1846, Letter 203, in *Universität und Ministerium,* ed. Felshow and Heuser, 267.

107. Liebig to Knapp, April 30, 1846, Letter 19b, in *Liebig und Hofmann,* ed. Zott and Heuser, 17, annotated with Knapp's confirmation of Linde's innocence. This is the same letter that Felshow and Heuser dated to May 1, 1846. This discrepancy might only be resolved by consulting the original source, which I have been unable to do.

108. Liebig to Hofmann, April 28, 1846, Letter 18, in *Liebig und Hofmann,* ed. Brock, 50–51.

109. Liebig to Linde, May 25, 1846, Letter 209, in *Universität und Ministerium,* ed. Felschow and Heuser, 277–278.

110. Liebig to Linde, May 1, 1846, Letter 203, *Universität und Ministerium,* ed. Felschow and Heuser, 267–268.

111. Liebig to Hofmann, May 25, 1846, Letter 21, in *Liebig und Hofmann,* ed. Brock, 52–54. Bullock's English patent (BP 1846, No. 11,204) was granted on May 12. In August 1846, Bullock also applied for, and received, a Scottish patent.

112. Liebig, "On Amorphous Quinine," appeared in the May 23 issue; Wakley's editorial endorsement appeared on the issue's opening page.

113. Brock, *Gatekeeper,* 97–98, cited Liebig to Berzelius, November 26, 1837, in *Berzelius und Liebig,* ed. Carrière, 134, on this point.

114. Theophilus Redwood, "The Progress of Pharmaceutical Science," *Pharmaceutical Journal* 5 (1846): 103–106.

115. The exchange between Redwood and Bullock appeared in *Lancet* 1 and 2 (1846), *London Medical Gazette* 38 (1846), and *Pharmaceutical Journal* 5 and 6 (1846–1847).

116. Anon., "Ueber Chinoidin," *Augsburger Allgemeine Zeitung* 169 (June 7, 1846): 1349. Liebig to Linde, June 11, 1846, Letter 211, in *Universität und Ministerium,* ed. Felshow and Heuser, 281, indicated that Liebig delayed the contract with Bullock (see *Liebig und Hofmann,* ed. Brock, 55–56, Document 22a), awaiting the effect of this anonymous publication on the price of quinoidine.

117. Ferdinand L. Winckler, "Ueber die chemischen Zusammensetzung des Chinoidins und das zweckmäßigste Verfahren dasselbe zu Reinigen," *Jahrbuch für praktische Pharmacie* 7 (1843): 69, proposed the name "amorphous quinine," for his most highly purified extract of quinoidine.

118. Ferdinand L. Winckler, "Ueber die chemischen Zusammensetzung des käuflichen Chinoidins und die Bedeutung desselben als Arzneimittel," *Jahrbuch für praktische Pharmacie* 10 (1846): fn. on 382.

119. Justus Liebig, "Ueber die Zusammensetzung und die medicinische Wirksamkeit des Chinoidins," *Großherzoglich Hessische Zeitung* 1 (June 12, 1846): 823–824; Ferdinand L. Winckler, "Einige Worte ueber den mutmaßlichen Werth des Chinoidins als Heilmittel," *Großherzoglich Hessische Zeitung* 1 (June 19, 1846): 859–860.

120. Redwood's publications make no mention of Winckler—as seems entirely consistent with the very different causes of their disagreements with Liebig. Though they shared similar responses to Liebig's chemistry, Redwood's concern was moral and practical—whereas Winckler's was intensely personal.

121. Liebig to Hofmann, June 22, 1846, Letter 24, in *Liebig und Hofmann*, ed. Brock, 56–57, acknowledged the realization of his fears.

122. Ferdinand L. Winckler, "Ueber die chemischen Zusammensetzung des käuflichen Chinoidins und die Bedeutung desselben als Arzneimittel," *Großherzoglich Hessische Zeitung* 1 (June 30, 1846), included Liebig's extensive fn. on 919.

123. Winckler referred to Geiger, *Handbuch der Pharmacie,* 1185–1186.

124. In his entry on cinchona bark, for example, Henry Watts, *A Dictionary of Chemistry,* vol. 1 (London: Longman, 1863), 970, tabulated the alkaloid content of a range of barks produced by different species of cinchona, giving the attribution of these data. Winckler's name appeared on numerous occasions, Liebig's not at all.

125. Liebig to Hofmann, April 29, 1846, Letter 19, in *Liebig und Hofmann*, ed. Brock, 51, provides the only contrary evidence. Asked by Hofmann to supply one of Winckler's articles on quinoidine, Liebig claimed not to have the journal in which it appeared, the *Jahrbuch [für praktische Pharmacie]*. This is entirely consistent, of course, with the idea that Liebig had once possessed copies of Winckler's publications, passing these to Bullock while developing their plans for amorphous quinine. Indeed, it is possible that Liebig—anticipating exactly the challenge that Winckler had just made—asked Bullock to suggest the name "amorphous quinine" to Hofmann precisely to retain plausible deniability in these circumstances.

126. Winckler, "käuflichen Chinoidins," 382, referring to Ferdinand L. Winckler, "Briefliche Notiz über die wahrscheinliche Entdeckung eines neuen Chinarinden-Alkaloids (Chinidin)," *Repertorium für die Pharmacie* 85 (1844): 392–397. This was when Winckler noted for the first time that white crystals were present that resembled another cinchona alkaloid he had previously called "quinidine." Brock's identification of amorphous quinine as quinidine—and hence his choice to denote this episode as the "Quinidine-Quinine Conspiracy"—thus perpetuates widespread confusion surrounding the history of the cinchona alkaloids. As Hornix, "Tales of Hofmann," 521, explained, what Winckler called "quinidine" in 1846 (not 1848, as both Brock and Hornix state) was no such thing. Winckler's quinidine was, in fact,

the alkaloid now known as cinchonine. As Theophilus Redwood, "On Amorphous Quinine," *Pharmaceutical Journal* 6 (1846): 129, made clear, contemporary pharmaceutical chemists were well aware that Liebig's procedure could not produce a single, pure alkaloid. Indeed, this was the major scientific (rather than moral) criticism that Redwood raised against Bullock—and, by implication, Liebig.

127. Winckler, "käuflichen Chinoidins," 384–385.

128. The fact that some of Winckler's work had been reprinted in the *Pharmaceutical Journal* makes a connection with Redwood highly plausible.

129. Winckler, "käuflichen Chinoidins," 386–388.

130. Jacob Bell and Theophilus Redwood, "The Rise and Progress of a Philosopher," *Pharmaceutical Journal* 6 (1847): 141–142, criticized Gardner. Having been threatened by Gardner, the journal published an "Apology to Dr. Gardner," on 148–149. Gardner, meanwhile, presented his "Vindication of Dr. Gardner," on 149–151. In their "Counter Statement," on 151–160, Bell and Redwood now made public chapter and verse of recent events at the RCC, reproducing an "Extract from the report of the Sub-Committee for Accounts," together with other documents leaked to them by "a member of the College." Bell and Redwood (on 157) also deprecated "the appointment of a foreigner (no matter how distinguished he might be)" to what was supposed to be a "NATIONAL Institution."

131. Gardner, "Address." Cf. *Liebig und Hofmann*, ed. Brock, 55–56, Document 22a. For the Pharmaceutical Society's view, see Redwood, "On Amorphous Quinine," 161.

132. Redwood, "On Amorphous Quinine," 162.

133. Anon., "Editor's note," *London Medical Gazette* 38 (1846): 256, cf. Editor's note, "Amorphous Quinine," *Pharmaceutical Journal* 6 (1846–1847): 55–56, which quoted Thomas Wakley, "Editorial," *The Lancet* (July 25, 1846). See also Thomas Wakley, "Communication," *Pharmaceutical Journal* 6 (1847): 144.

134. John Grove, "Amorphous Quinine," *Lancet* 2 (1846): 399–401. One of the key objections to Bullock's actions was that he sought to establish a monopoly over the supply of amorphous quinine—which Liebig's letter to Hofmann (February 6, 1846) confirms was indeed the intention. In October, responding to Grove, John L. Bullock, "Amorphous Quinine," *Lancet* 2 (1846): 436–467, called on a lawyer friend to refute this notion.
No letter between Liebig and Hofmann survives between September and December 1846.
It is noteworthy that, although almost all of Liebig's letters to Hofmann survive, many of Hofmann's do not. Several of Liebig's letters refer to communications from Hofmann that Brock and his collaborators were unable to locate—including one sent on June 1, 1846 (*Liebig und Hofmann*, ed. Brock, 56) as an enclosure within a letter Hofmann wrote to his mother in Giessen. Hornix, "Tales of Hofmann," 519, noted this highly uneven survival pattern, suggesting that, whereas Liebig deliberately

destroyed potentially damning evidence contained in Hofmann's letters, Hofmann kept Liebig's letters out of carelessness. A more reasonable inference, it seems to me, is that Hofmann preserved letters from Liebig that would have helped mitigate his complicity, had the full extent of the matter come to light.

135. Liebig to Hofmann, July 15, 1847, Letter 36, in *Liebig und Hofmann,* ed. Brock, 75, confirmed Liebig's purchase of Linde's quinoidine, leaving Liebig to carry an estimated loss of 950 thalers. Liebig was glad it was not more.

136. Royal College of Chemistry, Minutes of Committees, "Committee of Management appointed 9th September 1846" (AICL C4/568).

137. Roberts, "Establishment," 473.

138. August Wilhelm Hofmann, "A Page of Scientific History: Reminiscences of the Early Days of the Royal College of Chemistry," *Quarterly Journal of Science* 8 (1871): 151, framed this decision as a gesture of solidarity toward Sir James Clark.

139. Hofmann had initially sought Liebig's advice and deferred to his opinions, especially concerning arrangements at the RCC, but he displayed much greater independence when resuming their correspondence in February 1847.

140. Hofmann, *Reports;* Hofmann, *Heilmittellehre.*

141. Hofmann, *Life-Work of Liebig,* 8.

142. Hofmann, "Page," 146.

143. Jacob Bell, "Laboratories for Practical Instruction," *Pharmaceutical Journal* 6 (1847): 196.

144. For example, Bud and Roberts, *Science versus Practice.*

145. Indeed, the RCC's later merger with the Royal School of Mines and its emergence in 1907 as one of the three constituent institutions forming Imperial College (now Britain's elite university devoted to science, technology and medicine) encourage the view that the RCC was an essentially academic institution. This tendency has been especially persistent in historical literature intended for chemists. See, for example, John J. Beer, "A. W. Hofmann and the Founding of the Royal College of Chemistry," *Journal of Chemical Education* 37 (1960): 248–251; Brigitte Osterath, "August Hofmann and the Chemists Factory," *Chemistry World* (August 10, 2017). https://www.chemistryworld.com/feature/hofmanns-chemistry-factory/3007787.article.

146. Sir James Clark to Liebig, August 23, 1845, Letter 8, in *Liebig und Hofmann,* ed. Brock, 33–34.

147. Sir James Clark to Liebig, August 23, 1845, Letter 8, in *Liebig und Hofmann,* ed. Brock, 33–34.

148. Hofmann's academic and institutional accomplishments have attracted relatively little modern historical interest yet featured heavily in contemporary accounts of his life and work, including the memorial lectures delivered in England and Germany by Abel, Playfair, Perkin, Armstrong, and Tiemann.

Following his return to his homeland in the mid-1860s, Hofmann became Germany's leading chemist: founder and repeated president of the German Chemical Society, and professor at Berlin University. For more on Hofmann and chemical laboratories, see chapter 8.

149. See, for example, Gay, *History of Imperial College*, 19–20. Also see Meinel and Scholz, *Allianz*.

CHAPTER 5

1. Klein, "Techniques of Modelling," 146–167, explained Dumas's use of Berzelian formulae in formulating type theory.

2. For an analysis of the distinction between atoms and equivalents, and the debates this caused, see Rocke, *Nationalizing Science*, 97ff.

3. Jean Baptiste Dumas, "Mémoire sur la Loi des Substitutions et la Théorie des Types," *Comptes Rendus* 10 (1840): 149–178.

4. The concept of electronegativity was fundamental to Berzelian dualistic theory, because it related chemical and electrical properties. However, as William B. Jensen, "Electronegativity from Avogadro to Pauling. Part 1. The Origins of the Electronegativity Concept," *Journal of Chemical Education* 73 (1996): 11–20, explained, it is not the case that the electrochemical dualistic theory relied on simple electrostatic attraction between positive and negative radicals—an oversimplification at the root of other interpretations that see this theory as a forerunner to ionic bonding.

5. Hofmann, "Metamorphosen des Indigo's," Liebig's fn. on 1.
It is difficult to believe that Liebig—though undoubtedly partisan—was actually persuaded by such limited experimental evidence. Instead, it seems likely that he sought to promote Hofmann's work in Paris. Cf. Keas, "Nature."

6. Liebig, "neuen Apparat," 13.

7. Liebig, "organischen Säuren," 186.

8. Liebig, "Stickstoff-Verbindungen." Melamine was produced by dry distillation of ammonium thiocyanate (*Schwefelcyanammonium*).

9. Liebig, "organischen Säuren," 186–187. Liebig described "M" as a radical—a term that originated in dualistic theory but later lost a definitive association with this theory.

10. Liebig, "Basen, Organische."

11. Liebig's inference here mirrored Dumas's earlier suggestion that alcohol contained hydrogen in two distinct forms. Klein, "Techniques of Modelling," showed how Dumas reached this conclusion by paper-based manipulation of Berzelian formulae.

12. Rocke, *Chemical Atomism*, 54 ff., explained how the lack of agreement concerning molecular weight determination resulted in competing four- and two-volume formulae for organic compounds. Two-volume formulae, standard in inorganic chemistry, used vapor density measurements as the basis for molecular weights.

Thus, a substance's molecular weight was the weight of vapor occupying the same volume as two unit weights of hydrogen (measured at the same temperature and pressure). Berzelius, however, did not accept that vapor density determined molecular size, instead assigning formulae so as to accommodate his commitment to electrochemical dualistic theory. Sustaining whole numbers of atoms in dualistic formulae for certain classes of organic compounds, such as acid anhydrides, entailed an effective doubling of molecular weight. Thus, many of the formulae that Berzelius assigned were four-volume formulae, e.g., $C_4H_6O_3$ for acetic acid anhydride. Berzelius's formula for acetic acid anhydride is identical to the modern molecular formula. But it led Berzelius to assign $C_4H_8O_4$ (four-volume) rather than $C_2H_4O_2$ (two-volume) for acetic acid, and in this case his molecular formula was double that which we accept today. This example illustrates why historical formulae of organic compounds expressed using the four-volume system frequently contain double the number of atoms compared with their modern counterparts.

Readers may also notice that many otherwise familiar formulae, e.g., Hofmann's formula for aniline ($C_{12}H_7N$), contain twice as many carbon atoms as are assigned today (C_6H_7N). This difference is caused by Hofmann's use of a relative atomic mass for carbon that was roughly half its present-day value.

13. Liebig, "Herr Gerhardt," expressed Liebig's exasperation with his former pupil. Rocke, *Quiet Revolution,* 86–93, related Gerhardt's fortunes during this tumultuous period.

14. Charles Gerhardt, "Sur la Classification Chimique des Substances Organiques," *Revue Scientifique et Industrielle* 14 (1843): 588.

15. See, for example, Carl Remigius Fresenius, "Ueber die Constitution der Alkaloide," *Annalen der Chemie und Pharmacie* 61 (1847): 149–156.

16. Rocke, *Image and Reality,* 9. See also Rocke, *Quiet Revolution,* 95–96.

17. Holmes, "Liebig," 343–344, described Liebig's disillusion with theory and the mounting tensions between him and Berzelius.

18. Klein, "Shifting Ontologies," noted this important transition. Illustrated in chapter 2 by the Giessen studies of camphor and oil of turpentine, the equivalence of substances from artificial and natural sources was fundamental to Liebig's definition of organic bases.

19. According to Armstrong, "Hofmann Memorial Lecture," 654, Hofmann continued using aniline from both indigo and coal tar as late as 1849.

20. Once established, Hofmann's London laboratory provided a refuge for many Giessen-trained chemists—among whom Peter Griess, pioneer of azo dye chemistry, is perhaps the best known—at a time when opportunities in Germany remained limited. See Keas, "Structure and Philosophy," and Hannah Gay, "'Pillars of the College': Assistants at the Royal College of Chemistry," *Ambix* 67 (2000): 135–169.

21. Theodore Gervaise Chambers, *Register of the Associates and Old Students of the Royal College of Chemistry, the Royal College of Mines and the Royal College of Science;*

with Historical Introduction and Biographical Notes and Portraits of Past and Present Professors (London: Hazell, 1896). Hofmann to Liebig, Letter 11, Okt–Dez 1845, in *Liebig und Hofmann*, ed. Brock, 38, reported the skillfulness of his new assistants. Nicholson went on to become a partner in the London dye company Simpson, Nicholson and Maule; while Abel succeeded Michael Faraday as professor of chemistry at the Woolwich Arsenal, London's military training academy.

22. Abel, "Hofmann Memorial Lecture," 591, explained the assistants' responsibility to prepare aniline from indigo.

23. August Wilhelm Hofmann, "Eine sichere Reaction auf Benzol," *Annalen der Chemie und Pharmacie* 55 (1845): 200–205. Hofmann first used this test-tube procedure in his previous study of styrole. See Blyth and Hofmann, "On Styrole."

24. Charles Blachford Mansfield, "Researches on Coal Tar," *Quarterly Journal of the Chemical Society* 1 (1849), 244–268, 264–265, reported a new process (patented the previous year) for extracting pure benzene from coal tar. Mansfield's process relied on his earlier isolation of pure benzene by a series of painstaking small-scale distillations carried out in glass retorts and a crystallization that made innovative use of a French-press coffee maker. Having purified benzene by this means, Mansfield showed that it boiled at a constant 175 °F (just over 80 °C), more than 10 degrees below the boiling points previously reported by Faraday and Mitscherlich. Equipped with this new and reliable way of identifying pure benzene, Mansfield was able to draw on existing technical knowledge concerning the industrial rectification of alcohol (alcohol with specific gravity 0.825 boiled at almost exactly the same temperature) to scale up the procedure using industrial apparatus made of metal. This was the procedure he patented. Benzene, once obtained, was converted into aniline using Zinin's reaction. However, this transformation did not become feasible on a large scale until Perkin's development of an industrial process in 1856.

25. According to Perkin, "Hofmann Memorial Lecture," 604, aniline remained "costly" at the time of his discovery of mauve. Travis, *Rainbow Makers*, 39, reported Hofmann's request in early October 1856 that Perkin begin supplying him with aniline. See also Travis, "Perkin's Mauve," for an account of how Perkin translated his discovery into a viable industrial process.

26. Hofmann, "Metamorphoses of Indigo," 299–300.

27. Hofmann, "New Researches," 26–28. As chapter 4 explained, this publication was delayed by Hofmann's priority dispute with Gerhardt over quinoline.

28. Muspratt and Hofmann, "On Toluidine."

29. James S. Muspratt and August Wilhelm Hofmann, "On Nitraniline, A New Decomposition Product of Dinitrobenzol," *Memoirs and Proceedings of the Chemical Society* 3 (1845): 110–125, produced nitroaniline from benzene in two steps: Mitscherlich's nitration followed by Zinin's reduction. What Muspratt and Hofmann called "peroxide" of nitrogen corresponds to today's nitro group.

30. Muspratt and Hofmann, "On Nitraniline," 124.

31. Fresenius, "Constitution der Alkaloide," 149–150.

32. See chapter 4 for a discussion of how Liebig handled the priority dispute between Hofmann and Gerhardt.

33. August Wilhelm Hofmann, "Researches on the Volatile Organic Bases. I: On the Action of Cyanogen on Aniline, Toluidine, and Cumidine," *Quarterly Journal of the Chemical Society* 1 (1849): 159–173.

34. See chapter 8. See also Catherine M. Jackson, "Re-Examining the Research School: August Wilhelm Hofmann and the Re-Creation of a Liebigian Research School in London," *History of Science* 44 (2006): 281–319.

35. Hofmann used "electronegative radical" to indicate that cyanogen, like oxide of nitrogen and the halogens, sat high in Berzelius's table of electronegativity—a rank order that roughly corresponds to the modern electronegativity scale, though it was derived from very different principles, evidence, and reasoning.

36. Hofmann, "Volatile Organic Bases. I."

37. Hofmann, "Volatile Organic Bases. I," 171.

38. Hofmann, "Volatile Organic Bases. I," 173.

39. It is a mark of the difficulty of this nitrogen determination that even Will and Varrentrapp's method did not provide useful results.

40. Hofmann, "Volatile Organic Bases. I," 171 ff.

41. Hofmann, "Volatile Organic Bases. I," 171–172. See also Liebig, "organischen Säuren;" Liebig, "Stickstoff-Verbindungen."

42. Both revisionist points are contra Keas, "Nature." In fact, Hofmann made no reference to type theory, nor to any change in his theoretical convictions, in this paper.

43. Hofmann, "Volatile Organic Bases. I," 170–171.

44. Hofmann, "Volatile Organic Bases. I," 173.

45. August Wilhelm Hofmann, "Researches on the Volatile Organic Bases. II: On the Action of Iodine on Aniline," *Quarterly Journal of the Chemical Society* 1 (1849): 269–281.

46. Hofmann, "Volatile Organic Bases. II," 280.

47. Hofmann, "Volatile Organic Bases. 1," 173, adduced evidence of both kinds in the case of cyaniline, but clearly gave primacy to the results of synthesis.

48. August Wilhelm Hofmann, "Researches on the Volatile Organic Bases. III: Action of Chloride, Bromide, and Iodide of Cyanogen on Aniline. Melaniline, a New Conjugated Alkaloid," *Quarterly Journal of the Chemical Society* 1 (1849): 285–317, 313–314.

49. Hofmann, "Uebersicht," 241–343.

50. Hofmann, "Volatile Organic Bases. III," 290. Hofmann also isolated melaniline from the reaction between aniline and bromide of cyanogen.

51. Hofmann, "Volatile Organic Bases. III," 298 (various agents); 310 (reagents); 311 (synoptical table).

52. Liebig, "organischen Säuren," 187.

53. Hofmann, "Volatile Organic Bases. III," 313–314.

54. Hofmann, "Volatile Organic Bases. III," 314.

55. Hofmann, "Volatile Organic Bases. III," 314–317.

56. Hofmann, "Volatile Organic Bases. III," 317.

57. Hofmann, "Volatile Organic Bases. III," 317.

58. August Wilhelm Hofmann, "Researches on the Volatile Organic Bases. V. On the Action of Acids and Bases upon Cyaniline. VI. Metamorphoses of Dicyanomelaniline. Formation of the Aniline-Term Corresponding to Cyanic Acid. VII. Action of Anhydrous Phosphoric Acid on Various Aniline-Salts and Anilides," *Quarterly Journal of the Chemical Society* 2 (1850): 300–335, 311–312.

59. Hofmann, "Volatile Organic Bases. V," 312.

60. Hofmann, "Volatile Organic Bases. V," 313–314.

61. Hofmann, "Volatile Organic Bases. V," 313–314.

62. Hofmann, "Volatile Organic Bases. V," 315.

63. Hofmann, "Volatile Organic Bases. V," 310.

64. Hofmann, "Volatile Organic Bases. V," 315.
Klein, "Techniques of Modelling," 147, introduced the term "paper-tools" to describe how paper-based manipulation of Berzelian formulae enabled Dumas to produce "traditional [i.e., nonatomistic] models of the constitution of organic compounds and their reactions," subsequently demonstrating how this approach led him to propose novel atomistic explanations of constitution and reactivity—namely, his law of substitution and theory of types. Klein, *Experiments, Models, Paper Tools,* developed this argument, showing that paper tools were productive in early nineteenth-century organic chemistry, despite the opposition of chemists, including Berzelius. In 1829, as Klein explained, Berzelius dismissed such paper-based manipulations (this time by Dumas and Félix-Polydore Boullay) as "theoretical speculations" that were not in accord with available experimental evidence.
The present study helps clarify the continuities and discontinuities in chemists' use of Berzelian formulae ca. 1850. As one might expect, given Klein's demonstration that paper tools produced a novel, atomistic understanding of substance and reactivity, Hofmann did not use Berzelian formulae in the same way as Dumas had more than a decade earlier. Hofmann's formulae represented the outcome of chemical reactions, and they incorporated the constitution of novel products only where this was indicated by experimental evidence. This process mobilized concepts including substitution, which had been suggested by Dumas's earlier work with paper tools. Unlike Dumas, however,

Hofmann's formula models were increasingly atomistic, and his laboratory reasoning built on experimentally derived knowledge of properties and reactivity rather than on constitutional inferences produced by manipulating formulae.

65. Adolphe Wurtz, "Note sur la Formation de l'Uréthane par l'Action du Chlorure de Cyanogène Gazeux sur l'Alcool," *Comptes Rendus* 22 (1846): 503–505.

August Wilhelm Hofmann, "Erinnerungen an Adolph Wurtz," *Berichte der deutschen chemischen Gesellschaft* 20 (1887): 815–996. Reprinted in *Zur Erinnerung an vorangegangene Freunde,* vol. 3 (Braunschweig: Vieweg, 1888). Hofmann described meeting Wurtz at one of Liebig's regular Sunday lunch parties. The meal—accompanied by what Hofmann described as "unbelievable quantities" (p. 835) of good Rhine wine—was followed by several hours spent discussing the lab's current research. Wurtz investigated the chemistry of hypophosphoric acid—a curious inorganic acid—while in Giessen.

66. Friedrich Wöhler and Justus Liebig, "Cyansäures Aethyl- und Methyloxyd," *Annalen der Chemie und Pharmacie* 54 (1845): 370–371, developed a reaction introduced by Liebig and Wöhler, "Cyansäuren."

67. Hofmann, "Adolph Wurtz," 219–220; Rocke, *Nationalizing Science,* 102, noted Wurtz's presentation of his Giessen "credentials" to Dumas.

68. Adolphe Wurtz, "Sur une Série d'Alcalis Organiques Homologues avec l'Ammonique," *Comptes Rendus* 28 (1849): 223, termed his study a "sort of transition between mineral and organic chemistry." Although Wurtz perhaps sought to promote the unification of chemistry, it is equally the case that this work situated Wurtz at the intersection of organic and inorganic chemistry. See also Adolphe Wurtz, "Note sur l'Éther Cyanurique et sur le Cyanurate de Methylene," *Comptes Rendus* 26 (1848): 368–370.

Wurtz's papers revealed his relative inexperience as an organic chemist. His analyses certainly did not conform to Giessen standards. As was then common in French publications, he provided limited raw experimental data and none of the results of combustion analysis. Time spent in Liebig's laboratory had not persuaded Wurtz to do organic chemistry the Giessen way.

69. Hofmann, "Volatile Organic Bases. V," 318–319.

70. Hofmann, "Volatile Organic Bases. VI," 327–339.

71. Hofmann, "Volatile Organic Bases. VI," 327.

72. Hofmann, "Volatile Organic Bases. VII," 332–333.

73. Hofmann, "Volatile Organic Bases. VII," 334.

74. The formulae shown here are taken from Hofmann, "Volatile Organic Bases. VII," "synoptical table" facing 331, which organized more than 50 compounds belonging to the ammonia and aniline series in Hofmann's four-part, dualistic taxonomy. Although anilocyanic acid provided an example of an "anilimide" or "anile" (the aniline analog of imides or imidogen compounds), there was no entry in the category of "anilo-nitrile."

It is worth noting that Hofmann here used formulae intended to clarify how each substance fitted his proposed taxonomy—even though elsewhere he formulated the same substances in different ways, e.g., oxanilide as $C_{12}H_6N,C_2O_2$ rather than $NH_2,C_2O_2,C_{12}H_4$. Hofmann certainly used formulae to represent constitutional interpretations of the substances he studied. His uncommented-on use of different formulations, however, indicates that formulae were not decisive in directing Hofmann's experimental program or in his interpretation of its results. In other words, Hofmann's laboratory reasoning was not driven by the paper-based manipulation of formulae.

CHAPTER 6

1. Wurtz, "Note sur l'Éther Cyanurique;" Adolphe Wurtz, "Recherches sur les Éthers Cyaniques et leurs Dérivés," *Comptes Rendus* 27 (1848): 241–243; Wurtz, "Sur une Série."

2. Wurtz, "Sur un Série," 223.

3. Jean Baptiste Dumas, "Rapport sur un Mémoire de M. Wurtz, Relative à des Composés Analogues à l'Ammoniaque," *Comptes Rendus* 29 (1849): 203–205.

4. Rocke, *Quiet Revolution,* 96–97, which cited Liebig to Hofmann, April 23, 1849, in *Liebig und Hofmann,* ed. Brock, 84. As Rocke, *Nationalizing Science,* 166–167, fn. 22, explained, this referred to an earlier letter from Hofmann to Liebig, dated March 24, 1849, that is no longer extant. In his letter, Hofmann evidently pointed out that Wurtz had done no more than restate Liebig's previous proposals concerning the constitution of the organic bases.

5. Armstrong, "Hofmann Memorial Lecture," 659.

6. Wurtz, "Recherches sur les Éthers Cyaniques," seems to have prepared cyanate esters with relative ease, for he gave no explanation of how he arrived at his published procedure—distillation of potassium cyanide with potassium sulphovinate (ethyl potassium sulfate)—nor any experimental detail. Yet the reaction was apparently reliable, since he was also able to prepare methyl cyanate.

7. Wurtz, "Recherches sur les Éthers Cyaniques," 243.

8. Wurtz, "Recherches sur les Éthers Cyaniques." Examination of Wurtz's published work clearly reveals this distinction: Whereas Hofmann's papers were long and detailed, Wurtz's seldom extended beyond a few pages.

9. Hofmann, "Adolph Wurtz," 932, explained that Wurtz expected the product to be ammonia by analogy with the recently published Frankland-Kolbe hydrolysis of ethyl cyanide. Thus, as Rocke, *Quiet Revolution,* 96, noted, reasoning from the copula (i.e., dualistic) theory was partly responsible for Wurtz's error.

10. Wurtz, "Sur un Série," 223. According to Keas, "Nature," 106; and Rocke, *Nationalizing Science,* 166–167, fn. 22, Hofmann expressed this concern in a letter to Liebig

that is no longer extant. Liebig's response was reassuring. Hofmann used the term "ether" to refer to compounds now called "esters."

11. Hofmann, "Volatile Organic Bases. V," 319.

12. Hofmann, "Volatile Organic Bases. V,"322, 328.

13. Though he did not say so, Hofmann's classification organized these compounds as members of homologous series, providing Wurtz's earlier suggestion with a more concrete empirical foundation.

14. Hofmann, "Volatile Organic Bases. VI," 328–339.

15. Hofmann, "Volatile Organic Bases. VI," 330.

16. Hofmann, "Volatile Organic Bases. VII," 334–335.

17. Cf. Keas, "Nature," 106.

18. Hofmann, "Volatile Organic Bases. VII," 334–335.

19. Hofmann, "Volatile Organic Bases. VII," 335.

20. The primary instance of this divergence concerns Hofmann's account of his route to chloraniline—which I interpret as a justifiable simplification serving clear disciplinary goals; cf. Keas, "Nature."

21. Hofmann, "Researches Regarding," 96. Hofmann's categories were derived from his four-part taxonomy and correspond respectively to today's primary, secondary, and tertiary amines.

22. Hofmann, "Researches Regarding," 96.

23. As Keas, "Nature," 102, pointed out, Hofmann had in fact previously established the impossibility of this method.

24. Hofmann, "Researches Regarding," 98.

25. Hofmann, "Researches Regarding," 98–99.

26. John J. Griffin, *Chemical Handicraft: A Classified and Descriptive Catalogue of Chemical Apparatus, Suitable for the Performance of Class Experiments, for Every Process of Chemical Research, and for Chemical Testing in the Arts* (London: Griffin, 1866), 204–206, described and illustrated various Liebig condensers—including the now-familiar version in which both inner tube and outer sleeve are made of glass.

27. See, for example, the numerous occurences in the *Chemisches Zentralblatt* for 1869. There are also some 80 occurrences of the term in Karl Elbs, *Synthetische Darstellungsmethoden der Kohlenstoff-Verbindungen*, vol. 1. *Synthesen mittels Metallorganischer und mittels Cyanverbindungen; Synthesen durch molekulare Umlagerung und durch Addition* (Leipzig: Barth, 1889).

28. This observation begins to provide an explanation of organic chemistry's unusually strong emphasis on genealogies of training.

29. Hofmann, "Researches Regarding," 116.

30. Hofmann, "Researches Regarding," 100.

31. Ursula Klein, *Technoscience in History: Prussia, 1750–1850* (Cambridge, MA: MIT Press, 2020), chapter 3, highlighted the close relationship between pharmaceutical and chemical laboratories during the eighteenth century and the movement of apparatus between these sites. Klein's volume developed many themes connecting artisanal expertise and academic science, including chemistry's material dependence on pharmacy ca. 1800.

32. Batka, *Verzeichniss* (1832), 5.

33. See, for example, Batka, *Verzeichniss* (1832), item V61 *Heberpipette* (lifting pipette), priced at 12 kr., i.e., 1/10th the cost of Batka's filter- and separating funnels.

34. For example, the catalogs of London supplier John Joseph Griffin did not offer separatory funnels until after 1850.

35. AICL, item 1005, RCC *Musterbuch* (undated), is a hardbacked, foolscap notebook of some 50 pages. It includes a wide variety of glassware, including Liebig's Kaliapparat (No. 312), Geissler's modified Kaliapparat (No. 332), and Will and Varrentrapp's apparatus for nitrogen (in No. 333).

36. Minutes of Council Meeting, 23 September 1845 (AICL C3/566). The Council Minutes record purchases of German glassware amounting to some £150 on the following dates: January 6, 1846; July 14, 1847, December 13, 1848; and January 15, 1850.

37. Making taps is one of the most technically demanding tasks undertaken by scientific glassblowers.

38. Hans Schindler, "Notes on the History of the Separatory Funnel," *Journal of Chemical Education* 34 (1957): 528–530.

39. See also the similar mention of "tap-funnel" in James Smith Brazier and G. Gossleth, "Contributions towards the History of Caproic and Œnanthylic Acids," *Quarterly Journal of the Chemical Society* 3 (1851): 210–229. According to August Wilhelm Hofmann, "Researches into the Molecular Constitution of the Organic Bases," *Philosophical Transactions of the Royal Society of London* 141 (1851): 357–398, Smith Brazier ably assisted Hofmann in his later studies of the organic bases.

40. Hofmann, "Researches Regarding," 100.

41. Hofmann, "Researches Regarding," 131.

42. Hofmann, "Researches Regarding," 124.

43. Hofmann, "Researches Regarding," 107.

44. Hofmann, "Researches Regarding," 97. Hofmann here used the term "alkaloids" to refer to mainly artificial bases. Although (as previously noted) the term is now used almost exclusively in reference to natural products, this was not always the case in the period covered by this book.

45. Both Gerhardt and Wurtz—who grew up in Strasbourg around the same time—spent their careers primarily in France. Rocke (*Quiet Revolution*, 94–95) rightly

cautioned against inferring from their similar trajectories that Wurtz shared Gerhardt's views on chemistry. Yet there are definite similarities in how each offered theoretical speculations whose experimental basis was limited.

46. Hofmann, "Researches Regarding," 123 ff.

47. Hofmann, "Researches Regarding," 125. When comparing cumidine and methyl ethyl aniline, Hofmann relied on the boiling point regularities established by Kopp's work (see chapter 7) to argue for a difference of at least 11 degrees in their boiling points.

48. Hofmann, "Researches Regarding," 126.

49. Hofmann, "Researches Regarding," 127.

50. Hofmann, "Researches Regarding," 128.

51. Hofmann, "Researches Regarding," 129.

52. Hofmann, "Researches Regarding," 130.

53. Hofmann, "Researches Regarding," 130.

54. Cf. Beer, "A. W. Hofmann."

55. Garfield, *Mauve*, 36, dismissed Perkin's plan as "naïve." Even among scholarly accounts, Travis, *Rainbow Makers*, 36, is unusual in admitting Hofmann's notion that naphthalene might be converted into quinine was "not unreasonable." Yet Perkin's modified attempt to produce quinine by oxidizing allyl toluidine was— for Travis—the result of "using no more than the chemical formulae as a guide." See also Travis, "Perkin's Mauve." Indeed, the implausibility of Perkin's proposed conversion continues to fascinate historically inclined chemists, as reflected in "Mauveine—the Final Word?"—a series of occasional articles published by the Royal Society of Chemistry's Historical Group.

56. Tiemann, "Gedächtnisrede," 3390.

57. August Kekulé, *Lehrbuch der organischen Chemie, oder der Chemie der Kohlenstoffverbindungen* (Erlangen: Enke, 1861), 91.

58. Charles M. Gerhardt, *Traité de Chimie Organique* (Paris: Didot, 1853), 129–142.

59. Given widespread historical references to what is now called the "Williamson ether synthesis," it is worth pointing out that Williamson did not in fact refer to what he was doing as synthesis. Nor did he interpret his results as demonstrating the existence of the "water type." Despite the subsequent appearance of similarities between their achievements, comparing this account of Hofmann's work with prior accounts focused on Williamson establishes their distinct motivations and interpretational frameworks—as well as the divergent vocabulary they used to describe what they were doing. Cf. Rocke, *Image and Reality*, 22–30, 32–33.

60. Rocke, *Image and Reality*, 18–30, explained how Williamson's findings lent support to Gerhardt and Laurent's theory and argued that Williamson's use of visual

aids underpinned an approach that led Williamson to speculate on the existence of atomic and molecular motion.

61. This work appeared in Liebig's *Annalen* between 1848 and 1851 as a series of 10 papers, culminating with August Wilhelm Hofmann, "Beiträge zur Kenntniss der flüchtigen organischen Basen. X (fortsetzung)," *Annalen der Chemie und Pharmacie* 79 (1851): 11–39. See also, e.g., August Wilhelm Hofmann, "Recherches sur la Constitution Moléculaire des Bases Organiques Volatiles," *Annales de Chimie et de Physique* 30 (1850): 87–118.

62. Justus Liebig, Johann Christian Poggendorff, and Friedrich Wöhler, eds. *Handwörterbuch der reinen und angewandten Chemie, Supplement* (Braunschweig: Vieweg und Sohn, 1850), 469.

63. Gerhardt, *Traité*, 8–9.

64. Kekulé, *Lehrbuch*, 91.

65. Hermann Kolbe, *Ausführliches Lehrbuch der organischen Chemie*, vol. 1 (Braunschweig: Vieweg, 1854), 53.

66. William Gregory, *A Handbook of Organic Chemistry*, 4th edition (London: Walton and Maberly, 1856), 12–16.

67. Armstrong, *Introduction*, 329.

68. Kolbe, *Ausführliches Lehrbuch*, 265; Kekulé, *Lehrbuch*, 465. Both textbooks focused on principle rather than on practice. Neither provided experimental details, and Kekulé recommended using ethyl iodide rather than Hofmann's preferred ethyl bromide.

69. George Fownes, *A Manual of Elementary Chemistry: Theoretical and Practical*, 9th edition (London: Churchill, 1863), 644.
As chapter 8 explains, George Fownes, *A Manual of Elementary Chemistry: Theoretical and Practical* (London: Churchill, 1844) was adopted by Hofmann's laboratory at the RCC, its revision taken on by Henry Bence Jones following Fownes's death in 1849.

70. Ludwig Gattermann, *Die Praxis des organischen Chemikers*, (Leipzig: Veit, 1894), 52–58. I have chosen to use a literal translation of Gattermann's title, rather than the looser translation applied to English language versions of his text, e.g., Ludwig Gattermann, *The Practical Methods of Organic Chemistry*, trans. William B. Schover (New York: Wiley, 1896).

71. Gattermann, *Praxis*, 135–138.

72. Keas, "Nature," 106; Rocke, *Nationalizing Science*, 168–169.

73. Beer, "A. W. Hofmann," 250, 248; Aaron J. Ihde, *The Development of Modern Chemistry* (New York: Harper and Row, 1964), 210. Both Armstrong, "Hofmann Memorial Lecture," 656–658, and Tiemann, "Gedächtnisrede," 3389, examined this question, finding firmly in Hofmann's favor.

74. The origins of this view might be traced to, e.g., Armstrong, "Hofmann Memorial Lecture," 643.

75. Tiemann, "Gedächtnisrede," 3389.

76. Tiemann, "Gedächtnisrede," 3389–3390.

77. Tiemann, "Gedächtnisrede," 3389–3390.

78. Meinel, "Regierender Oberchemiker," 36.

CHAPTER 7

1. C. M. Jackson, "Chemical Identity Crisis."

2. Frank Wrigglesworth Clarke, *The Constants of Nature* (Washington, DC: Smithsonian Institution, 1873). Clarke's was the first volume in a Smithsonian series of that name, and as discussed below, the title was more aspirational than descriptive of its contents.

3. Klein, "Shifting Ontologies."

4. It is important here to distinguish between "synthetic" and "artificial." Whereas synthetic camphor (i.e. camphor produced by laboratory synthesis) is identical to the natural molecule, artificial camphor is a quite different substance.

5. Hofmann, "Steinkohlen-Teeröl."

6. Wurtz, "Sur une Série." See also Hofmann, "Adolph Wurtz," 932.

7. Hofmann, "Researches Regarding," 100.

8. For example, Hofmann, "Researches Regarding," 110.

9. Wöhler to Berzelius, 1 December 1831, in *Briefwechsel,* ed. Wallach, 385–387, 386, described experiments using a *"Kugelröhre* [filled] with caustic potash . . . in the same way as in organic analysis." These experiments produced highly variable results, but according to Wallach, they prompted Wöhler to suggest a method of measuring the carbon dioxide content of atmospheric air that was subsequently developed by Carl Brunner.
John J. Griffin, *Descriptive Catalogue of Chemical Apparatus Manufactured and Imported by John J. Griffin.* (London: Griffin, 1850), part 1 (dated July 1841), 37, supplied "Liebig's Apparatus for the absorption of Carbonic Acid in a solution of caustic potash."

10. The most recent biographical studies of Geissler are: Simón Reif-Acherman, "Heinrich Geissler: Pioneer of Electrical Science and Vacuum Technology," *Proceedings of the IEEE* 103 (2015): 1672–1684; and Günter Dörfel, "Der Meister und seine Schule—zur Biographie und Wirkung des Instrumentenbauers Heinrich Geißler," *Sudhoff's Archiv* 98 (2014): 91–108. Both emphasize Geissler's foundational work in physics. Dörfel built on earlier studies, including: Karl Eichhorn, "Heinrich Geissler: Leben und Werk eines Pioniers der Vakuumtechnik," *Schriftenreihe Deutsches Röntgen-Museum* 6 (1984); Karl Eichhorn, "Heinrich Geissler (1814–1879): His Life, Times, and Work," trans. Heidi Collins, *Bulletin of Scientific Instruments* 27 (1990): 17–19; Hans Kangro, "Geissler, Johann Heinrich Wilhelm," in *Complete Dictionary of Scientific Biography*, vol. 5 (Detroit: Scribner, 2008), 340–341.

11. Dr. H. Geisslers Nachfolger (Franz Müller), *Gedenkblatt zur Erinnerung an Heinrich Geissler, Dr. Phil., Glastechniker* (Bonn, 1890), dated Geissler's firm to 1840. Cf. Kangro, "Geissler," which dated the firm's foundation to "1852 or earlier," following Eichhorn, "Heinrich Geissler: Leben und Werk," 6. Dörfel, "Der Meister," 92, fn. 4, accused Müller of having "thoughtlessly circumvented" dates concerning Geissler and his firm, while Eichhorn's study was "truly sound." In fact, although it is plausible that Müller overstated both the firm's longevity and his own contribution to its growth, neither Dörfel nor Eichhorn definitively established that Geissler founded his firm as late as the 1850s. Indeed, Geissler's work with Plücker on the magnetic properties of gases—which none of these earlier studies cite—indicates that Geissler was well established in Bonn and connected to the university by the start of 1851.

12. Julius Plücker, "Ueber das magnetische Verhalten der Gase," *Annalen der Physik und Chemie* 83 (1851): 90. See also Julius Plücker and Heinrich Geissler, "Studien über Thermometrie und verwandte Gegenstände," *Annalen der Physik und Chemie* 162 (1852): 238–279.

13. Plücker, "magnetische Verhalten," 88, noted that access to substances as "chemically pure as possible" was critical to the accuracy of his experiments.

14. Griffin, *Chemical Handicraft*, 400, advertised Liebig's original at a cost of 1s 6d, while Geissler's modified device was a shilling more expensive.

15. Julius Plücker, "Ueber das magnetische Verhalten der Gase (Zweite Mittheilung)," *Annalen der Physik und Chemie* 84 (1851): 162, explained Geissler's modification of Liebig's Kaliapparat.

16. Although Kopp's (1838) Marburg PhD was awarded for a dissertation concerning highly theoretical work in inorganic chemistry, he had previously constructed a novel differential barometer. His Giessen training included a study of the action of nitric acid on ethyl sulfonic acid.
Kopp's biographical details are taken from August Wilhelm Hofmann, "Sitzung vom 22. Februar 1892," *Berichte der deutschen chemischen Gesellschaft* 25 (1892): 505–523; Thomas Edward Thorpe, "The Life Work of Hermann Kopp," *Journal of the Chemical Society, Transactions* 63 (1893): 775–815; and the abridged Thomas Edward Thorpe, "Obituary Notices of Fellows Deceased: Hermann Kopp," *Proceedings of the Royal Society of London* 60 (1896–1897): 1–35.
See also Alan Rocke's translation of Hermann Kopp's (1882) *Aus der Molecular-Welt*: Alan J. Rocke and Hermann Kopp, *From the Molecular World: A Nineteenth-Century Science Fantasy* (Berlin: Springer, 2012), 1–28, esp. 10–13.

17. Hermann Kopp, "Ueber die Vorausbestimmung einiger physikalischen Eigenschaften bei mehreren Reihen organischer Verbindungen," *Annalen der Chemie und Pharmacie* 41 (1842): 79–89 (part 1) and 169–189 (part 2), 88.

18. See, e.g., Kopp, "Vorausbestimmung," 87.

19. Kopp, "Vorausbestimmung," Liebig's fn. on 188.

20. This problem is confirmed by the boiling-point data that Kopp gathered, as well as by the divergent boiling points reported for coniine in chapter 9.

21. Thorpe, "Life Work," 786.

22. Kopp, "Vorausbestimmung," 89, 170.

23. Kopp, "Vorausbestimmung," 170, mapped out organic taxonomy in terms of horizontal and vertical series.

24. Kopp, "Vorausbestimmung," 188.

25. Kopp, "Vorausbestimmung," 87.

26. Hermann Kopp, "Ueber die Siedepunkte einiger isomerer Verbindungen, und über Siedepunktsregelmässigkeiten überhaupt," *Annalen der Chemie und Pharmacie* 55 (1845): 166–200, 178–179.

27. Kopp, "Vorausbestimmung," 87, 89, 188. Liebig's fn. on 188 makes it clear that Kopp's choice of the word "control" was indeed due to Liebig.

28. Muspratt and Hofmann, "On Toluidine," 376.

29. Arthur Church, "On the Benzole Series—Determination of Boiling-Points," *Philosophical Magazine* 9 (1855): 256–260. See also Hermann Kopp, "On the Relation between Boiling-Point and Composition in Organic Compounds," *Philosophical Transactions of the Royal Society* 150 (1860): 258

30. Arthur Church, "Notes on Boiling Points," *Chemical News* 1 (1860): 206.

31. Kopp, "On the Relation," 272–275; see also Hofmann's fn. on 274. Hofmann's development of the ammonia type is the subject of chapters 5 and 6.

32. Kopp, "On the Relation," 275–276.

33. For example, Eugen Sell, *Grundzüge der modernen Chemie: Organische Chemie*, vol. 2 (Berlin: Hirschwald, 1870), 554; William A Miller, *Elements of Chemistry: Theoretical and Practical, Part 3: Organic Chemistry* (London: Longman, 1869), 942.

34. Kopp, "On the Relation," 258.

35. Joseph-Louis Gay-Lussac, "Sur la Déliquescence des Corps," *Annales de Chimie et de Physique* 82 (1812):171–177, cited by Chang, *Inventing Temperature,* 21.

36. In other words, Kopp sought actively to manage what Chang, *Inventing Temperature,* 21, called the "complex and unruly phenomenon" of the boiling point.

37. Kopp, "Vorausbestimmung," 182.

38. Kopp, "Vorausbestimmung," 182. Hermann Kopp, "On a Great Regularity in the Physical Properties of Analogous Organic Compounds," *Philosophical Magazine* 20 (1842): 187–197, 195, noted the difficulty of obtaining the products of substitution uncontaminated by neighboring compounds in the series, but this difficulty did not prevent Kopp from identifying regularities in physical data. Cf. Kopp, "Ueber die

Siedepunkte," 179, where Kopp explained the need to measure boiling point immediately after purification to avoid decomposition.

39. Kopp, "Vorausbestimmung," Liebig's fn. on 188.

40. Chang, *Inventing Temperature*, 21.

41. Kopp, "Ueber die Siedepunkte," 169–172.

42. Kopp, "Ueber die Siedepunkte," 171.

43. Kopp, "Ueber die Siedepunkte," 172.

44. Kopp, "Ueber die Siedepunkte," 174. See also: Hermann Kopp, *Geschichte der Chemie*, vol. 1 (Braunschweig: Vieweg, 1843), 33ff.

45. According to August Wilhelm Hofmann, *Zur Erinnerung an Gustav Magnus* (Berlin: Dümmler, 1871), 69, this practice had been largely abandoned by 1870.

46. Kopp, "Ueber die Siedepunkte," 174.

47. Liebig to Wöhler, 1 May 1832, in "Liebig-Wöhler Briefwechsel," ed. Meinel and Steinhauser, for example, recommended Collardeau's thermometers as "indispensable for organic analysis," because "each degree is divided into 5 parts."

48. Kopp, "Ueber die Siedepunkte," 172–174, 179–180.

49. Kopp, "Ueber die Siedepunkte," 171–172, explained how this measure avoided the boiling point elevation otherwise observed in glass vessels of ca. 1 °C compared with metal. The presence of residual sulfuric acid was also known to elevate boiling points measured in glass.

50. Kopp, "Ueber die Siedepunkte," 172–174.

51. Kopp, "Ueber die Siedepunkte," 179. Chang, *Inventing Temperature*, 21, explained how Gay-Lussac determined "100.000 °C exactly" as the boiling point of water, thereby helping to stabilize an important fixed point in thermometry.

52. Ludwig Gattermann, *Die Praxis des organischen Chemikers* (Leipzig: Veit, 1894), 15–33. Reduced-pressure distillation followed from the mid-century improvements in vacuum technology, in which glassblowers, including Geissler, played a major part.

53. As discussed in relation to coniine, chemists learned to situate the thermometer bulb in the vapor near the still head during distillation. A. Siwoloboff, "Ueber die Siedepunktbestimmung kleiner Mengen Flüssigkeiten," *Berichte der deutschen chemischen Gesellschaft* 19 (1886), 795–796, introduced the still-classic method of measuring the boiling point of small samples.

54. Kopp, "Vorausbestimmung." Regnault's results came in for particular criticism on p. 176.

55. Regnault's biography is drawn from Sébastien Poncet and Laurie Dahlberg, "The Legacy of Henri Victor Regnault in the Arts and Science," *International Journal of Arts*

and Sciences 4 (2011): 377–400; and Simón Reif-Acherman, "Henri Victor Regnault: Experimentalist of the Science of Heat," *Physics in Perspective* 12 (2010): 396–442.

56. Victor Regnault, "Ueber die Zusammensetzung des Chlorkohlenwasserstoffs (Oel des ölbildenden Gases)," *Annalen der Pharmacie* 14 (1835): 22–38. Liebig's fn. is on 22.

57. Fox, *Caloric Theory*, chapter 8, presented a most informative overview of Regnault's career, including his many contributions to the study of heat.

58. Regnault, "Ueber die Zusammensetzung," 27.

59. *Encyclopaedia Britannica*, 7th edition, vol. 20 (Edinburgh: Black, 1842), 602.

60. Much data from Regnault, "Ueber die Zusammensetzung," was tabulated by Kopp, "Vorausbestimmung."

61. Otto H. Sibum, "Reworking the Mechanical Value of Heat: Instruments of Precision and Gestures of Accuracy in Early Victorian England," *Studies in History and Philosophy of Science* 26 (1995): 73–106, 76, explained that Joule's thermometers "could not be rebuilt because of a lack of sufficient information and possibly the skill to do so."

62. Auguste A. de la Rive, "Notices Respecting New Books: H. Regnault, Relation des Expériences pour Déterminer les Principales Lois Physiques, et les Données Numériques qui Entrent dans le Calcul des Machines à Vapeur," *Philosophical Magazine* 36 (1850): 41–62, 54.

63. Victor Regnault, *Relation des Expériences Entreprises par Ordre de Monsieur le Ministre des Travaux Publics, et sur la Proposition de la Commission Centrale des Machine à Vapeur, pour Déterminer les Principales Lois Physiques, et les Données Numériques qui Entrent dans le Calcul des Machines à Vapeur*, vol. I (Paris: Didot Freres, 1847).
Hasok Chang, "Spirit, Air, and Quicksilver: The Search for the 'Real' Scale of Temperature," *Historical Studies in the Physical and Biological Sciences* 31 (2001): 249–284; and Chang, *Inventing Temperature*, 74–84, described Regnault's foundational contributions to thermometry. Jules Jamin, *Cours de Physique de l'École Pólytechnique* (Paris: Gauthier-Villars, 1886) provided a helpful nineteenth-century account of Regnault's experimental work.

64. Regnault, *Relation des Expériences*, 165.

65. Regnault, *Relation des Expériences*, 165–166.

66. Rive, "Notices," 55.

67. Rive, "Notices," 54. See also, Regnault, *Relation des Expériences*, 120, 166.

68. Regnault, *Relation des Expériences*, 167.

69. Regnault, *Relation des Expériences*, 168 ff.

70. Rive, "Notices," 54, 59.

71. Chang, "Spirit, Air, and Quicksilver," 273.

72. Rive, "Notices," 58–59; see also Regnault, *Relation des Expériences*, 190 ff.

73. Rive, "Notices," 60–61.

74. Plücker and Geissler, "Studien über Thermometrie."

75. Plücker and Geissler, "Studien über Thermometrie," 238.

76. Plücker and Geissler, "Studien über Thermometrie," 239.

77. Dr. H. Geisslers Nachfolger (Franz Müller), "Gedenkblatt."

78. Dr. H. Geisslers Nachfolger (Franz Müller). "Gedenkblatt."

79. Hermann Kopp, "Investigations of the Specific Heat of Solid Bodies," *Philosophical Transactions of the Royal Society* 155 (1865): 71–202, 91. Kopp's paper was received and read (by Thomas Graham) during the spring of 1864.

80. August Wilhelm Hofmann, "Noch einiges über die Amine der Methyl- und Aethylreihe," *Berichte der deutschen chemischen Gesellschaft* 22 (1889): 699–705, 700–701.

81. For example, Geissler's normal thermometers were used aboard H.M.S. Challenger. See C. W. Thomson, John Murray, George S. Nares, and Frank T. Thomson, *Report on the Scientific Results of the Voyage of H.M.S. Challenger during the Years 1873–76 under the Command of Captain George S. Nares and the Late Captain Frank Tourle Thomson, R.N.* (Edinburgh: H.M.S.O., 1885), 110, 425.

82. Louis P. Casella, *An Illustrated and Descriptive Catalogue of Philosophical, Meteorological, Mathematical, Surveying, Optical and Photographic Instruments* (London: Lane, 1861), 15.

83. British Association proposal for a Physical Observatory at Kew, accepted June 1842, cited in Robert Henry Scott, "The History of Kew Observatory," *Proceedings of the Royal Society of London* 38 (1885): 37–86, 50.

84. For example, Casella, *Illustrated and Descriptive Catalogue,* 16, item 54: Standard Thermometer. Casella's catalog also offered "Chemical Thermometers" (p. 20), which typically operated over a much wider range, up to several hundred degrees.

85. See, e.g., Thomas Thomson, *A System of Chemistry*, vol. 2 (London: Baldwin, Craddock & Joy, 1817), 381, reporting an 8-degree difference in reported melting temperatures for oleic acid.

86. Hofmann, "Researches Regarding," contained numerous such references.

87. Peter J. Ramberg, "Wilhelm Heintz (1817–1880) and the Chemistry of the Fatty Acids," *Bulletin for the History of Chemistry* 38 (2013): 19–28, is a notable exception.

88. Michel E. Chevreul, *Recherches Chimiques sur les Corps Gras d'Origine Animale* (Paris: Levrault, 1823), fn. on 27, reported stearic acid's melting point as 70 °C, significantly higher than margaric acid's.

89. Michel E. Chevreul, *Considérations Générales sur l'Analyse Organique et sur ses Applications* (Paris: Levrault, 1824), 202, and Chevreul, *Recherches,* 61, cf. fn. on 82.

90. Thomson, *System,* 378, reported "134°" (F equiv. 56.7 °C); cf. Edward Turner, *Elements of Chemistry,* 4th ed (London: John Taylor, 1833), 882, reported "140°" (F equiv. 60 °C).

91. Jöns Jacob Berzelius, *Traité de Chimie*, vol. 2, trans. by A. J. L. Jourdan and Esslinger (Brussels: Wahlen, 1838), 493.

92. Chevreul, *Considerations*, 33, listed melting point as a characteristic property.

93. Ramberg, "Wilhelm Heintz." Recrystallization to constant melting point did not reliably identify pure fatty acids because of the formation of what are now called eutectic mixtures, i.e., mixtures of similar compounds that appear to be pure, because they melt at a defined temperature below the melting point of either pure component.

94. Chemists today relate optical activity to three-dimensional molecular structure or stereochemistry. This connection between optical rotation and stereochemistry was first proposed in 1874 by Jacobus H. van't Hoff and Joseph A. le Bel. In the 1880s, however, optical activity remained for most organic chemists an empirical phenomenon unconnected to the arrangement of atoms in three-dimensional space. For more on how this connection was established in the 1890s, see C. M. Jackson, "Emil Fischer."

95. Ladenburg used the method of resolution using tartrate salts developed by Louis Pasteur during the 1850s. For more on this method, see Gerald L. Geison, *The Private Science of Louis Pasteur* (Princeton, NJ: Princeton University Press, 1995).

96. As today's chemists know, the melting point of an optically inactive, racemic mixture may differ from that of each pure enantiomer, due to the formation of a eutectic mixture (discussed in relation to margaric acid above).

97. C. F. Roth, "Ein neuer Apparat zur Bestimmung von Schmelzpunkten," *Berichte der deutschen chemischen Gesellschaft* 19 (1886): 1970–1973, 1973, gave the provenance of his apparatus, which by that time Roth had been using for at least a year. Dr. H. Geisslers Nachfolger (Franz Müller), *Preisverzeichnis*, item 4445, 389, image on 390.

98. Roth, "neuer Apparat," 1972.

99. Roth, "neuer Apparat," 1970–1971.

100. See, e.g., Hans Landolt and Richard Börnstein, *Physikalisch-Chemischen Tabellen* (Berlin: Springer, 1883), 195.

101. Roth, "neuer Apparat," 1972.
I have been unable definitively to identify which maker supplied the thermometer used by Roth and Ladenburg, but good circumstantial evidence makes Geissler, or his successor Müller, by far the most likely candidates—beginning with the fact that Müller made Roth's melting point apparatus.
In addition, Ladenburg's prior training and experience surely predisposed him to patronize Geissler's firm. Wilhelm Körner, Ladenburg's former colleague in August Kekulé's Ghent laboratory during the 1860s, had since established melting points as key criteria for distinguishing and counting isomeric aromatic compounds, and therefore a key method for establishing benzene's symmetry properties. Ladenburg's continuing hostility to Kekulé's theory of aromatic structure focused on its inability adequately to explain benzene. Wilhelm Körner, "Studj sull'Isomeria delle Cosi dette

Sostanze Aromatiche a Sei Atomi di Carbonio," *Gazzetta Chimica Italiana* 4 (1874): 305–446, 335 fn, specified the use of "H. Geisler's [sic] normal thermometer," lending additional support to my inference.

102. Roth's formula was 0.000156α; here α is the extended length of the rising thread in degrees, t is the measured temperature, and t_0 is the ambient temperature.

103. Albert Ladenburg, "Synthese der activen Coniine," *Berichte der deutschen chemischen Gesellschaft* 19 (1886): 2578–2583, 2582.

104. Kopp, "Ueber die Siedepunkte," 175–176, made a very similar point, drawing an analogy between his investigation of boiling point regularities and chemists' earlier development of stoichiometry.

105. Klein and Lefèvre, *Materials,* offered a rare and major contribution to this history. Built on Klein, "Shifting Ontologies," it examined chemists' changing experimental production and understanding of materials in the period up to about 1830. Cf. Joachim Schummer, "The Impact of Instrumentation on Chemical Species Identity from Chemical Substances to Molecular Species," in *From Classical to Modern Chemistry: The Instrumental Revolution,* ed. Peter J. T. Morris (Cambridge: Royal Society of Chemistry, 2002), 188–211. Because Schummer considered that "we have good reasons to believe that the standard methods for purifying liquids and solids remained much the same from the mid-eighteenth century onwards," he was led to "assume that our concept of purity was already (implicitly) well established in the eighteenth century" (191).

106. Reif-Acherman, "Heinrich Geissler," summarized these developments.

107. August Wilhelm Hofmann, "Sitzung vom. 27. Januar 1879" *Berichte der deutschen chemischen Gesellschaft* 12 (1879): 147–148.

108. J. P. Kuenen and W. W. Randall, "The Expansion of Argon and of Helium as Compared with That of Air and Hydrogen," *Proceedings of the Royal Society of London* 59 (1896): 60–65, 61.

109. Clarke, *Constants of Nature.* Henry amended Clarke's original and more prosaic title, *Table of Specific Gravities, Boiling Points, and Melting Points of Solids and Liquids.*

110. Clarke, *Constants of Nature,* 2–3.

111. Thomas Carnelley, *Melting and Boiling Point Tables,* vol. 1 (London: Harrison, 1885), vi, identified just four previous collections of physical constants, the earliest of which was Clarke's *Constants of Nature.* Landolt and Börnstein, *Physikalisch-Chemische Tabellen.* Known simply as "Landolt and Börnstein," these tables reached their 6th edition in 1950.

CHAPTER 8

1. This chapter develops arguments first presented in Catherine M. Jackson, "Chemistry as the Defining Science: Discipline and Training in Nineteenth-Century

Chemical Laboratories," *Endeavour* 35 (2011): 55–62; and Catherine M. Jackson, "The Laboratory," in *Companion to the History of Science*, ed. Bernard Lightman (Oxford: Blackwell-Wiley, 2016), 296–309.

2. The most recent monograph concerned with the history of chemical laboratories, Peter J. T. Morris, *The Matter Factory: A History of the Chemistry Laboratory* (London: Reaktion, 2015), does not improve our understanding of these key issues.

3. David Cahan, "The Institutional Revolution in German Physics, 1865–1914," *Historical Studies in the Physical Sciences* 15 (1985): 1–65; David Cahan, *An Institute for an Empire: The Physikalische-Technische Reichsanstalt, 1871–1918* (Cambridge: Cambridge University Press, 1989).
See also, e.g., Arleen Marcia Tuchman, *Science, Medicine, and the State in Germany: The Case of Baden, 1815–1871* (Oxford: Oxford University Press, 1993).

4. August Wilhelm Hofmann, *Die Frage der Theilung der philosophischen Facultät* (Berlin: Dümmler, 1881), 17.

5. Originally highlighted in the context of physics by Cahan, "The Institutional Revolution," the laboratory's increasing scale is a persistent (if unattributed) theme of Morris, *Matter Factory*.

6. Rocke, "Giessen Model;" Margaret Rossiter, *The Emergence of Agricultural Science: Justus Liebig and the Americans, 1840–1880* (New Haven, CT: Yale University Press, 1975).

7. The laboratory's role in shaping chemists' social hierarchy was examined by Christoph Meinel, "Chemische Laboratorien: Funktion und Disposition," *Berichte zur Wissenschaftsgeschichte* 23 (2000): 287–302.

8. Brock, *Gatekeeper*, 47. Volhard, *Liebig*, 57–85, described Liebig's laboratory before 1839.

9. Liebig, *Anleitung*, 9 ff., described the use of "glowing coals" in combustion analysis.

10. Volhard, *Liebig*, 63.

11. J. Paul Hofmann, *Chemische Laboratorium;* J. Paul Hofmann, *Acht Tafeln zur Beschreibung des chemischen Laboratoriums zu Giessen* (Heidelberg: Winter, 1842).

12. J. Paul Hofmann, *Chemische Laboratorium*, x–xii.

13. J. Paul Hofmann, *Chemische Laboratorium*, 9.

14. Hofmann, "Adolph Wurtz," described the close friendship that developed between Hofmann and Wurtz during the latter's time in Giessen.

15. These enclosures, visible on Trautschold's woodcut, were described by J. Paul Hofmann, *Chemische Laboratorium*, 3, 11–12; and Carl Wilhelm Bergemann's (1840) report to the Prussian Minister, included as Appendix I in Brock, *Gatekeeper*, 333–341. Cf. Morris, *Matter Factory*, 99, and see notes 78 and 83 below.

J. Paul Hofmann, *Chemische Laboratorium,* 18, explained that the pharmaceutical laboratory was not equipped to allow operations producing dangerous gases to be safely performed.

16. J. Paul Hofmann, *Chemische Laboratorium,* 42–46.

17. Volhard, *Liebig,* 65–67.

18. J. Paul Hofmann, *Chemische Laboratorium,* 2, 11, 18.

19. J. Paul Hofmann, *Chemische Laboratorium,* x.

20. See, for example, Cremer, *Das neue chemische Laboratorium,* discussed below.

21. Muspratt and Hofmann, "On Toluidine," 367.

22. Robins, *Technical School and College Building,* 3.

23. Historical studies include Christine Nawa, "A Refuge for Inorganic Chemistry: Bunsen's Heidelberg Laboratory," *Ambix* 61 (2014): 115–140; and Tuchman, *Science, Medicine, and the State in Germany.* Numerous nineteenth-century surveys, produced by visitors from Britain, France, and Italy, documented Germany's rapidly increasing investment in institutional laboratories for scientific and, especially, chemical training: Festing, *Report;* Robins, *Technical School and College Building;* Roster, *Delle Scienze Sperimentali;* Wurtz, *Les Hautes Études Pratiques dans les Universités Allemandes;* and Wurtz, *Les Hautes Études Pratiques dans les Universités d'Allemagne et d'Autriche-Hongue.*

24. August Wilhelm Hofmann, *Report on the Chemical Laboratories in Process of Building in the Universities of Bonn and Berlin* (London: Clowes and Son, 1866). As discussed below, Hofmann also toured existing German chemical laboratories as preparation for designing these new buildings.

25. Correspondence between Hofmann and Liebig initially focused heavily on arrangements for the new laboratory. Sir James Clark to Liebig, 23 August 1845, Letter 8, in *Liebig und Hofmann,* ed. Brock, 33–34, also consulted Liebig about the new laboratory. Minutes of the Laboratory Committee Meeting, 6 December 1845 (AICL C3/566), approved an initial purchase of glassware and other apparatus from Germany costing £69 7s 6d. Further purchases continued into the 1850s.

26. Minutes of Committees, 18 November 1845, records the establishment of a Laboratory Committee (AICL C4/568). Hofmann's appointment to the Building Committee took place at a Council Meeting, 19 October 1845 (AICL C3/566 Rough Minute Book).

27. Edward Walford, "Hanover Square and Neighbourhood," *Old and New London* 4 (1878): 314–326. *British History Online,* accessed November 19, 2021, http://www .british-history.ac.uk/old-new-london/vol4/pp314-326. On Bullock and Gardner's plans to incorporate a school for chemistry at the Royal Institution, see Bentley, "Chemical Department," 161.

28. Jacob Volhard and Emil Fischer, *August Wilhelm von Hofmann: Ein Lebensbild* (Berlin: Springer, 1902), 32. Volhard's paraphrasing of Hofmann, "Page," 150–151,

made explicit the earlier implicit reference to the then-popular German panoptica as places of entertainment (rather than to Jeremy Bentham's 1843 Panopticon, inspired by the Parisian École Polytechnique).
Hofmann, *Reports,* xi, confirmed his commitment to "pure," academic research.

29. The new Metropolitan School of Science and Arts, later the Royal School of Mines, was one of the three colleges merged to form Imperial College in 1907. Gay, *History of Imperial College.*

30. According to Roberts, "Establishment," 473, more than 70% of the RCC's funding derived from these sources.

31. Hofmann, "New Researches," was read to London's Chemical Society on December 15, 1845.

32. R. A. Smith, *The Life and Works of Thomas Graham, DCL, FRS: Illustrated by 64 Unpublished Letters,* edited by Joseph J. Coleman (Glasgow: Smith, 1884), letter to John Graham, 29 October 1844, cited in Bentley, "Chemical Department," 161. Liebig had asked Graham whether he would be willing to host the new college at London's University College.

33. August Wilhelm Hofmann, "Remarks on the Importance of Cultivating Experimental Science from a National Point of View," in Hofmann, *Reports,* xxiii–xliv.

34. James Campbell Brown, *Essays and Addresses* (London: Churchill, 1914), 66. Armstrong, "Hofmann Memorial Lecture," 588, described Hofmann's occasional impromptu lectures in the laboratory. Hofmann did not deliver a formal course of lectures at the RCC until the new laboratory in Oxford Street was completed. Lectures were therefore a desirable but not essential complement to the crucial laboratory-based training program.

35. Hofmann and Henry Bence Jones continued to produce revised editions of Fownes's (1844) *Manual of Elementary Chemistry* following Fownes' death in 1849 at the age of 33. The RCC's Library Catalogue (AICL 983 SC) lists every edition of Fownes's textbook up to and including the 11th, published in the mid-1870s.

36. Hofmann, *Reports,* xlvii.
For more on George Fownes, see J. S. Rowe, "The Life and Work of George Fownes, F.R.S. (1815–49)," *Annals of Science* 6 (1950): 422–435.

37. Abel, "Hofmann Memorial Lecture," 586, highlighted the extent of Hofmann's work in connection with the erection and equipment of the RCC's new Oxford Street laboratories.

38. Abel, "Hofmann Memorial Lecture," 589.

39. Hofmann to Liebig, Letter 10, undated [October?] 1845, in *Hofmann und Liebig,* ed. Brock, 36–37.

40. According to Minutes of General Meetings, 31 August 1846 (AICL C5/564), Hofmann was repeatedly asked to provide instructions for apparatus to remove sulphuretted hydrogen from the temporary laboratory.

41. Minutes of the Managing Committee, 7 October 1846 (AICL C4/568).

42. Minutes of the Building Committee, 18 February 1846 and 2 March 1846 (AICL C3/566). My study revealed only one occasion (June 1846) when Hofmann was absent from a meeting of the Building Committee.

43. Warren de la Rue, "On Cochineal," *Memoirs and Proceedings of the Chemical Society* 3 (1845): 454–480. See also, Warren de la Rue, "On a Modification of the Apparatus of Will and Varrentrapp for the Estimation of Nitrogen," *Memoirs and Proceedings of the Chemical Society* 3 (1847): 347–348, read to the Society on April 5, 1847.

44. Abel, "Hofmann Memorial Lecture," 588, contains the most complete description of the completed laboratory.

45. Hofmann's report, Minutes of General Meetings, 5 June 1848 (AICL C5/564). Hofmann's instructions for the removal of sulphuretted hydrogen were requested at Management Committee meetings in September and October 1846 (AICL C4/568).

46. Hofmann's report, Minutes of General Meetings, 3 June 1850 (AICL C5/564).

47. Minutes of Council, 18 December 1849 (AICL C3/567). Neil Arnott, MD, was a donor to the RCC.

48. Minutes of Council, 21 May 1850 (AICL C3/567).

49. Minutes of General Meetings, 3 June 1850 (AICL C5/564). Minutes of Council, 16 July, 15 October, 18 November, and 17 December 1850; and 21 January 1851 (AICL C3/567).

50. William H. Perkin, "Some of the Laboratories I Have Worked In," a manuscript dated 1906 and quoted in *Perkin Centenary, London: 100 years of Synthetic Dyestuffs*, edited by Anon (New York: Pergammon, 1958), 18.

51. Gersham Henry Winkles, "On the Existence of Trimethylamine in the Brine of Salted Herrings," *Quarterly Journal of the Chemical Society* 7 (1855): 63–68. See also, Jackson, "Chemical Identity Crisis," 196.

52. According to Perkin, "Some of the Laboratories," combustions at the RCC were performed using charcoal—even though Liebig, *Anleitung*, 9, used coal for this purpose.

53. August Hofmann, "On the Use of Gas as Fuel in Organic Analysis," *Quarterly Journal of the Chemical Society* 6 (1854): 209–216; see especially the discussion of improvements on 214.

54. Brown, *Essays and Addresses*, 65.

55. Council meeting, 11 November 1846 (AICL C3/566).

56. Mansfield, "Researches on Coal Tar."

57. Ferdinand Tiemann, "Erinnerung an das fünfundzwanzigjährige Bestehen der deutschen chemischen Gesellschaft und an ihren ersten Präsidenten: August Wilhelm von Hofmann veranstaltet Gedächtnissfeier," *Berichte der deutschen chemischen Gesellschaft* 25 (1892): 3369–3414, 3384–3385. See also Edward R. Ward, "Charles

Blatchford Mansfield, 1819–55: Coal Tar Chemist and Social Reformer," *Chemistry in Britain* 15 (1979), 297–304.

58. Perkin, "Hofmann Memorial Lecture," 602.

59. Rue, "On Cochineal."

60. Volhard, *Liebig*, 65–67.

61. Minutes of General Meetings, 31 August 1846, 5 June 1848 (AICL C5/564).

62. Hofmann, "Volatile Organic Bases. VI," 313.

63. Hofmann, "Volatile Organic Bases. VI," 328–339.

64. Hofmann, "Researches into the Molecular Constitution of the Organic Bases," 390; see also Smith Brazier and Gossleth, "Contributions."

65. Edward Frankland, "On the Isolation of the Organic Radicals," *Quarterly Journal of the Chemical Society* 3 (1850): 263–296, 265. Sealed tube reactions are still used and still hazardous.

66. Hofmann, "Researches Regarding." See also August Wilhelm Hofmann, "Contributions towards the History of Thialdine," *Quarterly Journal of the Chemical Society* 10 (1858): 193–202; and August Wilhelm Hofmann, "Contributions to the History of the Phosphorus-Bases," *Quarterly Journal of the Chemical Society* 13 (1861): 289–325.

67. Edward Frankland, *Experimental Researches in Pure, Applied, and Physical Chemistry* (London: van Voorst, 1877), 144–145.

68. Volhard, *Liebig*, 65–67.

69. Bentley, "Chemical Department," 173, confirmed that funding problems continued after the 1853 merger.

70. Hannah Gay, "Pillars of the College," 140–141.

71. Keas, "Structure and Philosophy," 216–234, introduced the term "inner ring" to describe the separation between Hofmann's German research assistants, who were not paid by the RCC, and the salaried English assistants who undertook most of the college's elementary teaching.

72. Hofmann, *Report on the Chemical Laboratories*, 9.

73. According to Hofmann, *Report on the Chemical Laboratories*, 48, this was one of the reasons that he was offered the job in Berlin.

74. Hofmann, *Report on the Chemical Laboratories*, 2.

75. Hofmann, *Report on the Chemical Laboratories*, 8–9, 52. Estimations of present-day cost were converted using the calculators available at https://www.measuringworth.com/calculators, last accessed March 17, 2022.

76. Hofmann, *Report on the Chemical Laboratories*, 13.

77. Hofmann, *Report on the Chemical Laboratories*, 36.

78. Hofmann, *Report on the Chemical Laboratories*, 37, cf. Morris, *Matter Factory*, 99.

79. Hofmann, *Report on the Chemical Laboratories*, 9.

80. Hofmann, *Report on the Chemical Laboratories*, 15, 21.

81. Hofmann, *Report on the Chemical Laboratories*, 14–15.

82. Hofmann, *Report on the Chemical Laboratories*, 14–15.

83. Hofmann, *Report on the Chemical Laboratories*, 18, 38–40; cf. Morris, *Matter Factory*, 99. See also Hermann Kolbe, *Das neue chemische Laboratorium der Universität Leipzig* (Leipzig: Brockhaus, 1868), xxii–xxiii.

84. Hofmann, *Report on the Chemical Laboratories*, 18.

85. Hofmann, *Report on the Chemical Laboratories*, 18.

86. Hofmann, *Report on the Chemical Laboratories*, 28, 55 (on Berlin).

87. Robins, *Technical School and College Building*, 49.

88. Hofmann, *Report on the Chemical Laboratories*, 24.

89. Minutes of the Library Committee (AICL SC 982).

90. Hofmann, *Report on the Chemical Laboratories*, 20.

91. On Beilstein, see Michael D. Gordin, "Beilstein Unbound: The Pedagogical Unravelling of a Man and His Handbuch," in *Pedagogy and the Practice of Science*, ed. David Kaiser, 11–40.

92. Hofmann, *Report on the Chemical Laboratories*, 48–52.

93. Hofmann, *Report on the Chemical Laboratories*, 48–52.

94. Anon., "Farewell Dinner to Dr Hofmann," *Chemical News* 11 (1865): 213.

95. August Wilhelm Hofmann, "Vereinsabend bei dem Präsidenten am 15ten Mai 1869," *Berichte der deutschen chemischen Gesellschaft* 2 (1869): 223–236.

96. Cremer, *Das neue chemische Laboratorium*, 4, detailed these installations. Festing, *Report*, 7.

97. Cremer, *Das neue chemische Laboratorium*, 5.

98. Hofmann, *Report on the Chemical Laboratories*, 58. Hofmann delivered an address on lecture demonstrations at the Berlin laboratory's opening exhibition in May 1869.

99. Hofmann, *Report on the Chemical Laboratories*, 53. These included the two main teaching laboratories, the director's private laboratory and the lecture theater, as well as several subsidiary smaller rooms (59–63).

100. Robins, *Technical School and College Building*, 49–50.

101. Hofmann, *Report on the Chemical Laboratories*, 55.

102. Hofmann, *Report on the Chemical Laboratories*, 61; Festing, *Report*, 7.

103. August Wilhelm Hofmann, "On a New Class of Bodies Homologous to Hydrocyanic Acid. II," *Proceedings of the Royal Society of London* 16 (1868): 148. A range of circumstantial evidence supports my inference that Hofmann was working at the RCC during late summer 1867, around the time he finally relinquished his London appointment. Publication in an English journal is indicative, but the provision of starting material by Nicholson seems decisive: Ethylamine is a volatile, flammable, hazardous substance, dangerous enough to transport across London, let alone across Europe. Hofmann's presence at the RCC, moreover, seems unlikely to have inconvenienced his successor Edward Frankland, who retained his professorial appointment at the Royal Institution (RI). Frankland, whose research activity during this period was limited, perhaps found the RI's recently renovated laboratory facilities adequate. The Berlin laboratory remained far from fully operational at this time, making it unlikely that Hofmann would do such dangerous, unpleasant work there.

104. Leopold Pebal, *Das chemische Institut der K.K. Universität Graz* (Wien: Faesy & Frick, 1880), discussed the issue of large-scale operation in relation to laboratory design.

105. Cremer, *Das neue chemische Laboratorium*, 5, Blatt 9, figs. 1–4.

106. Tiemann, "Gedächtnisrede," 3394.

107. Ladenburg's biographical details are taken from Frederic S. Kipping, "Ladenburg Memorial Lecture," *Journal of the Chemical Society, Transactions* 103 (1913):1871–1895; and Albert Ladenburg, *Lebenserinnerungen* (Breslau: Trewent und Granier, 1912).

108. Ludwig Carius, "Ueber die Elementaranalyse organischer Verbindungen," *Annalen der Chemie und Pharmacie* 116 (1860): 1–30, Taf. 1.

109. Carius, "Elementaranalyse."

110. Albert Ladenburg, "Eine neue Methode der Elementaranalyse," *Annalen der Chemie und Pharmacie* 135 (1865): 1–24.

111. Wolfgang von Hippel, "Becoming a Global Corporation—BASF from 1865 to 1900," in *German Industry and Global Enterprise: BASF: The History of a Company*, ed. Werner Abelshauser, Wolfgang von Hippel, Jeffrey Allan Johnson, and Raymond G. Stokes, 5–114 (Cambridge: Cambridge University Press, 2003), 15–16.

112. Ladenburg, *Lebenserinnerungen*, 25–26.

113. Albert Ladenburg, "Synthèse de l'Acide Anisique et de l'un de ses Homologues," *Bulletin de la Société Chimique de France* 5 (1866): 257–261; Albert Ladenburg and A. Fitz, "Sur Quelques Dérivés de l'Acide Paroxybenzoique," *Bulletin de la Société Chimique de France* 5 (1866): 414–423.

114. Kipping, "Ladenburg Memorial Lecture," 1882–1885, explains how experiments between 1869 and 1878 prompted and confirmed Ladenburg's rejection of Kekulé's benzene ring.

115. Albert Ladenburg and Carl Leverkus, "Sur la Constitution de l'Anéthol," *Comptes Rendus* 63 (1866): 89–91. Ladenburg's observation (*Lebenserinnerungen*, 28) was indeed surprising, because—as Rocke, *Image and Reality*, 199, noted—Wurtz had presented Kekulé's benzene ring to the Société Chimique de France the previous year.

116. Rocke, *Nationalizing Science*, 182–184. The first quotation is from Wurtz's letter to the dean, April 1, 1860, cited on p. 182. Wurtz was a committed advocate of the importance of good laboratory facilities in chemistry. Wurtz, *Les Hautes Études Pratiques dans les Universités Allemandes*, was the first of two reports commissioned by the French government.

117. James Mason Crafts, "Friedel Memorial Lecture," *Journal of the Chemical Society, Transactions* 77 (1900): 998.

118. Kipping, "Ladenburg Memorial Lecture," 1873–1874.

119. Perhaps that is why the first course Ladenburg offered in Heidelberg drew on his longstanding fascination with history. This was the origin of Ladenburg's lectures, which first appeared in English as Albert Ladenburg, *Lectures on the History of the Development of Chemistry*, trans. Leonard Dobbin (Edinburg: Alembic Club, 1900).

120. Ladenburg, *Lebenserinnerungen*, 42–46.

121. Albert Ladenburg, *Vorträge über die Entwicklungsgeschichte der Chemie in den letzten Hundert Jahren* (Braunschweig: Vieweg, 1869).

122. Ladenburg, *Lebenserinnerungen*, 42–46.

123. For more on chemistry's movement between medical and philosophical faculties, see Christoph Meinel, "Artibus Acadmicis Inserenda: Chemistry's Place in Eighteenth and Early Nineteenth Century Universities," *History of Universities* 7 (1988): 89–115.

124. Ladenburg, *Lebenserinnerungen*, 52; Hans Dieter Nägelke, *Hochschulbau im Kaiserreich: Historische Architektur im Prozess bürgerlicher Konsensbildung* (Kiel: Ludwig, 2000), 383. On the "old chemical laboratory," see Carl Himly's annual reports, e.g., Carl Himly, "Jahersbericht," *Schriften der Universität zu Kiel* 27 (1880), 70.

125. Ladenburg, *Lebenserinnerungen*, 52.

126. Hippel, "Becoming," 23–26. On Caro and Baeyer, see Reinhardt and Travis, *Heinrich Caro*.

127. Ladenburg *Lebenserinnerungen*, 53–54.

128. Ladenburg's interest in benzene's substitution isomers seems to have been a key step in this trajectory. See, for example, Albert Ladenburg, "Ueber isomere Bisubstitutionsderivate des Benzols," *Berichte der deutschen chemischen Gesellschaft* 8 (1875): 853–855.

129. Ladenburg *Lebenserinnerungen*, 63–66.

130. Anon., "Zusammenstellung im Jahre 1877 in Ausführung begriffen gewesener Staatsbauten," *Zeitschrift für Bauwesen* 28 (1878): 487; Nägelke, *Hochschulbau im Kaiserreich*, 383.

131. The large laboratory (Saal) was originally intended for 16 students. According to Josef W. Durm et al., eds., *Handbuch der Architektur,* 4. Teil, 6. Halbbd, Heft 2a (Darmstadt: Bergsträsser, 1905 [1889]), 338, the number of students working in this laboratory seems later to have increased to 24. This increase is corroborated by Ladenburg's observation in his report, "Jahresbericht 1881–82," *Schriften der Universität zu Kiel* 27 (1882), 30, that student enrollment in the laboratory had risen as high as 29.

132. Albert Ladenburg, "Jahresbericht 1878," *Schriften der Universität zu Kiel* 25 (1879), 74; Albert Ladenburg, "Jahresbericht 1879," *Schriften der Universität zu Kiel* 26 (1880), 68.

133. Ladenburg, "Jahresbericht 1878," 74.

134. Ladenburg, "Jahresbericht 1881–82." On the later building work, see L. Mecking and G. Jacob, *Kiel als Universitätsstadt* (Kiel: Mühlau, 1921), 28–30. This volume contains both a contemporary and an early twentieth-century city plan indicating that the chemical institute more than doubled its footprint during this period.

135. Ladenburg, "Jahresbericht 1881–82."

136. Roth, "neuer Apparat."

137. Albert Ladenburg and C. F. Roth, "Nachweis der Identität von synthetischen Piperidin mit dem aus Piperin gewonnenen," *Berichte der deutschen chemischen Gesellschaft* 17 (1884): 513–515. Albert Ladenburg, "Die Constitution des Atropins," *Annalen der Chemie und Pharmacie* 217 (1883): 74–149, note on 87.

138. Roth earned his PhD in Ladenburg's laboratory in 1883, remaining in Kiel as an apothecary. Roth, "neuer Apparat," was submitted from Ladenburg's Kiel institute, indicating that Roth continued to assist Ladenburg at the institute until at least 1886. According to *Berichte der deutschen chemischen Gesellschaft* 16 (1883): 3098, Roth's apothecary shop was at Schlossgarten 12.
Already Ladenburg's assistant in 1883, Stöhr's 1884 PhD "Über die Hydroparacumarsäure" was completed in Munich, presumably in Adolf Baeyer's laboratory. Returning to Kiel, Stöhr habilitated in 1888 to become a Privatdozent. In 1893, he applied for a US patent for dimethylpiperazine, a new treatment for gout to be manufactured by the Bayer chemical company. In 1897, Stöhr was appointed director of the Kiel torpedo laboratory. Kiel Directory of Scholars, accessed November 19, 2021, https://cau.gelehrtenverzeichnis.de/person/33baf937-ea2f-3248-1955-4d4c60188df7.
Following completion of his 1884 doctoral dissertation in Kiel, "Beiträge zur Kenntniss der Alkine," Laun left Ladenburg's laboratory, apparently joining Merck in Darmstadt by 1886. He died soon afterward. Carsten Burhop, Michael Kissener, Hermann Schäfer, and Joachim Scholtyseck, *Merck: From a Pharmacy to a Global Corporation, 1668–2018* (Munich: Beck, 2018), figure 40.

139. According to Johnson, "Academic Chemistry in Imperial Germany," 511, well over a third of all published papers in academic chemistry between 1879 and 1905 concerned organic chemistry.

140. Adolf Baeyer's laboratory is the best-known site of academic-industrial collaboration. But similar connections were fundamental in university laboratories elsewhere, including those of Hofmann and Ladenburg.

141. Ludwig Gattermann, *The Practical Methods of Organic Chemistry*, trans. from the 4th German edition by William B. Schover (New York: Macmillan, 1901), 60. Gattermann's was one of most famous manuals of practical organic chemistry, reaching its 16th German edition in 1921.

142. Brock, *Gatekeeper*, 51.

143. Hermann Kolbe, *Das chemische Laboratorium der Universität Marburg und die seit 1859 darin ausgeführten chemischen Untersuchungen nebst Ansichten und Erfahrungen über die Methode des chemischen Unterrichts* (Braunschweig: Vieweg, 1865), 7–8.

144. Cf. Cahan, "The Institutional Revolution;" Cahan, *Institute for an Empire;* Graeme Gooday, "Precision Measurement and the Genesis of Physics Teaching Laboratories in Victorian Britain," *British Journal for the History of Science* 23 (1990): 25–51.

CHAPTER 9

1. My chapter title refers to nineteenth-century organic chemists' self-conscious fashioning of synthesis as a powerful new science—a sensibility shared by their successors. Founded in 2000, the reference work *Science of Synthesis*, accessed November 19, 2021, https://www.thieme.de/en/thieme-chemistry/science-of-synthesis-54780.htm, seeks to provide a "dependable source of information on evaluated synthetic methods in organic chemistry." The editors review "synthetic methodology developed from the early 1800s to-date"—an approach that confirms synthetic organic chemists' long-established practice of appropriating reactions introduced prior to the turn to synthesis.

2. The prize was recorded in the Sitzungsprotokoll der physikalisch-mathematischen Klasse vom 1. Juli 1867: Archiv der BBAW, PAW (1812–1945), II-IX-16, Bl. 4 r, 4v. It was also announced in *Berichte der deutschen chemischen Gesellschaft* 2 (1869): 468–469. The original deadline of March 1, 1870, was extended to 1 March 1873, because no entries had been received. There is no record of any entry having been received at any time.

3. Muspratt and Hofmann, "On Toluidine," 367.

4. Target synthesis is the intentional laboratory preparation of a target molecule, which may be natural (found in nature) or artificial (not yet isolated from natural sources). For chemists, laboratory origin does not necessarily define a synthetic product as artificial. Natural product synthesis—the subset of total synthesis whose targets are naturally occurring substances, or natural products—by definition produces nature-identical (i.e., natural, not artificial) products. As chapter 7 showed, late-nineteenth-century chemists no longer considered origin in their identification

of organic compounds. This practice indicates the desirability of disentangling the widespread equivalence of synthetic and artificial in general usage from these terms' particular, distinct meanings in chemistry and related sciences.

5. Ladenburg, "Synthese der activen Coniine," 2583, used total synthesis to indicate coniine's production in principle from its constituent elements. Subsequently amended to denote preparation from commercially available starting materials, Ladenburg's definition of total synthesis seems to have been introduced by Marcellin Berthelot, "Synthèse de l'Esprit de Bois," *Comptes Rendus* 45 (1857): 920.

6. Hofmann, "Researches Regarding."

7. Edward Frankland and Benjamin Collins Brodie, "On a New Series of Organic Bodies Containing Metals," *Philosophical Transactions of the Royal Society* 142 (1852): 417–444. Frankland introduced "combining power" (440) and referred to the analogy between "organometallic bodies" and organic bases (442). Russell, *History of Valency*, 83–89, offered a detailed etymology of the term "valency," from Frankland's "combining power," via Hofmann's 1865 promotion of "quantivalence," "monovalent," "bivalent," etc.

8. For more on how chemists sought to address this problem through nomenclature, see Evan Hepler-Smith, "'Just as the Structural Formula Does': Names, Diagrams, and the Structure of Organic Chemistry at the 1892 Geneva Nomenclature Congress," *Ambix* 62 (2015): 1–28.

9. This approach was a significant development in the way chemists worked with formulae on paper, a practice whose nonvisual origins were examined by Klein, *Experiments, Models, Paper Tools*.

10. Jacobus H. van 't Hoff, *La Chimie dans l'Espace* (Rotterdam: Bazendijk, 1875); Joseph Achille le Bel, "Sur les Relations qui existent entre les Formules Atomiques des Corps Organiques et le Pouvoir Rotatoire de leurs Dissolutions," *Bulletin de la Société Chimique* 22 (1874): 337–347. Further discussion of organic chemists' skepticism toward this theory can be found in C. M. Jackson, "Emil Fischer."

11. Geison, *The Private Science of Louis Pasteur*, remains the classic account of Louis Pasteur's foundational work in this area.

12. August Wilhelm Hofmann, "On Insolinic Acid," *Proceedings of the Royal Society of London* 8 (1857): 3. Hofmann's paper was received December 20, 1855, read January 10, 1856, and published December 31, 1857.

13. The most recent monograph study of Kekulé's accomplishment is Rocke, *Image and Reality*.

14. Adolf Baeyer, "Ueber die Reduction aromatischer Kohlenwasserstoffe durch Jodphosphonium," *Annalen der Chemie und Pharmacie* 155 (1870): 266–281.

15. Peter Squire, *A Companion to the British Pharmacopoeia* (London: Churchill, 1866), 78–79. See also J. Schorm, "Beitrag zur Kenntniss des Coniins und seiner Verbindungen," *Berichte der deutschen chemischen Gesellschaft* 14 (1881): 1765–1769.

16. Archiv der BBAW, PAW (1812–1945), II-IX-16, Bl. 4 r, 4v; *Berichte der deutschen chemischen Gesellschaft* 2 (1869): 468–469.

17. Auguste Cahours, "Recherches sur un Nouvel Alcali Dérivé de la Pipérine," *Comptes Rendus* 34 (1852): 481–484; Auguste Cahours, "Recherches sur un Nouvel Alkali Dérivé de la Piperine," *Annales de Chimie et de Physique* 38 (1853): 76–103. Hans Christian Ørsted, "Über das Piperin, ein neues Pflanzenalkaloid," *Schweiggers Journal für Chemie und Physik* 29 (1820): 80–82, reported the isolation of piperine from black pepper. Piperidine was isolated from the same source ca. 1930.

18. The reader may recall that Hofmann also assigned formula, constitution, and identity without complete analytical information during his investigations of organic nitrogen. But Hofmann was operating in a dense network of secure empirical knowledge, whereas Schiff's proposal reflected a failure to grasp what corroborating evidence would be required to make a reliable claim of this kind.

19. Hugo Schiff, "Erste Synthese eines Pflanzenalkaloids (Synthese des Coniins)," *Berichte der deutschen chemischen Gesellschaft* 3 (1870): 946–947. Received December 4, 1870.

20. J.B., "Synthesis of Alkaloids," *Journal of the Chemical Society* 24 (1871): 143–144 (in the section "Abstracts of Chemical Papers," 117–154), referring to Hugo Schiff, "Synthesis of Alkaloids," *Pharmaceutical Journal, Transactions* 3rd series, 1 (1870–1871): 605. I have not so far identified J.B.

21. Schiff's biographical information is taken from H. Wichelhaus, "Hugo Schiff," *Berichte der deutschen chemischen Gesellschaft* 48 (1915): 1565–1569, and M. Betti, "Hugo Schiff," *Journal of the Chemical Society, Transactions* 109 (1916): 369–434.

22. His interpretation is nevertheless curious in light of his earlier development of a new class of compounds now known as Schiff's bases—a subset of the imine family. See Hugo Schiff, "Mittheilungen aus dem Universitätslaboratorium in Pisa: Eine neue Reihe organischer Basen," *Annalen der Chemie und Pharmacie* 131 (1864): 118–119.

23. Hugo Schiff, "Ueber die Synthese des Coniins," *Annalen der Chemie und Pharmacie* 157 (1871): 352–362.

24. Hugo Schiff, "Weiteres über das künstliche Coniin," *Berichte der deutschen chemischen Gesellschaft* 5 (1872): 42–44. The prefix "para" was then widely used to designate isomeric forms. It had not yet acquired its present-day meaning (to specify the 4-position in a monosubstituted benzene).

25. Hofmann, "Researches into the Molecular Constitution of the Organic Bases;" August Wilhelm Hofmann, "Beiträge zur Kenntniss der flüchtigen organischen Basen," *Annalen der Chemie und Pharmacie* 78 (1851): 253–286.

26. Adolf von Planta and August Kekulé, "Beiträge zur Kenntniss einiger flüchtigen Basen," *Annalen der Chemie und Pharmacie* 89 (1854): 129–156.

27. Hugo Schiff, "Ueber die Synthese des Coniins," *Annalen der Chemie und Pharmacie* 166 (1873): 97–98.

28. Schiff, "Ueber die Synthese des Coniins," *Annalen der Chemie und Pharmacie* 166 (1873): 98, suggested that his product, paraconiine, was an open chain enamino imine (N-but-1-enylbutan-1-imine) isomeric with his previously proposed constitution for coniine (see figure 9.1).

29. Cahours, "Recherches," 483.

30. August Wilhelm Hofmann, "Notes of Researches on the Poly-Ammonias," *Proceedings of the Royal Society of London* 9 (1857–1859): 151.

31. Hofmann knew that the reaction between an organic base and methyl or ethyl iodide proceeded in stages, each involving substitution of a reactive hydrogen by an alkyl group. This initial phase culminated in the formation of a quaternary alkyl ammonium salt, in which nitrogen was surrounded by four alkyl groups. Hofmann's simplest example was tetraethylammonium iodide. Treatment with silver oxide removed iodine, converting the salt to its hydroxide. When heated, tetraethylammonium hydroxide decomposed to triethylamine, liberating ethylene gas—Hofmann's original elimination.
On further investigation, Hofmann established both the anomalous behavior of tetramethylammonium iodide and—by looking for patterns in the way that a range of amines containing methyl, ethyl, amyl, and phenyl groups reacted—the basics of what became known as Hofmann's rules for elimination reactions.
The relevant literature—spanning some 30 years—is too vast to cite in its entirety, but some highlights are: Hofmann, "Researches into the Molecular Constitution of the Organic Bases;" August Wilhelm Hofmann, "Ueber die Einwirkung der Wärme auf die Ammoniumbasen," *Berichte der deutschen chemischen Gesellschaft* 14 (1881): 494–496, 659–669, 705–713.

32. August Wilhelm Hofmann, "Zur Kenntniss des Piperidins und Pyridins," *Berichte der deutschen chemischen Gesellschaft* 12 (1879): 984–990.

33. Wilhelm Koenigs, "Ueberführung von Piperidin in Pyridin," *Berichte der deutschen chemischen Gesellschaft* 12 (1879): 2341–2344.
See also Giulielmo Körner, "Synthese d'une Base Isomère a la Toluidine," *Comptes Rendus* 68 (1869): 824; Giulielmo Körner, "Synthese d'une Base Isomere à la Toluidine," *Giornale di Scienze Naturali ed Economiche* 5 (1869): 111–114. Note that Körner frequently published under the Italian version of his forename.
For more on Körner's pyridine formula and its relationship to James Dewar's proposal, see Alan J. Rocke, "Körner, Dewar and the Structure of Pyridine," *Bulletin for the History of Chemistry* 2 (1988): 4–6.

34. Hofmann, "Einwirkung der Wärme," 668.

35. Johann C. Poggendorff, *Biographisch-Literarisches Handwörterbuch,* vol. 4 (Leipzig: Barth, 1904): 1347. Schotten trained in Hofmann's Berlin laboratory, earning his PhD in 1878. Schotten's doctoral dissertation, published in Ferdinand Tiemann and Carl Schotten, "Ueber die mittelst der Chloroformreaction aus den drei isomeren Kresolen

darstellbaren Oxytoluylaldehyde und die zugehörigen Oxytoluylsäuren," *Berichte der deutschen chemischen Gesellschaft* 11 (1878): 767–784, made "a valuable contribution to the history of aromatic compounds" by applying the recently discovered Reimer-Tiemann reaction. See also Hofmann's report on Schotten's dissertation: Humboldt-Universität zu Berlin, Universitätsarchiv zu Berlin, LiHP No. 4, vol. 5A, Phil. Fak. 250, Bl. 216v-216r, 218v-218r. The new reaction was discovered in Hofmann's laboratory by Carl Reimer and Ferdinand Tiemann and was first published in Karl Reimer and Ferdinand Tiemann, "Ueber die Einwirkung von Chloroform auf alkalische Pheno-late," *Berichte der deutschen chemischen Gesellschaft* 9 (1876): 824; 1268; 1285.

36. Hofmann, "Einwirkung der Wärme."
Hofmann, "Einwirkung der Wärme," 710–711, identified the eliminated hydrocar-bon as conylene, which had first been isolated by Theodor Wertheim, "Beiträge zur Kenntniss des Coniins," *Annalen der Chemie und Pharmacie* 123 (1862): 157–186; and 130 (1864): 269–302. Wertheim was then Professor of Chemistry in Graz.

37. Hofmann, "Einwirkung der Wärme," 712.

38. Hofmann, "Einwirkung der Wärme," 707–708. Hofmann thanked Schotten on 713.

39. Even at 17, coniine's hydrogen content was lower than expected.

40. Hofmann, "Einwirkung der Wärme," 494.

41. Hofmann, "Einwirkung der Wärme," 665, 668.

42. Hofmann, "Einwirkung der Wärme," 713.

43. Hofmann, "Einwirkung der Wärme," 713.

44. Muspratt and Hofmann, "On Toluidine," 369–370.

45. Klein and Lefèvre, *Materials,* 230.

46. Hofmann, "Einwirkung der Wärme," 713.

47. August Wilhelm Hofmann, "Zur Kenntniss der Coniin-Gruppe," *Berichte der deutschen chemischen Gesellschaft* 18 (1885): 5–23; and 109–131 (part 2).

48. A. Krakau, "Zur Kenntniss des Chinolins und einiger anderen Alkaloïde," *Berichte der deutschen chemischen Gesellschaft* 13 (1880): 2316. Vyschnegradsky's colleague in St Petersburg, Alexander Krakau, posthumously summarized his friend's unpublished studies in alkaloid chemistry. Vyschnegradsky claimed that two major families of natural alkaloids were derived from quinoline and pyridine—a result usually attrib-uted to Ladenburg.

49. Carl Schotten, "Zur Kenntniss des Coniins," *Berichte der deutschen chemischen Gesellschaft* 15 (1882): 1947. Hofmann, "Zur Kenntniss der Coniin-Gruppe," 128, called it an "outspoken suggestion."

50. Hofmann seems to have regarded Mylius as a superior chemist, "a man of clear understanding, extensive knowledge, a rare degree of observational ability and unusual experimental talent." See also Hofmann's report on Mylius's dissertation:

Humboldt-Universität zu Berlin, Universitätsarchiv zu Berlin, LiHP No. 4, vol. 6A, Phil. Fak. 260, Bl. 5v-5r, 6, 13v-13r.

51. August Wilhelm Hofmann, "Noch einige Beobachtungen über Piperidin und Pyridin," *Berichte der deutschen chemischen Gesellschaft* 16 (1883): 587–590. Hofmann (on 591) thanked Mylius most warmly for his support.

52. Hofmann, "Einwirkung der Wärme," 668. Previously equivocal concerning Wilhelm Koenigs's proposal, Hofmann, "Noch einige Beobachtungen," 587, now accepted that piperidine could be oxidized to pyridine.

53. August Wilhelm Hofmann, "Zur Kenntniss des Coniins," *Berichte der deutschen chemischen Gesellschaft* 17 (1884): 829, 832.

54. Graebe, *Geschichte,* 372, regarded this step as crucial in the constitutional analysis and subsequent synthesis of coniine. Hofmann's failure to synthesize coniine, however, suggests a need to reconsider this judgment.

55. According to Hofmann, "Zur Kenntniss der Coniin-Gruppe," 128, making conyrine provided a "factual basis" for the earlier "outspoken suggestion" of A. Vyschnegradsky and W. Koenigs that coniine was related to pyridine.

56. Hofmann, "Zur Kenntniss des Coniins," 828–830, and "Zur Kenntniss der Coniin-Gruppe," 128.

57. Hofmann, "Zur Kenntniss des Coniins," 830.

58. Benzene's primary position was identified by an existing substituent group (e.g., hydroxyl in phenol); in pyridine (and by extension, piperidine and coniine) nitrogen occupied the primary position.

59. Hofmann, "Zur Kenntniss des Coniins," 833.

60. By the time both parts of Hofmann's article "Zur Kenntniss der Coniin-Gruppe" were submitted in January 1885, Ehestädt had taken over from Mylius in assisting Hofmann's legendary studies of the coniceïnes.

61. Tiemann, "Erinnerung," 3382–3383.

62. Ladenburg, "Isomere Bisubstitutionsderivate des Benzols," might be taken as a starting point for this work.

63. Ladenburg, "Polemisches und Theoretisches," *Berichte der deutschen chemischen Gesellschaft* 8 (1875): 1666–1670, exemplifies the increasingly disputatious nature of Ladenburg's involvement in this field.

64. As, for example, when Ferdinand Tiemann, "Bemerkung zu der Abhandlung des Hrn. Ladenburg: Constitution des Benzols in No. 15 Dieser Berichte," *Berichte der deutschen chemischen Gesellschaft* 8 (1875): 1344, noted that a compound Ladenburg claimed was new was, in fact, already known in Berlin.

65. Albert Ladenburg, "Künstliches Atropin," *Berichte der deutschen chemischen Gesellschaft* 12 (1879): 941.

66. Albert Ladenburg, "Ueber Pyridin und Piperidinbasen," *Annalen der Chemie und Pharmacie* 247 (1988): 1.

67. Albert Ladenburg, "Methode zur Synthese in der Pyridinreihe," *Berichte der deutschen chemischen Gesellschaft* 16 (1883): 1410. Albert Ladenburg, "Ueber die Synthese des γ-Aethylpyridins und die Beziehungen des Pyridins zum Benzol," *Berichte der deutschen chemischen Gesellschaft* 16 (1883): 2059–2063.

68. Hofmann, "Zur Kenntniss des Coniins," 832–833.

69. Albert Ladenburg, "Synthese des Piperidins," *Berichte der deutschen chemischen Gesellschaft* 17 (1884): 156.

70. Ladenburg, "Pyridin und Piperidinbasen," 2. Both compounds were aromatic—but nitrogen formed part of the aromatic kernel only in pyridine (C_5H_5N), whereas aniline was benzene with an amino group substituent ($C_6H_5NH_2$).

71. Ladenburg, "Synthese des γ-Aethylpyridins."

72. Ladenburg, "Pyridin und Piperidinbasen," 50, 54–55, presented experimental evidence for pyridine and piperidine having ring structures.

73. Ladenburg, "Synthese des Piperidins"; Wilhelm Koenigs, "Zur Constitution des Cinchonins," *Berichte der deutschen chemischen Gesellschaft* 14 (1881): 1856.

74. Albert Ladenburg, "Synthese des Piperidins und seiner Homologen," *Berichte der deutschen chemischen Gesellschaft* 17 (1884): 388–391.

75. Ladenburg, "Synthese des Piperidins und seiner Homologen," 391. In fact, Ladenburg would later ascertain that his first alkylation product was α-ethyl pyridine, and its coniine-like reduction product was therefore α-ethyl piperidine.

76. Albert Ladenburg, "Ueber synthetische Pyridin- und Piperidinbasen," *Berichte der deutschen chemischen Gesellschaft* 17 (1884): 772–773.
Falck was then extraordinary professor of pharmacology in Kiel, where he would remain until his death in 1926. Accessed November 19, 2021, https://cau.gelehrtenverzeichnis.de/c77d6e1e-a41b-dfc8-3513-4d4c60c12cc5.

77. Ladenburg, "Synthetische Pyridin- und Piperidinbasen," 774–775.

78. Albert Ladenburg and Ludwig Schrader, "Ueber Isopropylpyridine," *Berichte der deutschen chemischen Gesellschaft* 17 (1884): 1121–1123.

79. Ladenburg and Schrader, "Ueber Isopropylpyridine," fn. on 1123.

80. Albert Ladenburg, "Ueber das α-Isopropylpyridin." *Berichte der deutschen chemischen Gesellschaft* 17 (1884): 1676–1679. (Received July 18, 1884.)

81. Ladenburg, "α-Isopropylpyridin."

82. Ladenburg here acknowledged the observation that optically inactive mixtures of optical isomers generally melted at a slightly different temperature from the individual optical isomers.

83. Ladenburg, "α-Isopropylpyridin," 1679.

84. Albert Ladenburg, "Ueber synthetische Pyridin- und Piperidinbasen," *Berichte der deutschen chemischen Gesellschaft* 18 (1885): 1587–1590. Albert Ladenburg, "Ueber Aethylpyridine und Aethylpiperidine," *Berichte der deutschen chemischen Gesellschaft* 18 (1885): 2961–2967, corrected the same conflation in the ethyl series.

Ladenburg's error here confirms that his chemistry was not based on, or interpreted using, ring structures for pyridine or piperidine. Had this been the case, Ladenburg would surely have appreciated—as any modern chemist would—that γ-isopropyl piperidine's molecular symmetry excluded any possibility of optical activity. Indeed, the α- and γ-isomers of isopropyl piperidine are so similar in their physical properties that it is today precisely the γ-(4-) isomer's lack of optical activity that reliably distinguishes it from the α-(2-) isomer.

85. Albert Ladenburg, "Einfache Methode zur Ortsbestimmung in der Pyridinreihe," *Berichte der deutschen chemischen Gesellschaft* 18 (1885): 2967–2969, addressed the difficulty of reliably determining substitution position.

86. Ladenburg, "Synthetische Pyridin- und Piperidinbasen," 1589.

87. Albert Ladenburg and C. F. Roth, "Studien über das käufliche Picolin," *Berichte der deutschen chemischen Gesellschaft* 18 (1885): 47–54.

88. Albert Ladenburg, "Versuche zur Synthese des Coniin," *Berichte der deutschen chemischen Gesellschaft* 19 (1886): 441, referred to Otto Lange, "Ueber α- und Gamma-Picolin," *Berichte der deutschen chemischen Gesellschaft* 18 (1885): 3436–3441.

See also, William Ramsay, "On Picoline and Its Derivatives," *The London, Edinburgh, and Dublin Philosophical Magazine and Journal of Science* 2 (1876): 269–281. Marcellin Berthelot, "Synthèse de l'Acétylène par la Combinaison directe du Carbone avec l'Hydrogène," *Comptes Rendus* 54 (1862): 640–644.

89. Ladenburg, "Versuche zur Synthese des Coniin."

90. Finding that both propyl and isopropyl iodide reacted with pyridine to produce α-isopropyl pyridine implied that propyl iodide suffered a rearrangement during the alkylation.

91. Ladenburg, "Versuche zur Synthese des Coniin."

92. Travis, *Rainbow Makers*, described many such examples in the field of dye chemistry.

93. For example, Ladenburg, "Synthese des Piperidins und Seiner Homologen," 389. Whereas Koenigs had separated piperidine from unreacted pyridine by forming the nitro derivatives, Ladenburg and Stöhr's improved sodium reduction removed the need for this tiresome additional step, because virtually no pyridine withstood the reaction.

94. Ladenburg, "Synthese der activen Coniine," 2578. Ladenburg referred on 2583 to Wilhelm Laun's tragically early death.

95. Ladenburg, "Pyridin und Piperidinbasen," 3–4.

96. Ladenburg, "Pyridin und Piperidinbasen," 4–6. Fractional crystallization relies on differential solubility to achieve separation: less soluble compounds crystallize sooner than more soluble ones.

97. Chemists will note that this compound must, in fact, have been 2-(1-propenyl) pyridine rather than its unconjugated allyl isomer.

98. Ladenburg, "Synthese der activen Coniine," 2579. Ladenburg, "Pyridin und Piperidinbasen," 26–28, reported similar figures from a later attempt: 262 g pure α-picoline obtained from 600 g raw picoline produced 36 g crude allylpyridine, i.e., ca. 7% overall yield.

99. Ladenburg, "Synthese der activen Coniine," 2579.

100. Ladenburg, "Synthese der activen Coniine," 2580.

101. Ladenburg, "Synthese der activen Coniine."

102. Geison, *The Private Science of Louis Pasteur,* chapter 3.

103. Ladenburg, "Synthese der activen Coniine," 2582; Schorm, "Beitrag zur Kenntniss des Coniins und seiner Verbindungen."

104. Ladenburg, "Synthese der activen Coniine;" Roth, "neuer Apparat."

105. Ladenburg, "Synthese der activen Coniine," 2583; Ladenburg, "Pyridin und Piperidinbasen," 83–86. Ladenburg's work was well received by the 1886 German *Naturforscherversammlung* in Berlin.

106. On the Hofmann-Löffler reaction, developed from Hofmann's studies of the coniceïnes, see L. Stella, "Homolytic Cyclizations of N-Chloroalkenylamines," *Angewandte Chemie International Edition* 22 (1983): 337–422; M. Wolff, "Cyclization of N-Halogenated Amines (The Hofmann-Löffler Reaction)," *Chemical Review* 63 (1963): 55–64.

107. Ladenburg, "Methode zur Synthese in der Pyridinreihe," 1410.

108. Ladenburg, "Pyridin und Piperidinbasen," 2.

109. Ladenburg, "Pyridin und Piperidinbasen," 15, described the preparation of 667 g raw ethyl pyridine from 800 g pyridine in a reaction carried out in 267 sealed tubes!

110. Albert Ladenburg, "Ueber das spezifische Drehungsvermögen der Piperidinbasen," *Berichte der deutschen chemischen Gesellschaft* 19 (1886): 2584.

111. Ladenburg, "spezifische Drehungsvermögen."

112. Ladenburg, "spezifische Drehungsvermögen;" Albert Ladenburg, "Ueber das optische Drehungsvermögen der Piperidinbasen. II," *Berichte der deutschen chemischen Gesellschaft* 19 (1886): 2975–2977.

113. This standard measure was established by Hans Heinrich Landolt, subsequently appearing in the tables of physico-chemical data he produced with Richard Bornstein. Hans Landolt, *Das optische Drehungsvermögen organischer Substanzen und die*

praktische Anwendungen desselben: Für Chemiker, Physiker und Zuckertechniker (Braunschweig: Vieweg, 1879); Landolt and Börnstein, *Physikalisch-Chemischen Tabellen.*

CONCLUSION

1. Antoine Lavoisier, *Traité Élémentaire de Chimie,* vol. 1 (Paris: Chez Cuchet, 1793).

2. Because this study reveals analysis and synthesis as complementary, rather than opposing, methods, I am less inclined than others have been to attribute a significant role in establishing the value of synthesis to Lavoisier's countryman Marcellin Berthelot. See, for example, Henry E. Armstrong, "Marcellin Berthelot and Synthetic Chemistry. A Study and an Interpretation. 1827 (Oct. 25)–1907 (March 18)," *Journal of the Royal Society of Arts* 76 (1927): 145–171; Bensaude-Vincent and Simon, *Chemistry: The Impure Science,* chapter 6.

3. Hofmann, *Life-Work of Liebig,* 12.

4. Although derived by studying synthesis in academic rather than industrial settings, this revised understanding of the relationship between the artificial and natural may help develop aspects of the argument in Berenstein, "Global Sensation."

5. Cf. Schummer, "Impact of Instrumentation."

6. The classic study remains: Cahan, *Institute for an Empire.* Tuchman, *Science, Medicine, and the State in Germany,* offers a most instructive analysis of why Cahan's laboratory revolution does not fit the case of chemistry.

7. Speaking at the foundation of Germany's Kaiser Wilhelm Foundation in 1911, the chemist Emil Fischer characterized chemistry as "the true land of unlimited possibilities," whose "exploitation" was among the new foundation's major tasks. Emil Fischer, *Neuere Erfolge und Probleme der Chemie* (Berlin: Springer, 1911).

8. Emil Fischer and Ernst Beckmann, *Das Kaiser-Wilhelm-Institut für Chemie Berlin-Dahlem* (Braunschweig: Vieweg, 1913), listed the industrial sponsors of the first Kaiser Wilhelm Institute for chemistry.

9. Hofmann, *Reports,* lx.

10. Indeed, the development of biomimetic synthesis and increased reliance on room temperature, aqueous chemistry suggests that chemists are now considerably closer to overcoming this distinction. For a recent examination of this issue in relation to synthetic biology, see Francesco Bianchini, "A New Definition of 'Artificial' for Two Artificial Sciences," *Foundations of Science* (June 1, 2021). https://rdcu.be/cA4LW.

11. Meinel, "Regierender Oberchemiker," 36.

12. Ladenburg, "Synthese des Piperidins und seiner Homologen," 389.

13. Friedrich Konrad Beilstein, *Handbuch der organischen Chemie* (Hamburg: Voss, 1881); Gordin, "Beilstein Unbound."

14. Elbs, *Synthetischen Darstellungsmethoden;* Eugen Lellmann, *Principien der organischen Synthese* (Berlin: Oppenheim, 1887).

15. For more on chemists' deployment of digital technologies, see Evan Hepler-Smith, "'A Way of Thinking Backwards': Computing and Method in Synthetic Organic Chemistry," *Historical Studies in the Natural Sciences* 48 (2018): 300–337.

16. Graebe and Liebermann's patent, No. 3850, Carl Graebe and Carl Liebermann, "Farbstoff," *Berichte der deutschen chemischen Gesellschaft* 2 (1869): 505–506 (submitted December 18, 1868)—identified their product only as "yellow flakes of alizarin" that "could be used in the same way as various madder compounds."

17. Carl Graebe and Carl Liebermann, "Ueber Anthracen und Alizarin," *Berichte der deutschen chemischen Gesellschaft* 1 (1968): 49–51, reported the conversion of alizarin to anthracene using "Baeyer's method" (zinc dust reduction). Graebe and Liebermann consequently revised alizarin's molecular formula (from $C_{14}H_6O_3$ to $C_{14}H_8O_4$). Combined with Heinrich Limpricht's (1866) synthesis of anthracene (from benzoyl chloride), this result led them to conclude that anthracene contained three fused aromatic rings—suggesting alizarin might be made from the coal tar component, anthracene.

18. Thomas Anderson, "On the Constitution of Anthracene or Paranaphthaline, and Some of Its Products of Decomposition," *Transactions of the Royal Society of Edinburgh* 22 (1861): 681–690, described the oxidation of anthracene to anthraquinone. Thus, all that remained was to introduce the two hydroxyl (OH) groups that differentiated alizarin from anthraquinone. See also Carl Graebe and Carl Liebermann, "Ueber künstliches Alizarin," *Berichte der deutschen chemischen Gesellschaft* 2 (1969): 332–334.

19. Graebe and Liebermann, "Ueber künstliches Alizarin," 334. As Thomas Edward Thorpe, *A Dictionary of Applied Chemistry*, vol. 1 (London: Longmans, 1891), 49, explained, Graebe and Liebermann benefited from both alizarin's extremely high molecular symmetry and the unusually small number of steps its synthesis required.

20. Hans Rupe, *Adolf Baeyer als Lehrer und Forscher: Erinnerungen aus seinem Privatlaboratorium* (Stuttgart: Enke, 1932), 15 (*in Erscheinungen denken*).

21. Adolf Baeyer, "Synthese des Indigblaus," *Berichte der deutschen chemischen Gesellschaft* 11 (1878): 1296–1297. See also Travis, *Rainbow Makers;* Reinhardt and Travis, *Heinrich Caro.*

22. Adolf Baeyer, "Ueber die Verbindungen der Indigogruppe," *Berichte der deutschen chemischen Gesellschaft* 16 (1883): 2188–2204.

23. Kumar, *Indigo Plantations.*

24. As Kōstas Gavroglou and Ana Simões, *Neither Physics nor Chemistry: A History of Quantum Chemistry* (Cambridge, MA: MIT Press, 2012) explain, to some extent this matter remained unresolved until the twentieth-century development of quantum chemistry and molecular orbital theory.

25. Rocke, *Image and Reality*, chapter 5.

26. Rocke, *Quiet Revolution*, 299, described Hermann Kolbe's proposed constitution for benzene as equivalent to a six-membered ring when "structurally interpreted." This seems historically questionable in light of Kolbe's absolute rejection of structural theory. Kolbe, meanwhile, continued to promote his constitutional alternative to Kekulé's benzene ring until at least 1883. Thus, Rocke's original interpretation of Kolbe's constitution as not implying a ring strikes me as more plausible. See Alan J. Rocke, "Kekulé's Benzene Theory and the Appraisal of Scientific Theories," in *Scrutinizing Science: Empirical Studies of Scientific Change*, ed. A. Donovan and R. Laudan, 145–161 (Dordrecht: Kluwer, 1988).

27. See Rocke, *Image and Reality*, 216–217, on the general principles of isomer counting in the development of aromatic structure theory; and Rocke, *Quiet Revolution*, 305, on the specific case of Kolbe's search for a second isomeric phenol.

28. Albert Ladenburg, "Ueber Benzolformeln," *Berichte der deutschen chemischen Gesellschaft* 23 (1890): 1007–1011.

29. As discussed in chapter 9, Ladenburg suggested that neither β- nor γ-propyl pipderidine would be optically active, when in fact this is only the case for the γ-compound.

30. Gustav Schultz, "Feier der deutschen chemischen Gesellschaft zu Ehren August Kekulé's," *Berichte der deutschen chemischen Gesellschaft* 23 (1890): 1265–1312.

31. C. M. Jackson, "Emil Fischer."

32. In doing so, this study extends the discussion of theory in the history of chemistry, e.g., Rocke, "What Did 'Theory' Mean to Nineteenth-Century Chemists?" More broadly, it offers a historical perspective on the nature of chemical theory, a topic of mounting interest to philosophers of chemistry. According to Davis Baird, Eric Scerri, and Lee McIntyre, eds., *Philosophy of Chemistry: Synthesis of a New Discipline* (Berlin: Springer, 2006), 5, chemistry's unique experimental and material basis requires a specialized philosophy to take account of its difference from physics "with all its lovely unifying and foundational theory." For Baird, Scerri, McIntyre, and their contributors, a satisfactory philosophy of chemistry must encompass chemistry's methods and tools as well as its theories—a commitment developed in their successor volume: Eric Scerri and Lee McIntyre, *Philosophy of Chemistry: Growth of a New Discipline* (Berlin: Springer, 2014).

33. Ladenburg, *Vorträge*, 3–4. According to Rocke ("What Did 'Theory' Mean to Nineteenth-Century Chemists?," 146–147), Hermann Kopp, *Die Entwickelung der Chemie in der neueren Zeit* (Munich: Oldenbourg, 1873), 844, offered a similar explanation of chemical theory.

34. Baird, Scerri, and McIntyre, *Philosophy of Chemistry*; Scerri and McIntyre, *Philosophy of Chemistry*.

BIBLIOGRAPHY

Abel, Frederick A. "Hofmann Memorial Lecture." *Journal of the Chemical Society* 69 (1896): 580–596.

Anderson, Thomas. "On the Constitution of Anthracene or Paranaphthaline, and Some of Its Products of Decomposition." *Transactions of the Royal Society of Edinburgh* 22 (1861): 681–690.

Anon. "Editor's Note." *London Medical Gazette* 38 (1846): 256.

Anon. "Farewell Dinner to Dr Hofmann." *Chemical News* 11 (1865): 210–216.

Anon. "Preussische Staatsbauten, welche im Jahre 1877 in der Ausführung begriffen gewesen sind." *Atlas zur Zeitschrift für Bauwesen* 28 (1878): 52b.

Anon. "Zusammenstellung im Jahre 1877 in Ausführung begriffen gewesener Staatsbauten." *Zeitschrift für Bauwesen* 28 (1878): 487.

Anon. "Ueber Chinoidin." *Augsburger Allgemeine Zeitung* 169 (June 7, 1846): 1349.

Anon., ed. *Perkin Centenary, London: 100 years of Synthetic Dyestuffs*. New York: Pergammon, 1958.

Anschütz, Richard, and Robert Schulze. "Ueber einen einfachen Apparat zur bequemen Bestimmung hochliegender Schmelzpunkte." *Berichte der deutschen chemischen Gesellschaft* 10 (1877): 1800–1802.

Armstrong, Henry Edward. "Hofmann Memorial Lecture." *Journal of the Chemical Society* 69 (1896): 637–732.

Armstrong, Henry Edward. *Introduction to the Study of Organic Chemistry: The Chemistry of Carbon and Its Compounds*. London: Longmans, Green & Company, 1874.

Armstrong, Henry Edward. "Marcellin Berthelot and Synthetic Chemistry: A Study and an Interpretation. 1827 (Oct. 25)—1907 (March 18)." *Journal of the Royal Society of Arts* 76 (1927): 145–171.

Baeyer, Adolf. "Synthese des Indigblaus." *Berichte der deutschen chemischen Gesellschaft* 11 (1878): 1296–1297.

Baeyer, Adolf. *Ueber die chemische Synthese*. Munich: Akademie der Wissenschaften, 1878.

Baeyer, Adolf. "Ueber die Reduction aromatischer Kohlenwasserstoffe durch Jodphosphonium." *Annalen der Chemie und Pharmacie* 155 (1870): 266–281.

Baeyer, Adolf. "Ueber die Verbindungen der Indigogruppe." *Berichte der deutschen chemischen Gesellschaft* 16 (1883): 2188–2204.

Baird, Davis, Eric Scerri, and Lee McIntyre, eds. *Philosophy of Chemistry: Synthesis of a New Discipline*. Berlin: Springer, 2006.

Baldwin, Melinda. *Making "Nature": The History of a Scientific Journal*. Chicago: University of Chicago Press, 2015.

Ball, Philip. "Perkin, the Mauve Maker." *Nature* 440 (March 2006): 429.

Batka, Wenzel. *Verzeichniss der neuesten chemischen und pharmaceutischen Geraethschaften: mit Abbildungen*. Nuremberg: Schrag, 1829.

Batka, Wenzel. *Verzeichniss der neuesten chemischen und pharmaceutischen Gerätschaften mit Abbildungen: Herausgegeben bei Gelegenheit der Versammlung deutscher Naturforscher in Wien*. Leipzig: Barth, 1832.

Batka, Wenzel. *Verzeichniss der neuesten chemischen, physikalischen und pharmaceutischen Apparate, Geräthschaften und Instrumente*. Leipzig: Barth, 1857.

Beer, John J. "A. W. Hofmann and the Founding of the Royal College of Chemistry." *Journal of Chemical Education* 37 (1960): 248–251.

Beilstein, Friedrich Konrad. *Handbuch der organischen Chemie*. Hamburg: Voss, 1881.

Bel, Joseph Achille le. "Sur les Relations qui existent entre les Formules Atomiques des Corps Organiques et le Pouvoir Rotatoire de leurs Dissolutions." *Bulletin de la Société Chimique de France* 22 (1874): 337–347.

Bell, Jacob. "Laboratories for Practical Instruction." *Pharmaceutical Journal* 6 (1846–1847): 193–197.

Bell, Jacob, and Theophilus Redwood. "Apology to Dr. Gardner." *Pharmaceutical Journal* 6 (1846–1847): 148–149.

Bell, Jacob, and Theophilus Redwood. "Counter Statement." *Pharmaceutical Journal* 6 (1846–1847): 151–160.

Bell, Jacob, and Theophilus Redwood. "The Rise and Progress of a Philosopher." *Pharmaceutical Journal* 6 (1846–1847): 141–142.

Benfey, Otto Theodor, and Peter J. T. Morris. *Robert Burns Woodward: Architect and Artist in the World of Molecules*. Philadelphia: Chemical Heritage Foundation, 2001.

Bensaude-Vincent, Bernadette, and Jonathan Simon, eds. *Chemistry: The Impure Science*. London: Imperial College Press, 2008.

Bensaude-Vincent, Bernadette, and Isabelle Stengers. *A History of Chemistry*. Cambridge, MA: MIT Press, 1996.

Bentley, Jonathan. "The Chemical Department of the Royal School of Mines: Its Origins and Development under A. W. Hofmann." *Ambix* 17 (1970): 153–181.

Berenstein, Nadia. "Making a Global Sensation: Vanilla Flavor, Synthetic Chemistry, and the Meanings of Purity." *History of Science* 54 (2016): 399–424.

Beretta, Marco. "Between the Workshop and the Laboratory: Lavoisier's Network of Instrument Makers." *Osiris* 29 (2014): 197–214.

Berl, Ernst, ed. *Briefe von Justus Liebig: Nach neuen Funden*. Giessen: Liebig-Museum-Gesellschaft, 1928.

Berthelot, Marcellin. *La Chimie Organique Fondée sur la Synthèse*. Vol. 2. Paris: Mallet-Bachelier, 1860.

Berthelot, Marcellin. "Synthèse de l'Acétylène par la Combinaison Directe du Carbone avec l'Hydrogène." *Comptes Rendus* 54 (1862): 640–644.

Berthelot, Marcellin. "Synthèse de l'Esprit de Bois." *Comptes Rendus* 45 (1857): 916–920.

Berzelius, Jöns Jacob. "An Address to Those Chemists Who Wish to Examine the Laws of Chemical Proportions, and the Theory of Chemistry in General." *Annals of Philosophy* 5 (1815): 122–131.

Berzelius, Jöns Jacob. *Anwendung des Löthrohrs in der Chemie und Mineralogie*. Translated by Heinrich Rose. Nuremberg: Schrag, 1821.

Berzelius, Jöns Jacob. *Atomgewichts-Tabellen*. Braunschweig: Vieweg, 1845.

Berzelius, Jöns Jacob. *Chemische Operationen und Gerätschaften, nebst Erklärung chemischer Kunstworter, in alphabetischer Ordnung*. Translated by Friedrich Wöhler. Dresden: Arnold, 1831.

Berzelius, Jöns Jacob. "Essay on the Cause of Chemical Proportions, and on Some Circumstances Relating to Them: Together with a Short and Easy Method of Expressing Them." *Annals of Philosophy* 2 (1813): 443–454.

Berzelius, Jöns Jacob. "Essay on the Cause of Chemical Proportions, and on Some Circumstances Relating to Them: Together with a Short and Easy Method of Expressing Them. Part 2." *Annals of Philosophy* 3 (1813): 51–62.

Berzelius, Jöns Jacob. "Experiments to Determine the Definite Proportions in Which the Elements of Organic Nature Are Combined." *Annals of Philosophy* 4 (1814): 323–331.

Berzelius, Jöns Jacob. "Experiments to Determine the Definite Proportions in Which the Elements of Organic Nature Are Combined. Part 2." *Annals of Philosophy* 4 (1814): 401–409.

Berzelius, Jöns Jacob. *Jahres-Bericht über die Fortschritte der physischen Wissenschaften: Eingericht an die Schwedische Akademie der Wissenschaften, den 31. März 1831*. Vol. 11. Translated by Friedrich Wöhler. Tübingen: Laupp, 1832.

Berzelius, Jöns Jacob. *Lehrbuch der Chemie*. Vol. 6. 3rd edition. Translated by Friedrich Wöhler. Dresden: Arnold, 1837.

Berzelius, Jöns Jacob. *Lehrbuch der Chemie*. Vol. 7. 3rd edition. Translated by Friedrich Wöhler. Dresden: Arnold, 1838.

Berzelius, Jöns Jacob. "Nachtrag zum VI. Band: Ueber die Analyse organischer Körper durch Verbrennung." In *Lehrbuch der Chemie*. Vol. 7. 3rd edition, 610–630. Dresden: Arnold, 1838.

Berzelius, Jöns Jacob. *Traité de Chimie*. Vol. 2. Translated by A. J. L. Jourdan and Esslinger. Brussels: Wahlen, 1838.

Berzelius, Jöns Jacob. "Ueber das Atomgewicht des Kohlenstoffs." *Annalen der Pharmacie* 30 (1839): 241–249.

Berzelius, Jöns Jacob. "Ueber die Bestimmung der relativen Anzahl von einfachen Atomen in chemischen Verbindungen." *Annalen der Physik und Chemie* 7 (1826): 397–416.

Berzelius, Jöns Jacob, and Pierre Louis Dulong. "Nouvelles Déterminations des Proportions de l'Eau et de la Densité de quelques Fluides Élastiques." *Annales de chimie et de physique* 15 (1820): 386–395.

Betti, M. "Hugo Schiff." *Journal of the Chemical Society, Faraday Transactions* 109 (1916): 369–434.

Biagioli, Mario. *Galileo's Instruments of Credit: Telescopes, Images, Secrecy*. Chicago: University of Chicago Press, 2006.

Bianchini, Francesco. "A New Definition of 'Artificial' for Two Artificial Sciences." *Foundations of Science*, June 1, 2021. https://rdcu.be/cA4LW.

Blanchet, Rodolphe, and Ernst Sell. "Ueber die Zusammensetzung einiger organischer Substanzen." *Annalen der Pharmacie* 6 (1833): 259–308.

Blondel-Mégrelis, Marika. "Liebig or How to Popularize Chemistry." *HYLE—International Journal for Philosophy of Chemistry* 13 (2007): 43–54.

Blyth, John, and August Wilhelm Hofmann. "On Styrole, and Some of the Products of Its Decomposition." *Memoirs and Proceedings of the Chemical Society* 2 (1843): 334–358.

Borscheid, Peter. *Naturwissenschaft, Staat und Industrie in Baden, (1848–1914)*. Stuttgart: Klett, 1976.

Brock, William H. *Justus von Liebig: The Chemical Gatekeeper*. Cambridge: Cambridge University Press, 1997.

Brock, William H., ed. *Liebig und Hofmann in Ihren Briefen (1841–1873)*. Weinheim: VCH, 1984.

Brock, William H. "Liebig's Laboratory Accounts." *Ambix* 19 (1972): 47–58.

Bromeis, Johann Conrad. "Ueber das Chinolin." *Annalen der Chemie und Pharmacie* 51 (1845): 130–140.

Brooke, John H. "Laurent, Gerhardt and the Philosophy of Chemistry." *Historical Studies in the Physical Sciences* 6 (1975): 405–429.

Brooke, John H. "Organic Synthesis and the Unification of Chemistry: A Reappraisal." *British Journal for the History of Science* 5 (1971): 363–392.

Brooke, John H. "Wöhler's Urea and Its Vital Force: A Verdict from the Chemists." *Ambix* 15 (1968): 84–114.

Brooks, Nathan M. "Nikolai Zinin and Synthetic Dyes: The Road Not Taken." *Bulletin for the History of Chemistry* 27 (2002): 26–36.

Brown, James Campbell. *Essays and Addresses.* London: Churchill, 1914.

Brown, James Campbell. "Justus Liebig: An Autobiographical Sketch." In *Essays and Addresses,* 170–196. London: Churchill, 1914.

Brunner, Carl. "Beiträge zur organischen Analyse." *Annalen der Physik und Chemie* 44 (1838): 134–155.

Brunner, Carl. "Versuche über Stärkmehl und Stärkmehlzucker." *Annalen der Physik und Chemie* 4 (1835): 319–338.

Buckingham, John. *Chasing the Molecule.* Stroud: Sutton, 2004.

Bud, Robert, and Gerrylynn K. Roberts. *Science versus Practice: Chemistry in Victorian Britain.* Manchester: Manchester University Press, 1984.

Buff, Heinrich. "Ueber Indigsäure und Indigharz." *Jahrbuch der Chemie und Physik* 21 (1827): 38–59.

Buff, Heinrich, Hermann Kopp, and Friedrich Zamminer. *Lehrbuch der physikalischen und theoretischen Chemie.* Vol. 1. Braunschweig: Vieweg, 1857.

Bullock, John Lloyd. "Amorphous Quinine." *Lancet* 2 (1846): 436–467.

Burhop, Carsten, Michael Kissener, Hermann Schäfer, and Joachim Scholtyseck. *Merck: From a Pharmacy to a Global Corporation, 1668–2018.* Munich: Beck, 2018.

Cahan, David. *An Institute for an Empire: The Physikalische-Technische Reichsanstalt, 1871–1918.* Cambridge: Cambridge University Press, 1989.

Cahan, David. "The Institutional Revolution in German Physics, 1865–1914." *Historical Studies in the Physical Sciences* 15 (1985): 1–65.

Cahours, Auguste. "Recherches sur un Nouvel Alkali Dérivé de la Piperine." *Comptes Rendus* 34 (1852): 481–484.

Cahours, Auguste. "Recherches sur un Nouvel Alkali Dérivé de la Piperine." *Annales de Chimie et de Physique* 38 (1853): 76–103.

Carius, Ludwig. "Ueber die Elementaranalyse organischer Verbindungen." *Annalen der Chemie und Pharmacie* 116 (1860): 1–30.

Carnelley, Thomas. *Melting and Boiling Point Tables.* Vol. 1. London: Harrison, 1885.

Carrière, Justus, ed. *Berzelius und Liebig: Ihre Briefe von 1831–1845 mit erläuternden Einschaltungen aus gleichzeitigen Briefen von Liebig und Wöhler sowie wissenschaftlichen Nachweisen.* Munich: Lehmann, 1898.

Casella, Louis P. *An Illustrated and Descriptive Catalogue of Philosophical, Meteorological, Mathematical, Surveying, Optical and Photographic Instruments.* London: Lane, 1861.

Chambers, Theodore Gervaise. *Register of the Associates and Old Students of the Royal College of Chemistry, the Royal College of Mines and the Royal College of Science; with Historical Introduction and Biographical Notes and Portraits of Past and Present Professors.* London: Hazell, 1896.

Chang, Hasok. *Inventing Temperature: Measurement and Scientific Progress.* Oxford: Oxford University Press, 2004.

Chang, Hasok. "Spirit, Air, and Quicksilver: The Search for the 'Real' Scale of Temperature." *Historical Studies in the Physical and Biological Sciences* 31 (2001): 249–284.

Chevreul, Michel E. *Considérations Générales sur l'Analyse Organique et sur ses Applications.* Paris: Levrault, 1824.

Chevreul, Michel E. *Recherches Chimiques sur les Corps Gras d'Origine Animale.* Paris: Levrault, 1823.

Child, Ernest. *The Tools of the Chemist: Their Ancestry and American Evolution.* New York: Reinhold, 1940.

Church, Arthur. "Notes on Boiling Points." *Chemical News* 1 (1860): 205–206.

Church, Arthur. "On the Benzole Series—Determination of Boiling-Points." *Philosophical Magazine* 9 (1855): 256–260.

Clarke, Frank Wrigglesworth. *The Constants of Nature.* Washington, DC: Smithsonian Institution, 1873.

Collins, Harry. *Changing Order: Replication and Induction in Scientific Practice.* Chicago: University of Chicago Press, 1985.

Crafts, James Mason. "Friedel Memorial Lecture." *Journal of the Chemical Society, Transactions* 77 (1900): 993–1019.

Cremer, Albert. *Das neue chemische Laboratorium zu Berlin.* Berlin: Ernst und Korn, 1868.

Crosland, Maurice. *Gay-Lussac: Scientist and Bourgeois.* Cambridge: Cambridge University Press, 1978.

Crosland, Maurice. *The Society of Arcueil: A View of French Science at the Time of Napoleon.* Cambridge, MA: Harvard University Press, 1967.

Csiszar, Alex. *The Scientific Journal: Authorship and the Politics of Knowledge in the Nineteenth Century.* Chicago: University of Chicago Press, 2018.

Danger, Ferdinand P. *Die Kunst der Glasbläserei vor dem Lothrohre und an der Lampe oder Darstellung eines neuen Verfahrens, um alle physikalische und chemische Instrumente, welche in den Bereich dieser Kunst Gehören, als Barometer, Thermometer, Ureometer, Heber u.s.w. mit dem geringsten Kosten-Aufwande und auf die leichteste Art zu Verfertigen.* Quedlinburg, Leipzig: Basse, 1833.

Danger, Ferdinand P. "Supplément à l'Art de Souffler le Verre." *Journal des Connaissances Usueles et Pratiques* 17 (1833): 33–37.

Danger, Ferdinand P. *The Art of Glass-Blowing; or, Plain Instructions for Making the Chemical and Philosophical Instruments Which Are Formed of Glass; Such as Barometers, Thermometers, Hydrometers, Hour-Glasses, Funnels, Syphons, Tube Vessels for Chemical Experiments, Toys for Recreative Philosophy, & c.* London: Bumpus & Griffin, 1831.

Dechend, Hertha, ed. *Justus von Liebig in eigenen Zeugnissen und solchen seiner Zeitgenossen.* Weinheim: Chemie, 1953.

Dörfel, Günter. "Der Meister und Seine Schule—zur Biographie und Wirkung des Instrumentenbauers Heinrich Geißler." *Sudhoff's Archiv* 98 (2014): 91–108.

Dr. H. Geisslers Nachfolger (Franz Müller). *Gedenkblatt zur Erinnerung an Heinrich Geissler, Dr. Phil., Glastechniker.* Bonn: F. Müller, 1890.

Dr. H. Geisslers Nachfolger (Franz Müller). *Preisverzeichnis: Institut zur Anfertigung und Lager chemischer, bakteriologischer, physikalischer und meteorologischer Apparate.* 9th edition. Bonn: Georgi, 1904.

Dumas, Jean Baptiste. *Éloge Historique de Henri-Victor Regnault.* Paris: Académie des Sciences, 1881.

Dumas, Jean Baptiste. "Lettre de M. Dumas à M. Gay-Lussac, sur les Procédés de l'Analyse Organique." *Annales de Chimie et de Physique* 47 (1831): 198–213.

Dumas, Jean Baptiste. "Mémoire sur la Loi des Substitutions et la Théorie des Types." *Comptes Rendus* 10 (1840): 149–178.

Dumas, Jean Baptiste. "Mémoire sur les Substances Végétales qui se Rapprochent du Camphre, et sur Quelques Huiles Essentielles." *Annales de Chimie et de Physique* 50 (1832): 225–240.

Dumas, Jean Baptiste. "Rapport sur un Mémoire de M. Wurtz, Relative à des Composés Analogues à l'Ammoniaque." *Comptes Rendus* 29 (1849): 203–205.

Dumas, Jean Baptiste. "Recherches de Chimie Organique." *Annales de Chimie et de Physique* 53 (1833): 164–181.

Dumas, Jean Baptiste. "Sur l'Esprit Pyroacétique." *Annales de Chimie et de Physique* 49 (1832): 208–210.

Dumas, Jean Baptiste. "Sur l'Oxamide, Matière qui se Rapproche de Quelques Substances Animales." *Annales de Chimie et de Physique* 44 (1830): 129–143.

Dumas, Jean Baptiste. "Ueber das Oxamid, eine gewissen Thierstoffen verwandte Substanz." *Annalen der Physik und Chemie* 19 (1830): 474–487.

Dumas, Jean Baptiste, and Pierre Joseph Pelletier. "Recherches sur la Composition Élémentaire et quelques Propriétés Charactéristiques des Bases Salifiables Organiques." *Annales de Chimie et de Physique* 24 (1823): 163–191.

Durm, Josef W., et al., eds. *Handbuch der Architektur,* 4. Teil, 6. Halbbd, Heft 2a. Darmstadt: Bergsträsser, 1905 [1889].

Earman, John, and C. Glymour. "Relativity and Eclipses: The British Eclipse Expeditions of 1919 and Their Predecessors." *Historical Studies in the Physical and Biological Sciences* 11 (1980): 49–85.

Editor. "Amorphous Quinine." *Pharmaceutical Journal* 6 (1846–1847): 55–56.

Editor. "On Amorphous Quinine." *Pharmaceutical Journal* 6 (1846–1847): 160–163.

Eichhorn, Karl. "Heinrich Geissler (1814–1879): His Life, Times, and Work." Trans. Heidi Collins. *Bulletin of Scientific Instruments* 27 (1990): 17–19.

Eichhorn, Karl. "Heinrich Geissler: Leben und Werk eines Pioniers der Vakuumtechnik." *Schriftenreihe Deutsches Röntgen-Museum* 6 (1984): 1–14.

Elbs, Karl. *Die Synthetischen Darstellungsmethoden der Kohlenstoff-Verbindungen. Volume I. Synthesen mittels Metallorganischer und mittels Cyanverbindungen; Synthesen durch molekulare Umlagerung und durch Addition.* Leipzig: Barth, 1889.

Faraday, Michael. *Chemical Manipulation: Being Instructions to Students in Chemistry, on the Methods of Performing Experiments of Demonstration or of Research, with Accuracy and Success.* London: Phillips, 1827.

Felschow, Eva-Marie, and Emil Heuser, eds. *Universität und Ministerium im Vormärz: Justus Liebigs Briefwechsel mit Justin von Linde.* Vol. 3. Studia Giessensia. Giessen: Giessen University Press, 1992.

Festing, Edward. *Report of Visits to Chemical Laboratories at Bonn, Berlin, Leipzig, etc.* London: Eyre and Spottiswoode, 1871.

Feyerabend, Kurt. *Die Universität Kiel: Ihre Anstalten, Institute und Kliniken.* Dusseldorf: Lindner, 1929.

Fischer, Emil. *Neuere Erfolge und Probleme der Chemie.* Berlin: Springer, 1911.

Fischer, Emil, and Ernst Beckmann. *Das Kaiser-Wilhelm-Institut für Chemie Berlin-Dahlem.* Braunschweig: Vieweg, 1913.

Fownes, George. *A Manual of Elementary Chemistry: Theoretical and Practical.* London: Churchill, 1844.

Fownes, George. *A Manual of Elementary Chemistry: Theoretical and Practical.* 9th edition. Revised by Henry Bence Jones and August Wilhelm Hofmann. London: Churchill, 1863.

Fox, Robert. *The Caloric Theory of Gases: From Lavoisier to Regnault.* Oxford: Oxford University Press, 1971.

Fox, Robert. "The Rise and Fall of Laplacian Physics." *Historical Studies in the Physical Sciences* 4 (1974): 89–136.

Fox, Robert. *The Savant and the State: Science and Cultural Politics in Nineteenth-Century France.* Baltimore, MD: Johns Hopkins University Press, 2012.

Frankland, Edward. *Experimental Researches in Pure, Applied, and Physical Chemistry.* London: van Voorst, 1877.

Frankland, Edward. "On the Isolation of the Organic Radicals." *Quarterly Journal of the Chemical Society* 3 (1850): 263–96.

Frankland, Edward, and Benjamin Collins Brodie. "On a New Series of Organic Bodies Containing Metals." *Philosophical Transactions of the Royal Society* 142 (1852): 417–444.

Fresenius, Carl Remigius. *Anleitung zur quantitativen chemischen Analyse.* 5th extended and improved edition. Braunschweig: Vieweg, 1863.

Fresenius, Carl Remigius. "Ueber die Constitution der Alkaloide." *Annalen der Chemie und Pharmacie* 61 (1847): 149–156.

Fritzsche, Julius. "Bemerkung zu vorstehender Abhandlung des Herrn. Zinin." *Journal für praktische Chemie* 27 (1842): 153.

Fritzsche, Julius. "Ueber die Produkte der Einwirkung von Kali auf Indigblau." *Annalen der Chemie und Pharmacie* 39 (1841): 76–91.

Fruton, Joseph S. *Contrasts in Scientific Style: Research Groups in the Chemical and Biochemical Sciences.* Philadelphia: American Philosophical Society, 1990.

Fruton, Joseph S. *Proteins, Enzymes, Genes: The Interplay of Chemistry and Biology.* New Haven, CT: Yale University Press, 1999.

Fruton, Joseph S. "The Liebig Research Group—a Reappraisal." *Proceedings of the American Philosophical Society* 132 (1988): 1–66.

Galison, Peter. *How Experiments End.* Chicago: University of Chicago Press, 1987.

García-Belmar, Antonio. "Sites of Chemistry in the Nineteenth Century." *Ambix* 61 (2014): 109–114.

Gardner, John. "An Address Delivered in the Royal College of Chemistry, Hanover Square, on Wednesday Evening, June 3." *Lancet* 1 (1846): 637–641.

Gardner, John. "Vindication of Dr. Gardner." *Pharmaceutical Journal* 6 (1846–1847): 149–151.

Garfield, Simon. *Mauve: How One Man Invented a Colour That Changed the World.* London: Faber & Faber, 2001.

Gattermann, Ludwig. *Die Praxis des organischen Chemikers.* Leipzig: Veit, 1894.

Gattermann, Ludwig. *Die Praxis des organischen Chemikers.* 4th edition. Leipzig: Veit, 1900.

Gattermann, Ludwig. *The Practical Methods of Organic Chemistry.* Trans. William B. Schover. New York: Wiley, 1896.

Gattermann, Ludwig. *The Practical Methods of Organic Chemistry.* Trans. from the 4th German edition by William B. Schover. New York: Macmillan, 1901.

Gavroglou, Kōstas, and Ana Simões. *Neither Physics nor Chemistry: A History of Quantum Chemistry.* Cambridge, MA: MIT Press, 2012.

Gay, Hannah. "'Pillars of the College': Assistants at the Royal College of Chemistry." *Ambix* 67 (2000): 135–169.

Gay, Hannah. *The History of Imperial College London, 1907–2007: Higher Education and Research in Science, Technology, and Medicine.* London: Imperial College Press, 2007.

Gay-Lussac, Joseph-Louis. *Cours de Chimie.* Paris: Pichon et Didiers, 1828.

Gay-Lussac, Joseph-Louis. "Sur la Déliquescence des Corps." *Annales de Chimie et de Physique* 82 (1812): 171–177.

Gay-Lussac, Joseph-Louis. "Recherches sur l'Acide Prussique." *Annales de Chimie et de Physique* 95 (1815): 136–231.

Gay-Lussac, Jules, and Théophile-Jules Pelouze. "Sur la Composition de la Salicine." *Annales de Chimie et de Physique* 48 (1831): 111.

Gay-Lussac, Jules, and Théophile-Jules Pelouze. "Sur la Salicine." *Annales de Chimie et de Physique* 44 (1831): 220–221.

Geiger, Philipp Lorenz. *Handbuch der Pharmacie.* Vol. 1, Part II. 5th edition. Rev. by Justus Liebig. Heidelberg: Winter, 1843.

Geiger, Philipp Lorenz. "Pharmacognostische Notizen." *Magazin für Pharmacie* 33 (1831): 134–136.

Geison, Gerald L. *The Private Science of Louis Pasteur*. Princeton, NJ: Princeton University Press, 1995.

Gerhardt, Charles. "Sur la Classification Chimique des Substances Organiques." *Revue Scientifique et Industrielle* 14 (1843): 580–609.

Gerhardt, Charles M. "Quinoléine, Produit de Décomposition de la Quinine et de la Cinchonine." *Comptes Rendus* 1 (1845): 30–31.

Gerhardt, Charles M. *Traité de Chimie Organique*. Paris: Didot, 1853.

Gerhardt, Charles M. "Untersuchungen über die organischen Basen." *Annalen der Chemie und Pharmacie* 42 (1842): 310–313.

Gooday, Graeme. "Precision Measurement and the Genesis of Physics Teaching Laboratories in Victorian Britain." *British Journal for the History of Science* 23 (1990): 25–51.

Gordin, Michael D. "Beilstein Unbound: The Pedagogical Unravelling of a Man and His Handbuch." In *Pedagogy and the Practice of Science: Historical and Contemporary Perspectives*, edited by David Kaiser, 11–40. Cambridge, MA: MIT Press, 2005.

Götz, Wolfgang. "Johann Baptist Batka, 'Arznei-Waarenhändler in Prag.'" *Geschichte der Pharmazie* 46 (1994): 1–12.

Graebe, Carl. *Geschichte der organischen Chemie*. Vol. 1. Berlin: Springer, 1972 [1920].

Graebe, Carl, and Carl Liebermann. "Farbstoff." *Berichte der deutschen chemischen Gesellschaft* (1869): 505–506.

Graebe, Carl, and Carl Liebermann. "Ueber Anthracen und Alizarin." *Berichte der deutschen chemischen Gesellschaft* 1 (1868): 49–51.

Graebe, Carl, and Carl Liebermann. "Ueber künstliches Alizarin." *Berichte der deutschen chemischen Gesellschaft* 2 (1869): 332–234.

Graebe, Carl, and Carl Liebermann. "Ueber künstliche Bildung von Alizarin." *Berichte der deutschen chemischen Gesellschaft* 2 (1869): 14.

Graham, Thomas, and Friedrich Julius Otto. *Ausführliches Lehrbuch der Chemie*. Vol. 1. 3rd edition. Braunschweig: Vieweg, 1857.

Gregory, William. *A Handbook of Organic Chemistry*. 3rd edition. London: Taylor: Walton and Maberly, 1856.

Griffin, John J. *Chemical Handicraft: A Classified and Descriptive Catalogue of Chemical Apparatus, Suitable for the Performance of Class Experiments, for Every Process of Chemical Research, and for Chemical Testing in the Arts*. London: Griffin, 1866.

Griffin, John J. *Descriptive Catalogue of Chemical Apparatus Manufactured and Imported by John J. Griffin*. New edition. Corrected to March 1850 (London, 1850), Part 1 (first published July 1841). London: Griffin, 1850.

Grove, John. "Amorphous Quinine." *Lancet* 2 (1846): 399–401.

Hannaway, Owen. "Laboratory Design and the Aim of Science: Andreas Libavius versus Tycho Brahe." *Isis* 77 (1986): 585–610.

Hepler-Smith, Evan. "'A Way of Thinking Backwards': Computing and Method in Synthetic Organic Chemistry." *Historical Studies in the Natural Sciences* 48 (2018): 300–337.

Hepler-Smith, Evan. "'Just as the Structural Formula Does': Names, Diagrams, and the Structure of Organic Chemistry at the 1892 Geneva Nomenclature Congress." *Ambix* 62 (2015): 1–28.

Hess, Hermann. "Berichtigung zu meinem Aufsatze über die Constitution der Zuckersäure." *Annalen der Chemie und Pharmacie* 33 (1840): 116–17.

Hess, Hermann. "Ueber die Constitution der Zuckersäure." *Annalen der Pharmacie* 30 (1839): 302–13.

Hess, Hermann. "Ueber die Wasserstoffbestimmung bei der Analyse organischer Substanzen." *Annalen der Pharmacie* 26 (1838): 189–94.

Hess, Hermann. "Ueber die Zusammensetzung der Zuckersäure." *Annalen der Pharmacie* 26 (1938): 1–9.

Himly, Carl. "Jahresbericht." *Schriften der Universität zu Kiel* 27 (1880): 70.

Hippel, Wolfgang von. "Becoming a Global Corporation—BASF from 1865 to 1900." In *German Industry and Global Enterprise: BASF: The History of a Company*, edited by Werner Abelshauser, Wolfgang von Hippel, Jeffrey Allan Johnson and Raymond G. Stokes, 5–114. Cambridge: Cambridge University Press, 2003.

Hoff, Jacobus H. van't. *La Chimie dans l'Espace*. Rotterdam: Bazendijk, 1875.

Hofmann, August Wilhelm. "A Page of Scientific History: Reminiscences of the Early Days of the Royal College of Chemistry." *Quarterly Journal of Science* 8 (1871): 145–153.

Hofmann, August Wilhelm, ed. *Aus Justus Liebig und Friedrich Wöhler's Briefwechsel in den Jahren 1829–1873*. Braunschweig: Vieweg, 1888.

Hofmann, August Wilhelm. "Beiträge zur Kenntniss der flüchtigen organischen Basen. X (fortsetzung)." *Annalen der Chemie und Pharmacie* 79 (1851): 11–39.

Hofmann, August Wilhelm. "Chemische Untersuchung der organischen Basen in Steinkohlen-Teerol." *Annalen der Chemie und Pharmacie* 47 (1843): 37–87.

Hofmann, August Wilhelm. "Contributions to the History of the Phosphorus-Bases." *Quarterly Journal of the Chemical Society* 13 (1861): 289–325.

Hofmann, August Wilhelm. "Contributions towards the History of Thialdine." *Quarterly Journal of the Chemical Society* 10 (1858): 193–202.

Hofmann, August Wilhelm. *Die Frage der Theilung der philosophischen Facultät*. Berlin: Dümmler, 1881.

Hofmann, August Wilhelm. *Die organische Chemie und die Heilmittellehre*. Berlin: Hirschwald, 1871.

Hofmann, August Wilhelm. "Eine sichere Reaction auf Benzol." *Annalen der Chemie und Pharmacie* 55 (1845): 200–205.

Hofmann, August Wilhelm. "Erinnerungen an Adolph Wurtz." *Berichte der deutschen chemischen Gesellschaft* 20 (1887): 815–996. Reprinted in *Zur Erinnerung an Vorangegangene Freunde*. Vol. 3. Braunschweig: Vieweg, 1888.

Hofmann, August Wilhelm. "Metamorphosen des Indigo's: Erzeugung organischer Basen, welche Chlor und Brom enthalten." *Annalen der Chemie und Pharmacie* 53 (1845): 1–57.

Hofmann, August Wilhelm. "New Researches upon Aniline." *Memoirs and Proceedings of the Chemical Society* 3 (1845): 26–28.

Hofmann, August Wilhelm. "Noch einige Beobachtungen über Piperidin und Pyridin." *Berichte der deutschen chemischen Gesellschaft* 16 (1883): 586–591.

Hofmann, August Wilhelm. "Noch einiges über die Amine der Methyl- und Aethylreihe." *Berichte der deutschen chemischen Gesellschaft* 22 (1889): 699–705.

Hofmann, August Wilhelm. "Notes of Researches on the Poly-Ammonias." *Proceedings of the Royal Society of London* 9 (1857–1859): 150–156.

Hofmann, August Wilhelm. "On a New Class of Bodies Homologous to Hydrocyanic Acid, II." *Proceedings of the Royal Society of London* 16 (1868): 148–150.

Hofmann, August Wilhelm. "On Insolinic Acid." *Proceedings of the Royal Society of London* 8 (1857): 1–3.

Hofmann, August Wilhelm. "On the Metamorphoses of Indigo: Production of Organic Bases which contain Chlorine and Bromine." *Memoirs and Proceedings of the Chemical Society* 2 (1843): 266–300.

Hofmann, August Wilhelm. "On the Use of Gas as a Fuel in Organic Analysis." *Quarterly Journal of the Chemical Society* 6 (1854): 209–216.

Hofmann, August Wilhelm. "Recherches sur la Constitution Moléculaire des Bases Organiques Volatiles." *Annales de Chimie et de Physique* 30 (1850): 87–118.

Hofmann, August Wilhelm. *Report on the Chemical Laboratories in Process of Building in the Universities of Bonn and Berlin*. London: Clowes and Son, 1866.

Hofmann, August Wilhelm. *Reports of the Royal College of Chemistry, and Researches Conducted in the Laboratories in the Years 1845-6-7*. London: Schulze, 1849.

Hofmann, August Wilhelm. "Researches into the Molecular Constitution of the Organic Bases." *Philosophical Transactions of the Royal Society* 141 (1851): 357–398.

Hofmann, August Wilhelm. "Researches on the Volatile Organic Bases. I: On the Action of Cyanogen on Aniline, Toluidine, and Cumidine." *Quarterly Journal of the Chemical Society* 1 (1849): 159–73.

Hofmann, August Wilhelm. "Researches on the Volatile Organic Bases. II: On the Action of Iodine on Aniline." *Quarterly Journal of the Chemical Society* 1 (1849): 269–281.

Hofmann, August Wilhelm. "Researches on the Volatile Organic Bases. III: Action of Chloride, Bromide, and Iodide of Cyanogen on Aniline. Melaniline, a New Conjugated Alkaloid." *Quarterly Journal of the Chemical Society* 1 (1849): 285–317.

Hofmann, August Wilhelm. "Researches on the Volatile Organic Bases. V. On the Action of Acids and Bases upon Cyaniline. VI. Metamorphoses of Dicyanomelaniline. Formation of the Aniline-Term Corresponding to Cyanic Acid. VII. Action of Anhydrous Phosphoric Acid on Various Aniline-Salts and Anilides." *Quarterly Journal of the Chemical Society* 2 (1850): 300–335.

Hofmann, August Wilhelm. "Researches Regarding the Molecular Constitution of the Volatile Organic Bases." *Philosophical Transactions of the Royal Society* 140 (1850): 93–131.

Hofmann, August Wilhelm. "Sitzung vom 22. Februar 1892." *Berichte der deutschen chemischen Gesellschaft* 25 (1892): 505–523.

Hofmann, August Wilhelm. "Sitzung vom 29. Januar 1879." *Berichte der deutschen chemischen Gesellschaft* 12 (1879): 147–152.

Hofmann, August Wilhelm. *The Life-Work of Liebig: The Faraday Lecture for 1875.* London: Macmillan, 1876.

Hofmann, August Wilhelm. "Ueber die Einwirkung der Wärme auf die Ammonium-basen." *Berichte der deutschen chemischen Gesellschaft* 14 (1881): 494–496, 659–669, 705–713.

Hofmann, August Wilhelm. "Uebersicht der in der letzten Zeit unternommenen Forschungen über den Indigo und seine Metamorphosen." *Annalen der Chemie und Pharmacie* 48 (1843): 241–343.

Hofmann, August Wilhelm. "Vereinsabend bei dem Präsidenten am 15ten Mai 1869," *Berichte der deutschen chemischen Gesellschaft* 2 (1869): 223–236.

Hofmann, August Wilhelm. *Zur Erinnerung an Gustav Magnus.* Berlin: Dümmler, 1871.

Hofmann, August Wilhelm. "Zur Kenntniss der Coniin-Gruppe." *Berichte der deutschen chemischen Gesellschaft* 18 (1885): 5–23 (Part 1) and 109–131 (Part 2).

Hofmann, August Wilhelm. "Zur Kenntniss des Coniins." *Berichte der deutschen chemischen Gesellschaft* 17 (1884): 825–833.

Hofmann, August Wilhelm. "Zur Kenntniss des Piperidins und Pyridins." *Berichte der deutschen chemischen Gesellschaft* 12 (1879): 984–990.

Hofmann, Paul. J. *Acht Tafeln zur Beschreibung des chemischen Laboratoriums zu Giessen.* Heidelberg: Winter, 1842.

Hofmann, Paul. J. *Das chemische Laboratorium der Ludwigs-Universität zu Giessen.* Heidelberg: Winter, 1842.

Holmes, Frederic Lawrence. *Antoine Lavoisier: The Next Crucial Year: Or, the Sources of His Quantitative Method in Chemistry.* Princeton, NJ: Princeton University Press, 1997.

Holmes, Frederic Lawrence. "Justus Liebig." In *Dictionary of Scientific Biography.* Vol. 8. Edited by Charles Gillespie, 329–350. New York: Scribner, 1973.

Holmes, Frederic Lawrence. "Justus Liebig and the Construction of Organic Chemistry." In *Essays on the History of Organic Chemistry,* edited by James G. Traynham, 119–134. Baton Rouge: Louisiana State University Press, 1987.

Holmes, Frederic Lawrence. "The Complementarity of Teaching and Research in Liebig's Laboratory." *Osiris,* 2nd series, 5 (1989): 121–164.

Holton, Gerald. "Subelectrons, Presuppositions, and the Millikan–Ehrenhaft Dispute." In *The Scientific Imagination: Case Studies,* 25–83. Cambridge: Cambridge University Press, 1978.

Homburg, Ernst. "The Rise of Analytical Chemistry and Its Consequences for the Development of the German Chemical Profession (1780–1860)." *Ambix* 46 (1999): 1–32.

Hornix, W. J. "Tales of Hofmann." *Annals of Science* 44 (1987): 519–224.

Houseman, Percy A. "Camphor, Natural and Synthetic." *Science Progress in the Twentieth Century (1906–1916)* 3 (1908): 60–68.

Ihde, Aaron J. *The Development of Modern Chemistry*. New York: Harper and Row, 1964.

J. B. "Synthesis of Alkaloids." *Journal of the Chemical Society* 24 (1871): 143–144.

Jackson, Catherine M. "Chemical Identity Crisis: Glass and Glassblowing in the Identification of Organic Compounds." *Annals of Science* 72 (2015): 187–205.

Jackson, Catherine M. "Chemistry as the Defining Science: Discipline and Training in Nineteenth-Century Chemical Laboratories." *Endeavour* 35 (2011): 55–62.

Jackson, Catherine M. "Emil Fischer and the 'Art of Chemical Experimentation'." *History of Science* 55 (2017): 86–120.

Jackson, Catherine M. "Re-Examining the Research School: August Wilhelm Hofmann and the Re-Creation of a Liebigian Research School in London." *History of Science* 44 (2006): 281–319.

Jackson, Catherine M. "Synthetical Experiments and Alkaloid Analogues: Liebig, Hofmann and the Origins of Organic Synthesis." *Historical Studies in the Natural Sciences* 44 (2014): 319–363.

Jackson, Catherine M. "The Laboratory." In *Companion to the History of Science*, ed. Bernard Lightman, 296–309. Oxford: Blackwell-Wiley, 2016.

Jackson, Catherine M. "The 'Wonderful Properties of Glass': Liebig's Kaliapparat and the Practice of Chemistry in Glass." *Isis* 106 (2015): 43–69.

Jackson, Catherine M. "Visible Work: The Role of Students in the Creation of Liebig's Giessen Research School." *Notes and Records of the Royal Society* 62 (2008): 31–49.

Jackson, Myles W. *Spectrum of Belief: Joseph von Fraunhofer and the Craft of Precision Optics*. Cambridge, MA: MIT Press, 2000.

Jamin, Jules. *Cours de Physique de l'École Pólytechnique*. Paris: Gauthier-Villars, 1886.

Jensen, William B. "Electronegativity from Avogadro to Pauling. Part 1. The Origins of the Electronegativity Concept." *Journal of Chemical Education* 73 (1996): 11–20.

Johnson, Jeffrey A. "Academic Chemistry in Imperial Germany." *Isis* 76 (1985): 500–524.

Jorpes, J. Erik. *Jac Berzelius: His Life and Work*. Trans. Barbara Steele. Stockholm: Almquist and Wiksells, 1966.

Kaiser, David, ed. *Pedagogy and the Practice of Science: Historical and Contemporary Perspectives*. Cambridge, MA: MIT Press, 2005.

Kangro, Hans. "Geissler, Johann Heinrich Wilhelm." In *Complete Dictionary of Scientific Biography*. Vol. 5, 340–341. Detroit: Scribner, 2008.

Keas, Michael N. "The Nature of Organic Bases and the Ammonia Type." In *Die Allianz von Wissenschaft und Industrie August Wilhelm Hofmann (1818–1892): Zeit, Werk, Wirkung*. Ed. Christoph Meinel and Hartmut Scholz, 101–118. Weinheim: VCH, 1992.

Keas, Michael N. "The Structure and Philosophy of Group Research: August Wilhelm Hofmann's Research Program in London (1845–1865)." Unpublished PhD diss., University of Oklahoma, 1992.

Keen, Robin. *The Life and Work of Friedrich Wöhler (1800—1882)*. Ed. Johannes Büttner. Nordhausen, Germany: Bautz, 2011.

Kekulé, August. *Lehrbuch der organischen Chemie, oder der Chemie der Kohlenstoffverbindungen*. Erlangen: Enke, 1861.

Kipping, Frederic S. "Ladenburg Memorial Lecture." *Journal of the Chemical Society, Transactions* 103 (1913): 1871–1895.

Klein, Ursula. *Experiments, Models, Paper Tools: Cultures of Organic Chemistry in the Nineteenth Century*. Stanford, CA: Stanford University Press, 2003.

Klein, Ursula. "Shifting Ontologies, Changing Classifications: Plant Materials from 1700 to 1830." *Studies in History and Philosophy of Science* 36 (2005): 261–329.

Klein, Ursula. "Techniques of Modelling and Paper-Tools in Classical Chemistry." In *Models as Mediators: Perspectives on Natural and Social Science*, edited by Mary S. Morgan and Margaret Morrison, 146–67. Cambridge: Cambridge University Press, 1999.

Klein, Ursula. *Technoscience in History: Prussia, 1750–1850*. Cambridge, MA: MIT Press, 2020.

Klein, Ursula, and Wolfgang Lefèvre. *Materials in Eighteenth-Century Science: A Historical Ontology*. Cambridge, MA: MIT Press, 2007.

Kodweiss, Friedrich. "Ueber die Zusammensetzung der Harnsäure und über die Producte, welche durch ihre Zersetzung mit Salpetersäure erzeugt werden." *Annalen der Physik und Chemie* 19 (1830): 1–25.

Koenigs, Wilhelm. "Ueberführung von Piperidin in Pyridin." *Berichte der deutschen chemischen Gesellschaft* 12 (1879): 2341–2344.

Koenigs, Wilhelm. "Zur Constitution des Cinchonins." *Berichte der deutschen chemischen Gesellschaft* 14 (1881): 1852–1859.

Körner, Giulielmo. "Synthèse d'une Base Isomère a la Toluidine." *Comptes Rendus* 68 (1869): 824.

Körner, Giulielmo. "Synthèse d'une Base Isomère a la Toluidine." *Giornale di Scienze Naturali ed Economiche* 5 (1869): 111–114.

Körner, Wilhelm. "Studj sull'Isomeria delle Cosi dette Sostanze Aromatiche a Sei Atomi di Carbonio." *Gazzetta Chimica Italiana* 4 (1874): 305–446.

Kohler, Robert E. *From Medical Chemistry to Biochemistry*. Cambridge: Cambridge University Press, 1982.

Kohler, Robert E. "Lab History: Reflections." *Isis* 99 (2008): 761–768.

Kolbe, Hermann. *Ausführliches Lehrbuch der organischen Chemie*. Vol. 1. Braunschweig: Vieweg, 1854.

Kolbe, Hermann. *Das chemische Laboratorium der Universität Marburg und die seit 1859 darin ausgeführten chemischen Untersuchungen nebst Ansichten und Erfahrungen über die Methode des chemischen Unterrichts.* Braunschweig: Vieweg, 1865.

Kolbe, Hermann. *Das neue chemische Laboratorium der Universität Leipzig.* Leipzig: Brockhaus, 1868.

Kopp, Hermann. *Die Entwickelung der Chemie in der neueren Zeit.* Munich: Oldenbourg, 1873.

Kopp, Hermann. *Geschichte der Chemie.* Vol 1. Braunschweig: Vieweg, 1843.

Kopp, Hermann. "Investigations of the Specific Heat of Solid Bodies." *Philosophical Transactions of the Royal Society* 155 (1865): 71–202.

Kopp, Hermann. "On a Great Regularity in the Physical Properties of Analogous Organic Compounds." *Philosophical Magazine* 20 (1842): 187–197.

Kopp, Hermann. "On the Relation between Boiling-Point and Composition in Organic Compounds." *Philosophical Transactions of the Royal Society* 150 (1860): 257–276.

Kopp, Hermann. "Ueber die Siedepunkte einiger isomerer Verbindungen, und über Siedepunktsregelmässigkeiten überhaupt." *Annalen der Chemie und Pharmacie* 55 (1845): 166–200.

Kopp, Hermann. "Ueber die Vorausbestimmung einiger physikalischen Eigenschaften bei mehreren Reihen organischer Verbindungen." *Annalen der Chemie und Pharmacie* 41 (1842): 79–89 (Part 1) and 169–189 (Part 2).

Körner, Friedrich. *Anleitung zur Bearbeitung des Glases an der Lampe, und zur vollständigen Verfertigung der, durch das Lampenfeuer darstellbaren, physikalischen und chemischen Instrumente und Apparate.* Jena: Schmidt, 1831.

Krakau, A. "Zur Kenntniss des Chinolins und einiger anderen Alkaloïde." *Berichte der deutschen chemischen Gesellschaft* 13 (1880): 2310–2319.

Kuenen, J. P., and W. W. Randall. "The Expansion of Argon and of Helium as Compared with That of Air and Hydrogen." *Proceedings of the Royal Society of London* 59 (1896): 60–65.

Kuhn, Thomas S. *The Essential Tension: Selected Studies in Scientific Tradition and Change.* Chicago: University of Chicago Press, 1977.

Kumar, Prakash. *Indigo Plantations and Science in Colonial India.* Cambridge: Cambridge University Press, 2012.

Ladenburg, Albert. "Die Constitution des Atropins." *Annalen der Chemie und Pharmacie* 217 (1883): 74–149.

Ladenburg, Albert. "Eine neue Methode der Elementaranalyse." *Annalen der Chemie und Pharmacie* 135 (1865): 1–24.

Ladenburg, Albert. "Einfache Methode zur Ortsbestimmung in der Pyridinreihe." *Berichte der deutschen chemischen Gesellschaft* 18 (1885): 2967–2969.

Ladenburg, Albert. "Jahresbericht 1878." *Schriften der Universität zu Kiel* 25 (1879): 74–75.

Ladenburg, Albert. "Jahresbericht 1879." *Schriften der Universität zu Kiel* 26 (1880): 68.

Ladenburg, Albert. "Jahresbericht 1881–82." *Schriften der Universität zu Kiel* 27 (1882): 30.

Ladenburg, Albert. "Künstliches Atropin." *Berichte der deutschen chemischen Gesellschaft* 12 (1879): 941–944.

Ladenburg, Albert. *Lectures on the History of the Development of Chemistry since the Time of Lavoisier.* Trans. Leonard Dobbin. Edinburgh: Alembic Club, 1900.

Ladenburg, Albert. *Lebenserinnerungen.* Breslau: Trewendt and Granier, 1912.

Ladenburg, Albert. "Methode zur Synthese in der Pyridinreihe." *Berichte der deutschen chemischen Gesellschaft* 16 (1883): 1410–1411.

Ladenburg, Albert. "Polemisches und Theoretisches." *Berichte der deutschen chemischen Gesellschaft* 8 (1875): 1666–1670.

Ladenburg, Albert. "Synthese der activen Coniine." *Berichte der deutschen chemischen Gesellschaft* 19 (1886): 2578–2583.

Ladenburg, Albert. "Synthèse de l'Acide Anisique et de l'un de ses Homologues." *Bulletin de la Société Chimique de France* 5 (1866): 257–261.

Ladenburg, Albert. "Synthese des Piperidins." *Berichte der deutschen chemischen Gesellschaft* 17 (1884): 156.

Ladenburg, Albert. "Synthese des Piperidins und seiner Homologen." *Berichte der deutschen chemischen Gesellschaft* 17 (1884): 388–391.

Ladenburg, Albert. "Ueber Aethylpyridine und Aethylpiperidine." *Berichte der deutschen chemischen Gesellschaft* 18 (1885): 2961–2967.

Ladenburg, Albert. "Ueber Benzolformeln." *Berichte der deutschen chemischen Gesellschaft* 23 (1890): 1007–1011.

Ladenburg, Albert. "Ueber das α-Isopropylpyridin." *Berichte der deutschen chemischen Gesellschaft* 17 (1884): 1676–1679.

Ladenburg, Albert. "Ueber das optische Drehungsvermögen der Piperidinbasen. II." *Berichte der deutschen chemischen Gesellschaft* 19 (1886): 2975–2977.

Ladenburg, Albert. "Ueber das spezifische Drehungsvermögen der Piperidinbasen." *Berichte der deutschen chemischen Gesellschaft* 19 (1886): 2584.

Ladenburg, Albert. "Ueber die Synthese des γ-Aethylpyridins und die Beziehungen des Pyridins zum Benzol." *Berichte der deutschen chemischen Gesellschaft* 16 (1883): 2059–2063.

Ladenburg, Albert. "Ueber isomere Bisubstitutionsderivate des Benzols." *Berichte der deutschen chemischen Gesellschaft* 8 (1875): 853–855.

Ladenburg, Albert. "Ueber Pyridin und Piperidinbasen." *Annalen der Chemie und Pharmacie* 247 (1888): 1–98.

Ladenburg, Albert. "Ueber synthetische Pyridin- und Piperidinbasen." *Berichte der deutschen chemischen Gesellschaft* 17 (1884): 772–775.

Ladenburg, Albert. "Ueber synthetische Pyridin- und Piperidinbasen." *Berichte der deutschen chemischen Gesellschaft* 18 (1885): 1587–1590.

Ladenburg, Albert. "Versuche zur Synthese des Coniin." *Berichte der deutschen chemischen Gesellschaft* 19 (1886): 439–441.

Ladenburg, Albert. *Vorträge über die Entwickelungsgeschichte der Chemie in den letzten Hundert Jahren.* Braunschweig: Vieweg, 1869.

Ladenburg, Albert, and A. Fitz. "Sur quelques Dérivés de l'Acide Paroxybenzoique." *Bulletin de la Société Chimique de France* 5 (1866): 414–423.

Ladenburg, Albert, and Carl Leverkus, "Sur la Constitution de l'Anéthol." *Comptes Rendus* 63 (1866): 89–91.

Ladenburg, Albert, and C. F. Roth. "Nachweis der Identität von synthetischem Piperidin mit dem aus Piperin gewonnenen." *Berichte der deutschen chemischen Gesellschaft* 17 (1884): 513–515.

Ladenburg, Albert, and C. F. Roth. "Studien über das käufliche Picolin." *Berichte der deutschen chemischen Gesellschaft* 18 (1885): 47–54.

Ladenburg, Albert, and L. Schrader. "Ueber Isopropylpyridine." *Berichte der deutschen chemischen Gesellschaft* 17 (1884): 1121–1123.

Lafond. "Mémoire sur l'Art de Souffler le Verre." *Journal des Connaissances Usueles et Pratiques* 16 (1832): 175–194.

Lafond. "Ueber die Kunst Glas zu Blasen, mit Verbesserungen von Danger." *Annalen der Pharmacie* 7 (1833): 298–313.

Lafond, with Danger's corrections. "Ueber die Kunst Glas zu Blasen." *Polytechnisches Journal* 48 (1833): 121–140.

Landolt, Hans. *Das optische Drehungsvermögen organischer Substanzen und die praktische Anwendungen desselben: Für Chemiker, Physiker und Zuckertechniker.* Braunschweig: Vieweg, 1879.

Landolt, Hans, and Richard Börnstein. *Physikalisch-Chemischen Tabellen.* Berlin: Springer, 1883.

Lange, Otto. "Ueber α- und Gamma-Picolin." *Berichte der deutschen chemischen Gesellschaft* 18 (1885): 3436–3441.

Latour, Bruno. "Give Me a Laboratory and I Will Raise the World." In *Science Observed: Perspectives on Social Studies of Science,* edited by Karin Knorr Cetina and Michael Mulkay, 141–170. London: Sage Publications, 1983.

Laurent, Auguste. "Recherches sur l'Indigo." *Annales de Chimie et de Physique* 3 (1840): 393–434.

Lavoisier, Antoine L. "Mémoire sur la Combinaison du Principe Oxygine, avec l'Esprit-de-vin, l'Huile et Différents Corps Combustibles." *Histoire et Mémoires de l'Académie Royal des Sciences* 98 (1784): 593–608.

Lavoisier, Antoine L. *Oeuvres de Lavoisier.* Vol. 1. Paris: Imprimerie Impériale, 1864.

Lavoisier, Antoine L. *Oeuvres de Lavoisier.* Vol. 3. Paris: Imprimerie Impériale, 1865.

Lavoisier, Antoine L. *Traité Élémentaire de Chimie*. Vol. 1. Paris: Chez Cuchet, 1793.

Lellmann, Eugen. *Principien der organischen Synthese*. Berlin: Oppenheim, 1887.

Lesch, John E. "Conceptual Change in an Empirical Science: The Discovery of the First Alkaloids." *Historical Studies in the Physical Sciences* 2 (1981): 305–328.

Lesch, John E. *Science and Medicine in France: The Emergence of Experimental Physiology, 1790–1855*. Boston: Harvard University Press, 1984.

Lespius, B. *Festschrift zur Feier des 50 jährigen Bestehens der deutschen chemischen Gesellschaft*. Berlin: Friedländer, 1919.

Lewicki, Wilhelm, ed. *Berzelius und Liebig: Ihre Briefe 1831–1845 mit gleichzeitigen Briefen von Liebig und Wöhler*. 3rd edition. Göttingen: Cromm, 1991.

Liebig, Justus. "Amorphous Quinine: On Amorphous Quinine as It Exists in the Substance Known in Commerce as Quinoidine." *Lancet* 47 (1846): 585–587.

Liebig, Justus. "Analyse der Harnsäure." *Annalen der Pharmacie* 10 (1834): 47–48.

Liebig, Justus. "Analyse, organische." In *Handwörterbuch der reinen und angewandten Chemie*. Vol. 1, edited by Justus Liebig, Johann C. Poggendorff, and Friedrich Wöhler, 357–400. Braunschweig: Vieweg, 1842.

Liebig, Justus. *Anleitung zur Analyse organischer Körper*. Braunschweig: Vieweg, 1837.

Liebig, Justus. "Basen, Organische." In *Handwörterbuch der reinen und angewandten Chemie*. Vol. 1, edited by Justus Liebig, Johann C. Poggendorff, and Friedrich Wöhler, 693–709. Braunschweig: Vieweg, 1842.

Liebig, Justus. "Bemerkungen der Redaktion zu der Abhandlung der Herren Pelletier und Couerbe." *Annalen der Physik und Chemie* 31 (1834): 321–326.

Liebig, Justus. "Bemerkungen der Redaktion zu der Abhandlung der Herren Pelletier und Couerbe." *Annalen der Pharmacie* 10 (1834): 203–210.

Liebig, Justus. "Bemerkungen zu vorstehenden Berichtigung." *Annalen der Chemie und Pharmacie* 33 (1840): 117–125.

Liebig, Justus. "Bemerkungen zu vorstehender Abhandlung." *Annalen der Pharmacie* 26 (1838): 41–60.

Liebig, Justus. "Bemerkungen zu vorstehender Abhandlung." *Annalen der Chemie und Pharmacie* 38 (1841): 195–216.

Liebig, Justus. "Bemerkungen zur vorhergehenden Abhandlung." *Annalen der Pharmacie* 2 (1832): 19–30.

Liebig, Justus. "Chinolein oder Chinolin." *Annalen der Chemie und Pharmacie* 44 (1842): 279–280.

Liebig, Justus. "Der Zustand der Chemie in Oestereich." *Annalen der Pharmacie* 25 (1838): 339–347.

Liebig, Justus. "Der Zustand der Chemie in Preussen." *Annalen der Chemie und Pharmacie* 34 (1840): 97–136.

Liebig, Justus. "Erklärung." *Annalen der Pharmacie* 9 (1834): 363.

Liebig, Justus. *Familiar Letters on Chemistry in Its Relations to Physiology, Dietetics, Agriculture, Commerce and Political Economy*. 4th edition. Edited by John Blyth. London: Walton and Maberly, 1859.

Liebig, Justus. "Herr Gerhardt und die organische Chemie." *Annalen der Chemie und Pharmacie* 57 (1846): 93–118.

Liebig, Justus. *Instructions for the Chemical Analysis of Organic Bodies*. Trans. William Gregory. Glasgow: Griffin, 1839.

Liebig, Justus. "Nachricht, das chemisch-pharmaceutische Institut zu Giessen betreffend." *Magazin für Pharmacie* 20 (1827): 98–99.

Liebig, Justus. "Nachricht, das chemisch-pharmaceutische Institut zu Giessen betreffend." *Jahrbuch der Chemie und Physik* 21 (1827): 376.

Liebig, Justus. "On Amorphous Quinine as It Exists in the Substance Known in Commerce as Quinoidine," *Lancet* 1 (May 1846): 585–587.

Liebig, Justus. "Sur l'Acide Contenu dans l'Urine des Quadrupèdes Herbivores." *Annales de Chimie et de Physique* 43 (1830): 188–198.

Liebig, Justus. "Sur l'Argent et le Mercure Fulminans." *Annales de Chimie et de Physique* 24 (1823): 294–317.

Liebig, Justus. "Ueber die Analyse organischer Substanzen." *Annalen der Pharmacie* 18 (1830): 357–367.

Liebig, Justus. "Ueber die Aethertheorie." *Annalen der Pharmacie* 23 (1837): 12–42.

Liebig, Justus. "Ueber die bittere Substanz, welche durch Behandlung des Indigs, der Seide und der Aloë mit Saltpetersäure erzeugt wird." *Annalen der Physik und Chemie* 13 (1828): 191–208.

Liebig, Justus. "Ueber die Constitution der organischen Säuren." *Annalen der Pharmacie* 26 (1838): 113–190.

Liebig, Justus. "Ueber die Darstellung und Zusammensetzung der Anthranilsaure." *Annalen der Chemie und Pharmacie* 39 (1841): 91–96.

Liebig, Justus. "Ueber die Säure welche in dem Harn der grasfressenden vierfüssigen Thiere enthalten ist." *Annalen der Physik und Chemie* 17 (1829): 389–399.

Liebig, Justus. "Ueber die vorstehende Notiz des Hrn. Akademikers Hess in Petersburg." *Annalen der Pharmacie* 30 (1839): 313–319.

Liebig, Justus. "Ueber die Zusammensetzung der Camphersäure und des Camphers." *Annalen der Physik und Chemie* 20 (1830): 41–47.

Liebig, Justus. "Ueber die Zusammensetzung der Gerbesäure (Gerbestoff) und der Gallussäuren." *Annalen der Pharmacie* 10 (1834): 172–179.

Liebig, Justus. "Ueber die Zusammensetzung der Hippursäure." *Annalen der Pharmacie* 12 (1834): 20–24.

Liebig, Justus. "Ueber die Zusammensetzung der Hippursäure." *Annalen der Physik und Chemie* 32 (1834): 573–574.

Liebig, Justus. "Ueber die Zusammensetzung und die medicinische Wirksamkeit des Chinoidins." *Annalen der Chemie und Pharmacie* 58 (1846): 348–356.

Liebig, Justus. "Ueber die Zusammensetzung und die medicinische Wirksamkeit des Chinoidins." *Großherzoglich Hessische Zeitung* 1 (June 1846): 823–824.

Liebig, Justus. "Ueber einen neuen Apparat zur Analyse organischer Körper, und über die Zusammensetzung einiger organischen Substanzen." *Annalen der Physik und Chemie* 97 (1831): 1–47.

Liebig, Justus. "Ueber einige Stickstoff-Verbindungen." *Annalen der Pharmacie* 10 (1834): 1–47.

Liebig, Justus. "Vorläufige Notiz über die Identität des Leucols mit Chinolin nach A. W. Hoffmann." *Annalen der Chemie und Pharmacie* 53 (1845): 427–428.

Liebig, Justus, and Joseph-Louis Gay-Lussac, "Analyse du Fulminate d'Argent," *Annales de Chimie et de Physique* 25 (1824): 285–311.

Liebig, Justus, Johann C. Poggendorff, and Friedrich Wöhler, eds. *Handwörterbuch der reinen und angewandten Chemie.* Vol. 1. Braunschweig: Vieweg, 1842.

Liebig, Justus, Johann C. Poggendorff, and Friedrich Wöhler, eds. *Handwörterbuch der reinen und angewandten Chemie.* Vol. 7. Braunschweig: Vieweg, 1859.

Liebig, Justus, Johann Christian Poggendorff, and Friedrich Wöhler, eds. *Handwörterbuch der reinen und angewandten Chemie: Supplement.* Braunschweig: Vieweg und Sohn, 1850.

Liebig, Justus, and Friedrich Wöhler. "Ueber die Natur der Harnsäure." *Annalen der Physik und Chemie* 117 (1837): 561–569.

Liebig, Justus, and Friedrich Wöhler. "Ueber die Zusammensetzung der Honigstein-säure." *Annalen der Pharmacie* 18 (1830): 161–164.

Liebig, Justus, and Friedrich Wöhler. "Untersuchungen über die Cyansäuren." *Annalen der Pharmacie* 20 (1830): 368–400.

Liebig, Justus, and Friedrich Wöhler. "Untersuchungen über die Natur der Harn-säure." *Annalen der Pharmacie* 26 (1838): 241–336.

Llanos, Eugenio J., Wilmer Leal, Duc H. Luu, Jürgen Jost, Peter F. Stadler, and Guillermo Restrepo. "Exploration of the Chemical Space and Its Three Historical Regimes." *Proceedings of the National Academy of Sciences* 116 (2019): 12660–12665.

Löwig, Carl. *Der Chemiker Dr. Justus Liebig in Giessen vor das Gericht der öffentlichen Meinung.* Zurich: Orell, Fössli, and Co., 1833.

Mansfield, Charles Blachford. "Researches on Coal Tar." *Quarterly Journal of the Chemical Society* 1 (1849): 244–268.

Marchand, Richard Felix. "Bemerkungen über die organische Analyse." *Journal für praktische Chemie* 13 (1838): 509–516.

Mecking, L., and G. Jacob. *Kiel als Universitätsstadt.* Kiel: Mühlau, 1921.

Meinel, Christoph. "Artibus Acadmicis Inserenda: Chemistry's Place in Eighteenth and Early Nineteenth Century Universities." *History of Universities* 7 (1988): 89–115.

Meinel, Christoph. "August Wilhelm Hofmann: 'Regierender Oberchemiker.'" In *Die Allianz von Wissenschaft und Industrie August Wilhelm Hofmann (1818–1892): Zeit,*

Werk, Wirkung, edited by Scholz Hartmut and Christoph Meinel, 27–64. Weinheim: VCH, 1992.

Meinel, Christoph. "Chemische Laboratorien: Funktion und Disposition." *Berichte zur Wissenschaftsgeschichte* 23 (2000): 287–302.

Meinel, Christoph, and Hartmut Scholz, eds. *Die Allianz von Wissenschaft und Industrie August Wilhelm Hofmann (1818–1892): Zeit, Werk, Wirkung*. Weinheim: VCH, 1992.

Meinel, Christoph and Thomas Steinhauser, eds. "Liebig-Wöhler Briefwechsel, 1829–1873." In preparation.

Miller, William A. *Elements of Chemistry: Theoretical and Practical. Part 3: Organic Chemistry*. London: Longman, 1869.

Milt, Clara de. "Auguste Laurent, Founder of Modern Organic Chemistry." *Chymia* 4 (1953): 85–114.

Mitscherlich, Alexander, ed. *Gesammelte Schriften von Eilhard Mitscherlich*. Berlin: Mittler, 1896.

Mitscherlich, Alexander, "Organische Analyse vermittelst Quecksilberoxyds." *Fresenius's Zeitschrift für analytische Chemie* 15 (1876): 371–417.

Mitscherlich, Eilhard. *Lehrbuch der Chemie*. Vol. 1. 2nd edition. Berlin: Mittler, 1834.

Mitscherlich, Eilhard. "Organische Analyse vermittelst Quecksilberoxyds." *Fresenius's Zeitschrift für analytische Chemie* 15 (1876): 371–417.

Mitscherlich, Eilhard. "Ueber das Benzol und die Säuren der Oel- und Talgarten." *Annalen der Pharmacie* 9 (1834): 39–48.

Mitscherlich, Eilhard, and Justus Liebig. "Ueber die Zusammensetzung der Milchsäure." *Annalen der Pharmacie* 7 (1833): 47–48.

Morrell, Jack B. "The Chemist Breeders: The Research Schools of Liebig and Thomas Thomson." *Ambix* 19 (1972): 1–46.

Morris, Peter J. T., ed. *From Classical to Modern Chemistry: The Instrumental Revolution*. Cambridge: Royal Society of Chemistry, 2002.

Morris, Peter J. T. *The Matter Factory: A History of the Chemistry Laboratory*. London: Reaktion, 2015.

Munday, Pat. "Social Climbing through Chemistry: Justus Liebig's Rise from the *niederer Mittelstand* to the *Bildungsbürgertum*." *Ambix* 37 (1990): 1–19.

Muspratt, James S., and August Wilhelm Hofmann. "On Certain Processes in Which Anilene is Formed." *Memoirs and Proceedings of the Chemical Society* 2 (1845): 249–254.

Muspratt, James S., and August Wilhelm Hofmann. "On Nitraniline, A New Decomposition Product of Dinitrobenzol." *Memoirs and Proceedings of the Chemical Society* 3 (1845): 110–125.

Muspratt, James S., and August Wilhelm Hofmann. "On Toluidine, a New Organic Base." *Memoirs and Proceedings of the Chemical Society* 2 (1845): 367–383.

Nägelke, Hans Dieter. *Hochschulbau im Kaiserreich: Historische Architektur im Prozess bürgerlicher Konsensbildung*. Kiel: Ludwig, 2000.

Nawa, Christine. "A Refuge for Inorganic Chemistry: Bunsen's Heidelberg Laboratory." *Ambix* 61 (2014): 115–140.

Nicolaou, K. C. "The Art and Science of Total Synthesis at the Dawn of the Twenty-First Century." *Angewandte Chemie International Edition* 39 (2000): 44–122.

Nicolaou, K. C. "The Emergence and Evolution of Organic Synthesis and Why It Is Important to Sustain It as an Advancing Art and Science for Its Own Sake." *Israel Journal of Chemistry* 58 (2018): 104–113.

Nicolaou, K. C. "The Emergence of the Structure of the Molecule and the Art of Its Synthesis." *Angewandte Chemie International Edition* 52 (2013): 131–146.

Novitski, Marya E. *Auguste Laurent and the Prehistory of Valence.* Chur, Switzerland: Harwood Academic, 1992.

Oppermann, Charles. "Ueber die Zusammensetzung des Terpenthinöls und einiger von demselben entstehenden Producte." *Annalen der Physik und Chemie* 22 (1831): 193–207.

Ørsted, Hans Christian. "Über das Piperin, ein neues Pflanzenalkaloid." *Schweiggers Journal für Chemie und Physik* 29 (1820): 80–82.

Osterath, Brigitte. "August Hofmann and the Chemists Factory." *Chemistry World,* August 10, 2017. https://www.chemistryworld.com/feature/hofmanns-chemistry-factory/3007787.article.

Partington, James R. *A History of Chemistry.* Vol. 3. London: Macmillan, 1962.

Partington, James R. *A History of Chemistry.* Vol. 4. London: Macmillan, 1964.

Pebal, Leopold. *Das chemische Institut der K.K. Universität Graz.* Wien: Faesy & Frick, 1880.

Perkin, William H. "Hofmann Memorial Lecture." *Journal of the Chemical Society* 69 (1896): 596–637.

Perkin, William H. "Some of the Laboratories I Have Worked In." A manuscript dated 1906 and quoted in *Perkin Centenary, London: 100 years of Synthetic Dyestuffs,* edited by Anon, 18. New York: Pergammon, 1958.

Planta, Adolf von, and August Kekulé. "Beiträge zur Kenntniss einiger flüchtigen Basen." *Annalen der Chemie und Pharmacie* 89 (1854): 129–156.

Playfair, Lyon. "Hofmann Memorial Lecture." *Journal of the Chemical Society* 69 (1896): 575–579.

Plücker, Julius. "Ueber das magnetische Verhalten der Gase." *Annalen der Physik und Chemie* 83 (1851): 87–108.

Plücker, Julius. "Ueber das magnetische Verhalten der Gase (Zweite Mittheilung)." *Annalen der Physik und Chemie* 84 (1851): 161–180.

Plücker, Julius, and Heinrich Geissler. "Studien über Thermometrie und verwandte Gegenstände." *Annalen der Physik und Chemie* 162 (1852): 238–279.

Poggendorff, Johann C. *Biographisch-Literarisches Handwörterbuch.* Vol. 4. Leipzig: Barth, 1904.

Poncet, Sébastien, and Laurie Dahlberg. "The Legacy of Henri Victor Regnault in the Arts and Science." *International Journal of Arts and Sciences* 4 (2011): 377–400.

Prout, William. "On the Ultimate Composition of Simple Alimentary Substances; with Some Preliminary Remarks on the Analysis of Organized Bodies in General." *Philosophical Transactions of the Royal Society* 117 (1827): 355–388.

Ramberg, Peter J. "The Death of Vitalism and the Birth of Organic Chemistry: Wöhler's Urea Synthesis in Textbooks of Organic Chemistry." *Ambix* 47 (2000): 170–195.

Ramberg, Peter J. "Wilhelm Heintz (1817–1880) and the Chemistry of the Fatty Acids." *Bulletin for the History of Chemistry* 38 (2013): 19–28.

Ramsay, William. "On Picoline and Its Derivatives." *The London, Edinburgh, and Dublin Philosophical Magazine and Journal of Science* 2 (1876): 269–281.

Redwood, Theophilus. "On Amorphous Quinine." *Pharmaceutical Journal* 6 (1846–1847): 128–132.

Redwood, Theophilus. "The Progress of Pharmaceutical Science." *Pharmaceutical Journal* 5 (1846): 103–106.

Regnault, Victor. "Neue Untersuchungen über die Zusammensetzung der organischen Basen." *Annalen der Pharmacie* 26 (1838): 10–41.

Regnault, Victor. *Relation des Expériences Entreprises par Ordre de Monsieur le Ministre des Travaux Publics, et sur la Proposition de la Commission Centrale des Machine à Vapeur, pour Déterminer les Principales Lois Physiques, et les Données Numériques qui Entrent dans le Calcul des Machines à Vapeur.* Vol. 1. Paris: Didot, 1847.

Regnault, Victor. "Ueber die Zusammensetzung des Chlorkohlenwasserstoffs (Oel des ölbildenden Gases)." *Annalen der Pharmacie* 14 (1835): 22–38.

Regnault, Victor. "Weiterer Beitrag über die Zusammensetzung der organischen Basen." *Annalen der Pharmacie* 29 (1839): 58–63.

Reif-Acherman, Simón. "Heinrich Geissler: Pioneer of Electrical Science and Vacuum Technology." *Proceedings of the IEEE* 103 (2015): 1672–1684.

Reif-Acherman, Simón. "Henri Victor Regnault: Experimentalist of the Science of Heat." *Physics in Perspective* 12 (2010): 396–442.

Reimer, Karl, and Ferdinand Tiemann. "Ueber die Einwirkung von Chloroform auf alkalische Phenolate." *Berichte der deutschen chemischen Gesellschaft* 9 (1876): 824, 1268, 1285.

Reinhardt, Carsten. *Shifting and Rearranging: Physical Methods and the Transformation of Modern Chemistry.* Sagamore Beach, MA: Science History Publications, 2006.

Reinhardt, Carsten, and Anthony S. Travis. *Heinrich Caro and the Creation of Modern Chemical Industry.* Berlin: Springer, 2000.

Rive, Auguste A. de la. "Notices Respecting New Books: H. Regnault, Relation des Expériences pour Déterminer les Principales Lois Physiques, et les Données Numériques qui Entrent dans le Calcul des Machines à Vapeur." *Philosophical Magazine* 36 (1850): 41–62.

Roberts, Gerrylynn K. "The Establishment of the Royal College of Chemistry: An Investigation of the Social Context of Early Victorian Chemistry." *Historical Studies in the Physical Sciences* 7 (1976): 437–485.

Roberts, Gerrylynn K. "The Royal College of Chemistry (1845–1853). A Social History of Chemistry in Early Victorian Britain." PhD dissertation, Johns Hopkins University, 1973.

Robins, Edward C. *Technical School and College Building: Being a Treatise on the Design and Construction of Applied Science and Art Buildings, and Their Suitable Fittings and Sanitation, with a Chapter on Technical Education.* London: Whittaker, 1887.

Robiquet, Pierre Jean. "Note sur la Narcotine." *Journal de Pharmacie* 17 (1831): 637–643.

Rocke, Alan J. *Chemical Atomism in the Nineteenth Century: From Dalton to Cannizzaro.* Columbus, OH: Ohio State University Press, 1984.

Rocke, Alan J. *Image and Reality: Kekulé, Kopp, and the Scientific Imagination.* Chicago: University of Chicago Press, 2010.

Rocke, Alan J. "Kekulé's Benzene Theory and the Appraisal of Scientific Theories." In *Scrutinizing Science: Empirical Studies of Scientific Change,* edited by Arthur Donovan and Larry Laudan, 145–161. Dordrecht: Kluwer, 1988.

Rocke, Alan J. "Körner, Dewar and the Structure of Pyridine." *Bulletin for the History of Chemistry* 2 (1988): 4–6.

Rocke, Alan J. *Nationalizing Science: Adolphe Wurtz and the Battle for French Chemistry.* Cambridge, MA: MIT Press, 2001.

Rocke, Alan J. "Organic Analysis in Comparative Perspective: Liebig, Dumas, and Berzelius, 1811–1837." In *Instruments and Experimentation in the History of Chemistry,* edited by Frederic L. Holmes and Trevor H. Levere. Cambridge, MA: MIT Press, 2000, 275–310.

Rocke, Alan J. "Origins and Spread of the 'Giessen Model' in University Science." *Ambix* 50 (2003): 90–115.

Rocke, Alan J. *The Quiet Revolution: Hermann Kolbe and the Science of Organic Chemistry.* Berkeley, CA: University of California Press, 1993.

Rocke, Alan J. "What Did 'Theory' Mean to Nineteenth-Century Chemists?" *Foundations of Chemistry* 15 (2013): 145–156.

Rocke, Alan J., and Hermann Kopp. *From the Molecular World: A Nineteenth-Century Science Fantasy.* Berlin: Springer, 2012.

Rockstroh, Heinrich. *Die Glasblasekunst im Kleinen oder mittelst der Docht- oder der Strahlflamme, oder Anweisung, wie aus Glas mittelst der Docht- oder der Strahlflamme mancherlei Gegenstände im Kleinen zu gestalten: nebst einer Anweisung, wie Mikroskope, Barometer, Thermometer und Aräometer, Mikrometer und noch manche andere Gegenstände, bei welchen Glas das vornehmliche Material ist, verfertigt oder bewerkstelliget werden; auch einem Anhange von Glaskunststückchen und Glaskünsteleien.* Leipzig: Günther, 1833.

Rossignol, Paul. "Les Travaux Scientifiques de Joseph Pelletier." *Revue d'Histoire de la Pharmacie* 77 (1989): 135–152.

Rossiter, Margaret. *The Emergence of Agricultural Science: Justus Liebig and the Americans, 1840–1880*. New Haven, CT: Yale University Press, 1975.

Roster, Giorgio. *Delle Scienze Sperimentali e in Particolare della Chimica in Germania*. Milan: Civelli, 1872.

Roth, C. F. "Ein neuer Apparat zur Bestimmung von Schmelzpunkten." *Berichte der deutschen chemischen Gesellschaft* 19 (1886): 1970–1973.

Rowe, J. S. "The Life and Work of George Fownes, F.R.S. (1815–49)." *Annals of Science* 6 (1950): 422–435.

Rue, Warren de la. "On a Modification of the Apparatus of Will and Varrentrapp for the Estimation of Nitrogen." *Memoirs and Proceedings of the Chemical Society* 3 (1847): 347–348.

Rue, Warren de la. "On Cochineal." *Memoirs and Proceedings of the Chemical Society* 3 (1845): 454–480.

Runge, F. F. "Ueber einige Produkte der Steinkohlendestillation." *Annalen der Physik und Chemie* 31 (1834): 65–78.

Rupe, Hans. *Adolf Baeyer als Lehrer und Forscher: Erinnerungen aus seinem Privatlaboratorium*. Stuttgart: Enke, 1932.

Russell, Colin A. "August Wilhelm Hofmann—Cosmopolitan Chemist." In *Die Allianz von Wissenschaft und Industrie: August Wilhelm von Hofmann (1818–1892): Zeit, Werk, Wirkung*, eds. Christoph Meinel and Hartmut Scholz, 65–75. Weinheim: VCH, 1992.

Russell, Colin A. "The Changing Role of Synthesis in Organic Chemistry." *Ambix* 34 (1987): 169–180.

Russell, Colin A. *The History of Valency*. Leicester: Leicester University Press, 1971.

Scerri, Eric, and Lee McIntyre. *Philosophy of Chemistry: Growth of a New Discipline*. Berlin: Springer, 2014.

Schierenberg, Kurt August. "Eilhard Mitscherlich und Justus von Liebig." *Giessener Universitaet* 2 (1977): 106–115.

Schiff, Hugo. "Erste Synthese eines Pflanzenalkaloids (Synthese des Coniins)." *Berichte der deutschen chemischen Gesellschaft* 3 (1870): 946–947.

Schiff, Hugo. "Mittheilungen aus dem Universitätslaboratorium in Pisa: Eine neue Reihe organischer Basen." *Annalen der Chemie und Pharmacie* 131 (1864): 118–119.

Schiff, Hugo. "Synthesis of Alkaloids." *Pharmaceutical Journal, Transactions*. 3rd series. 1 (1870–71): 605.

Schiff, Hugo. "Ueber die Synthese des Coniins." *Annalen der Chemie und Pharmacie* 157 (1871): 352–362.

Schiff, Hugo. "Ueber die Synthese des Coniins." *Annalen der Chemie und Pharmacie* 166 (1873): 88–108.

Schiff, Hugo. "Weiteres über das künstliche Coniin." *Berichte der deutschen chemischen Gesellschaft* 5 (1872): 42–44.

Schindler, Hans. "Notes on the History of the Separatory Funnel." *Journal of Chemical Education* 34 (1957): 528–530.

Schorlemmer, Carl. *The Rise and Development of Organic Chemistry*. Manchester: Cornish and London, 1879.

Schorm, J. "Beitrag zur Kenntniss des Coniins und seiner Verbindungen." *Berichte der deutschen chemischen Gesellschaft* 14 (1881): 1765–1769.

Schotten, Carl. "Zur Kenntniss des Coniins." *Berichte der deutschen chemischen Gesellschaft* 15 (1882): 1947–1951.

Schultz, Gustav. "Feier der deutschen chemischen Gesellschaft zu Ehren August Kekulé's." *Berichte der deutschen chemischen Gesellschaft* 23 (1890): 1265–1312.

Schummer, Joachim. "The Impact of Instrumentation on Chemical Species Identity from Chemical Substances to Molecular Species." In *From Classical to Modern Chemistry: The Instrumental Revolution,* edited by Peter J. T. Morris, 188–211. Cambridge: Royal Society of Chemistry, 2002.

Schütt, Hans-Werner. *Eilhard Mitscherlich: Prince of Prussian Chemistry*. Philadelphia: Chemical Heritage Foundation, 1997.

Scott, Robert Henry. "The History of the Kew Observatory." *Proceedings of the Royal Society of London* 39 (1886): 37–86.

Secord, James. "Science, Technology and Mathematics." In *The Cambridge History of the Book in Britain*. Vol. 5, 1830–1914, edited by D. McKitterick, 443–474. Cambridge: Cambridge University Press, 2009.

Seebach, Dieter. "Organic Synthesis—Where Now?" *Angewandte Chemie International Edition* 29 (1990): 1320–1367.

Seeman, Jeffrey I. "On the Relationship between Classical Structure Determination and Retrosynthetic Analysis/Total Synthesis." *Israel Journal of Chemistry* 58 (2018): 28–44.

Seeman, Jeffrey I. "R. B. Woodward's Letters: Revealing, Elegant and Commanding." *Helvetica Chimica Acta* 100 (2017): e1700183. https://doi.org/10.1002/hlca.201700183.

Sell, Eugen. *Grundzüge der modernen Chemie: Organische Chemie*. Vol. 2. Berlin: Hirschwald, 1870.

Shapin, Steven. "History of Science and Its Sociological Reconstructions" *History of Science* 20 (1982): 157–211.

Shapin, Steven. "The House of Experiment in Seventeenth-Century England." *Isis* 79 (1988): 373–404.

Sheibley, Fred E. "Carl Julius Fritzsche and the Discovery of Anthranilic Acid, 1841." *Journal of Chemical Education* 20 (1943): 115–117.

Sibum, Otto H. "Reworking the Mechanical Value of Heat: Instruments of Precision and Gestures of Accuracy in Early Victorian England." *Studies in History and Philosophy of Science* 26 (1995): 73–106.

Siwoloboff, A. "Ueber die Siedepunktbestimmung kleiner Mengen Flüssigkeiten." *Berichte der deutschen chemischen Gesellschaft* 19 (1886): 795–796.

Smith, R. Angus. *The Life and Works of Thomas Graham, DCL, FRS: Illustrated by 64 Unpublished Letters.* Edited by Joseph J. Coleman. Glasgow: Smith, 1884.

Smith Brazier, James, and G. Gossleth. "Contributions towards the History of Caproic and Œnanthylic Acids." *Quarterly Journal of the Chemical Society* 3 (1851): 210–229.

Söderbaum, H. G. *Correspondance entre Berzelius et C. L. Berthollet (1810–1822).* Uppsala: Académie Royale des Sciences de Suède, 1912.

Squire, Peter. *A Companion to the British Pharmacopoeia.* London: Churchill, 1866.

Stella, L. "Homolytic Cyclizations of N-Chloroalkenylamines." *Angewandte Chemie International Edition* 22 (1983): 337–422.

Szabadváry, Ferenc. *History of Analytical Chemistry.* Chemin de la Sallaz, Switzerland: Gordon and Breach, 1992 [1966].

Thaulow, M. C. J. "Ueber die Zuckersäure." *Annalen der Physik und Chemie* 44 (1838): 497–513.

Thenard, Louis-Jacques. "De l'Action des Acides Végétaux sur l'Alcool, sans l'Intermède et avec l'Intermède des Acides Minéraux." *Mémoires de Physique et de Chimie de la Société d'Arcueil* 2 (1809): 5–22.

Thenard, Louis-Jacques. *Traité de Chimie Élémentaire, Théorique et Pratique.* Vol. 4. Paris: Crochard, 1816.

Thomson, Anthony Todd. *The London Dispensatory.* London: Longman, Hurst, Rees, Orme and Brown, 1811.

Thomson, C. W., John Murray, George S. Nares, and Frank T. Thomson. *Report on the Scientific Results of the Voyage of H.M.S. Challenger During the Years 1873–76 under the Command of Captain George S. Nares and the Late Captain Frank Tourle Thomson. R.N.* Edinburgh: H.M.S.O., 1885.

Thomson, Thomas. *A System of Chemistry.* Vol. 2. London: Baldwin, Craddock & Joy, 1817.

Thorpe, Thomas Edward. *A Dictionary of Applied Chemistry.* Vol. 1. London: Longmans, 1891.

Thorpe, Thomas Edward. "Obituary Notices of Fellows Deceased: Hermann Kopp." *Proceedings of the Royal Society of London* 60 (1897): 1–35.

Thorpe, Thomas Edward. "The Life Work of Hermann Kopp." *Journal of the Chemical Society, Transactions* 63 (1893): 775–815.

Tiemann, Ferdinand. "Bemerkung zu der Abhandlung des Hrn. Ladenburg: Constitution des Benzols in No. 15 Dieser Berichte." *Berichte der deutschen chemischen Gesellschaft* 8 (1875): 1344.

Tiemann, Ferdinand. "Erinnerung an das fünfundzwanzigjährige Bestehen der deutschen chemischen Gesellschaft und an Ihren ersten Präsidenten August Wilhelm Hofmann veranstaltete Gedächtnissfeier." *Berichte der deutschen chemischen Gesellschaft* 25 (1892): 3369–3414.

Tiemann, Ferdinand. "Gedächtnisrede August Wilhelm Hofmann." *Berichte der deutschen chemischen Gesellschaft* 25 (1892): 3377–3398.

Tiemann, Ferdinand, and Carl Schotten. "Ueber die mittelst der Chloroformreaction aus den drei isomeren Kresolen darstellbaren Oxytoluylaldehyde und die zugehörigen Oxytoluylsäuren." *Berichte der deutschen chemischen Gesellschaft* 11 (1878): 767–784.

Tomic, Sacha. *Aux Origines de la Chimie Organique: Méthodes et Pratiques de Pharmaciens et des Chimistes (1785–1835)*. Rennes: University of Rennes Press, 2010.

Travis, Anthony S. "Perkin's Mauve: Ancestor of the Organic Chemical Industry." *Technology and Culture* 31 (1990): 51–82.

Travis, Anthony S. *The Rainbow Makers: The Origins of the Synthetic Dyestuffs Industry in Western Europe*. Bethlehem, PA: Lehigh University Press, 1993.

Travis, Anthony S. *The Synthetic Nitrogen Industry in World War I: Its Emergence and Expansion*. Berlin: Springer, 2015.

Tuchman, Arleen Marcia. *Science, Medicine, and the State in Germany: The Case of Baden, 1815–1871*. Oxford: Oxford University Press, 1993.

Turner, Edward. *Elements of Chemistry*. 4th edition. London: John Taylor, 1833.

Usselman, Melvyn C. "Liebig's Alkaloid Analyses: The Uncertain Route from Elemental Content to Molecular Formulas." *Ambix* 50 (2003): 71–89.

Usselman, Melvyn C., Christina Reinhart, Kelly Foulser, and Alan J. Rocke. "Restaging Liebig: A Study in the Replication of Experiments." *Annals of Science* 62 (2005): 1–55.

Varrentrapp, Franz, and Heinrich Will. "Neue Methode zur Bestimmung des Stickstoffs in organischen Verbindungen." *Annalen der Chemie und Pharmacie* 39 (1841): 257–296.

Volhard, Jacob, and Emil Fischer. *August Wilhelm von Hofmann: Ein Lebensbild*. Berlin: Springer, 1902.

Volhard, Jacob. *Justus von Liebig*. Vol. 1. Leipzig: Barth, 1909.

Vorländer, Daniel. "Jacob Volhard." *Berichte der deutschen chemischen Gesellschaft* 45 (1912): 1855–1902.

Wakley, Thomas. "Communication." *Pharmaceutical Journal* 6 (1846–7): 144.

Wakley, Thomas. "Editorial." *The Lancet* (July 25, 1846): 105–109.

Walford, Edward. "Hanover Square and Neighbourhood." *Old and New London* 4 (1878): 314–326.

Wallach, Otto, ed. *Briefwechsel zwischen J. Berzelius und F. Wöhler*. Vol. 1. Leipzig: Englmann, 1901.

Ward, Edward R. "Charles Blatchford Mansfield, 1819–55: Coal Tar Chemist and Social Reformer." *Chemistry in Britain* 15 (1979): 297–304.

Warwick, Andrew C. *Masters of Theory: Cambridge and the Rise of Mathematical Physics*. Chicago: University of Chicago Press, 2003.

Watts, Henry. *A Dictionary of Chemistry*. Vol. 1. London: Longman, 1863.

Wertheim, Theodor. "Beiträge zur Kenntniss des Coniins." *Annalen der Chemie und Pharmacie* 123 (1862): 157–186 and 130.

Wertheim, Theodor. "Beiträge zur Kenntniss des Coniins." *Annalen der Chemie und Pharmacie* 130 (1864): 269–302.

Wichelhaus, H. "Hugo Schiff." *Berichte der deutschen chemischen Gesellschaft* 48 (1915): 1565–1569.

Winckler, Ferdinand Ludwig. "Briefliche Notiz über die wahrscheinliche Entdeckung eines neuen Chinarinden-Alkaloids (Chinidin)." *Repertorium für die Pharmacie* 85 (1844): 392–397.

Winckler, Ferdinand Ludwig. "Einige Worte ueber den mutmaßlichen Werth des Chinoidins als Heilmittel." *Großherzoglich Hessische Zeitung* 1 (June 12, 1846): 859–860.

Winckler, Ferdinand Ludwig. "Ueber die chemischen Zusammensetzung des käuflichen Chinoidins und die Bedeutung desselben als Arzneimittel." *Großherzoglich Hessische Zeitung* 1 (June 30, 1846): 919–920.

Winckler, Ferdinand Ludwig. "Ueber die chemischen Zusammensetzung des Chinoidins und das zweckmäßigste Verfahren dasselbe zu Reinigen." *Jahrbuch für praktische Pharmacie* 7 (1843): 65–69.

Winckler, Ferdinand Ludwig. "Ueber die chemischen Zusammensetzung des käuflichen Chinoidins und die Bedeutung desselben als Arzneimittel." *Jahrbuch für praktische Pharmacie* 10 (1846): 361–389.

Winkles, Gersham Henry. "On the Existence of Trimethylamine in the Brine of Salted Herrings." *Quarterly Journal of the Chemical Society* 7 (1855): 63–68.

Wöhler, Friedrich. "Ueber künstliche Bildung des Harnstoffs." *Annalen der Physik und Chemie* 88 (1828): 253–256.

Wöhler, Friedrich, and Justus Liebig. "Cyansäures Aethyl- und Methyloxyd." *Annalen der Chemie und Pharmacie* 54 (1845): 370–371.

Wöhler, Friedrich, and Justus Liebig. "Ueber die Bildung des Bittermandelöls." *Annalen der Pharmacie* 22 (1837): 1–24.

Wöhler, Friedrich, and Justus Liebig. "Untersuchungen über das Radical der Benzoesäure." *Annalen der Pharmacie* 3 (1832): 249–287.

Wöhler, Friedrich, and Justus Liebig. "Untersuchungen über das Radical der Benzoesäure." *Annalen der Physik und Chemie* 26 (1832): 325–343.

Wolff, Emil T. *Vollständige Uebersicht der elementar-analytischen Untersuchungen organischer Substanzen*. Halle: Eduard Anton, 1846.

Wolff, M. "Cyclization of N-Halogenated Amines (The Hofmann-Löffler Reaction)." *Chemical Review* 63 (1963): 55–64.

Woodward, Robert B. "Art and Science in the Synthesis of Organic Compounds: Retrospect and Prospect." In *Pointers and Pathways in Research: Six Lectures in the Fields of Organic Chemistry and Medicine*, edited by Maeve O'Connor, 23–41. Bombay: CIBA of India, 1963.

Woodward, Robert B. "Synthesis." In *Perspectives in Organic Chemistry*, edited by Alexander R Todd, 155–184. New York: Interscience, 1956.

Woodward, Robert B. "The Total Synthesis of Vitamin B12." *Pure and Applied Chemistry* 33 (1973): 145–178.

Woodward, Robert B., and Roald Hoffmann. "Stereochemistry of Electrocyclic Reactions." *Journal of the American Chemical Society* 87 (1965): 395–397.

Wurtz, Adolphe. *Les Hautes Études Pratiques dans les Universités Allemandes*. Paris: Imprimerie Imperialé, 1870.

Wurtz, Adolphe. *Les Hautes Études Pratiques dans les Universités d'Allemagne et d'Autriche-Hongue*. Paris: Masson, 1882.

Wurtz, Adolphe. "Note sur la Formation de l'Uréthane par l'Action du Chlorure de Cyanogène Gazeux sur l'Alcool." *Comptes Rendus* 22 (1846): 503–505.

Wurtz, Adolphe. "Note sur l'Éther Cyanurique et sur le Cyanurate de Methylene." *Comptes Rendus* 26 (1848): 368–370.

Wurtz, Adolphe. "Recherches sur les Éthers Cyaniques et leurs Dérivés." *Comptes Rendus* 27 (1848): 241–243.

Wurtz, Adolphe. "Sur une Série d'Alcalis Organiques Homologues avec l'Ammonique." *Comptes Rendus* 28 (1849): 223–226.

Zeise, W. C. "Neue Untersuchungen über das entzündliche Platinchlorüs." *Annalen der Pharmacie* 23 (1837): 1–11.

Zinin, Nikolai. "Beschreibung einiger neuer organischer Basen, dargestellt durch die Einwirkung des Schwefelwasserstoffes auf Verbindungen der Kohlenwasserstoffe mit Untersaltpetersäure." *Journal für praktische Chemie* 27 (1842): 140–153.

Zott, Regine, and Emil Heuser, eds. *Justus von Liebig und August Wilhelm Hofmann in Ihren Briefen. Nachträge 1845–1869*. Mannheim: Bionomica-Verlag, 1988.

INDEX